MW00341517

BARRON'S

IB

BIOLOGY

Camilla Walck, Ph.D.
IB Biology Teacher
Princess Anne High School
Virginia Beach, Virginia

BARRON'S

ABOUT THE AUTHOR

Camilla Walck has taught pre-IB biology, SL biology and HL biology over the past fifteen years at Princess Anne High School in Virginia Beach, Virginia. In addition she teaches anatomy at Tidewater Community College and has taught secondary science methods at Virginia Wesleyan College. Camilla is a Nationally Board Certified Teacher (NBCT) in adult and adolescent science. She has received national recognition from the U.S Chamber of Commerce as the 2012 National Life Science Teacher of the Year and was selected as a Claes Nobel Top Ten Teacher of the Year for 2013 by the National Society of High School Scholars. Camilla was selected to attend grade awarding for IB biology in Cardiff, Wales, and has served as an IB Biology Exam grader for IBO for several years.

© Copyright 2014 by Barron's Educational Series, Inc.

All rights reserved.

No part of this publication may be reproduced or distributed in any form or by any means without the written permission of the copyright owner.

All inquiries should be addressed to:
Barron's Educational Series, Inc.
250 Wireless Boulevard
Hauppauge, New York 11788
www.barronseduc.com

ISBN: 978-1-4380-0339-9

Library of Congress Control Number: 2014944226

PRINTED IN THE UNITED STATES OF AMERICA
9 8 7 6 5 4 3 2 1

10%
POST-CONSUMER WASTE
Paper contains a minimum of 10% post-consumer waste (PCW). Paper used in this book was derived from certified, sustainable forestlands.

CONTENTS

OPTIONAL TOPICS

PRACTICE TESTS

APPENDIX

Introduction

Congratulations on your choice to study the incredible world of biology! No topic of study is more important than the one that involves understanding how your own body functions and how the world around you affects your life. This book is specifically designed to prepare you to successfully master the material required for the International Baccalaureate (IB) Biology Exam. International Baccalaureate Biology scores range from a low score of 1 to a top score of 7. Most universities will accept any score of 4 or higher when assigning college credits. However, students majoring in biology will often be required to achieve a score of 5 or higher. Many institutions publish the required scores on IB exams in order to receive course credit, and these can be seen online at each institution's website.

The questions on the IB Biology Exam all come from the IB Biology Course syllabus. The syllabus is very specific in what it requires the student to be able to respond to on the exam. For this reason, this review book focuses specifically on the syllabus material in order to ensure that students concentrate on concepts that will successfully prepare them for the IB Biology Exam. When needed, background information required to understand the concepts presented is included along with the required material.

To get the most out of this book, you should do the following:

- Read each section carefully, and focus on the main concepts presented.
- Take notes on material that is new to you.
- Make labelled sketches of diagrams or drawings you may be required to draw on the exam.
- Complete all practice questions, and read your answers thoroughly.
- Review the topic again when you miss a practice question or fail to receive full credit for your response.
- Skim through previous topics of study each time you begin to review a new topic.

The material presented in this review book follows the presentation of objectives in the IB syllabus. Individual topics are introduced with topic objectives listed according to objective number. Students preparing for the standard-level (SL) exam need to review topics 1 through 6. Students preparing for the higher-level (HL) exam will need to review all the topics presented (1 through 11). One optional topic of study is required for both SL and HL students. The optional topics follow the standard-level and higher-level material and are presented as they appear in the syllabus (topics A through D). Your teacher should have prepared you for responding to one of the optional topics. You need to review only the topic presented to you in class. The topic taught in class should be the one you will be best prepared to respond to on the IB exam. However, you are free to respond to any one of the presented topics. For this reason, it is wise to read over all of the questions before making your final choice. Respond to the topic you feel you can earn the most points from.

Practice questions and answers (IB calls these mark schemes) are included at the end of each topic covered. The mark schemes give the detail needed to earn points. Each statement or response that should be awarded a point is listed next to a bullet point. If multiple responses mean the same thing, they are listed as being worth 1 possible point and will have back slashes (/) or the word "or" separating the possible responses. Only 1 point is awarded for a correct statement. You do not have to have the exact wording to earn the point, but you must cover the required details.

THE NATURE OF SCIENCE (NOS)

The nature of science (NOS) is an overarching theme throughout the IB Biology syllabus. For this reason, each topic presented will begin with a brief list of NOS connections that relate to the material presented. Your teacher should have referred to and related course topics to the NOS strands as material was presented. The 5 NOS strands focus on but are not limited to the following:

- What science is and how scientific endeavours are carried out
  - Scientific method and variations of it
  - Scepticism and accidental discoveries
  - Observations, analogies and theoretical understanding
  - Collaboration in scientific communities

- How science is presented and understood

  - Hypotheses, theories and laws
  - Explanations of natural phenomena
  - Causation and correlation
  - Controlled experimentation (reliability of data)

- Science and objectivity

  - Multiple trials and reliability
  - Biases in science (personal as well as social)
  - Data analysis: patterns and relationships

- The human aspect of science

  - Collaboration among scientists
  - International collaboration and data sharing
  - Ethical issues of scientific discoveries and experimentation
  - Risk and benefit analysis of scientific research and discoveries

- Public understanding and scientific literacy of science

  - Public knowledge of the proper scientific language and methodology
  - The role of scientists in educating the public about scientific discoveries
  - The appropriate use of scientific terminology

This brief summary of the NOS strands is by no means complete. The full descriptions can be seen in the biology curriculum guide.

THE IB BIOLOGY EXAM

The IB Biology Exam is composed of 3 papers (exams) that are given over a two-day period.

SL Exam Papers

Paper	Format	Weight (%)	Duration (hours)
I (taken day 1)	30 multiple-choice questions	20	¾ hour
II (taken day 1)	Section A: Data-based questions (DBQs)/short answer Section B: Essay (1 chosen from 2)	40 (20 Part A and 20 Part B)	1¼ hours
III (taken day 2)	Optional material (Respond to 1 of the 4 options presented) Section A: Short-response questions based on experimental design, analysis and techniques related to optional material Section B: Short answer and extended response based on optional material	20	1 hour

HL Exam Papers

Paper	Format	Weight (%)	Duration (hours)
I (taken day 1)	40 multiple-choice questions	20	1 hour
II (taken day 1)	Section A: Data-based questions (DBQs)/short answer Section B: Essay (2 chosen from 3)	36 (18 Part A and 18 Part B)	2¼ hours
III (taken day 2)	Optional material (Respond to 1 of the 4 options presented) Section A: Short-response questions based on experimental design, analysis and techniques related to optional material Section B: Short answer and extended response based on optional material	24	1¼ hours

The remaining 20% of your final IB Biology score (1–7) is calculated from your score on the internal assessment (IA). The internal assessment score comes from the laboratory investigation you conducted.

The IB Biology Exam is created specifically for testing knowledge of IB standard-level (SL) or higher-level (HL) material. Standard level does not go into as much detail as HL, and the exams reflect these differences. Standard-level material will be referred to as core material. Papers I and II will focus on topics 1–6 (SL and HL) and on topics 7–11 (HL only). When preparing for the SL exam, review only the SL (core) material. Higher-level students will need to prepare to be tested on all SL (core) and HL material. In addition to the SL (core) and HL material, one optional topic of study must be chosen from those presented for both SL and HL exams. Material in Paper 3 will test your knowledge of material presented in the topic options. Four optional areas of study (A–D) are available for both SL and HL students. Both SL and HL students are required to prepare for and respond to only 1 of the 4 optional area questions. The questions presented for SL students will be based solely on the core topics presented in the optional area of study. The questions presented for HL students will cover the core topics presented in the optional area of study as well as questions based on the additional HL topics. Material from the core topics will not appear on Paper 3, and material from the topic options will not appear on either Paper 1 or Paper 2.

Each question will appear with a number in parentheses next to the question. The number refers to the number of statements, ideas or labels that you must include in your answers in order to achieve the maximum points possible. For example, you may see "Explain diffusion (2)." You must provide 2 correct statements to earn full credit (2 points) for this question. You can include more than 2, but you can receive only a maximum of 2 points. Including extra responses beyond those requested is a good idea in case one of your responses is incorrect. You will not lose points on the IB exam for incorrect responses; you earn points only for correct responses. However, you will not earn points if you make contradictory statements in your response!

COMMAND TERMS

The single most common reason students fail to earn points is they fail to answer the question appropriately or completely. In order to receive full credit for your answers, you must make sure that you understand the detail that IB expects you to cover for each type of question. IB uses specific command terms for Paper 2 and Paper 3 that indicate the depth of understanding you must show in your responses in order to earn points. In order to make sure you understand what is required of each response in Paper 2 and Paper 3, you must understand each command term. There are three objectives, each with their own command. **It is vital that you understand each of these command terms and the detail they require.**

Objective 1: Command Terms and Their Meanings

Term	Meaning
Define	Give the precise meaning of a word, phrase or physical quantity. Give the exact meaning of a term, a statement or a quantity.
Draw	Use pencil lines to represent structures.
Label	Add labels to a diagram. Identify structures by labelling a given diagram.
List	Give a sequence of names of other brief answers with no explanation. Give brief sequences of events or identifying statements without giving specific details.
Measure	Find a value for a quantity. Quantify the value.
State	Give a specific name, value or other brief answer without explanation or calculation. Identify the specific name, value or structure without in depth details or calculations.

Objective 2: Command Terms and Their Meanings

Term	Meaning
Annotate	Add brief notes to a diagram or graph. Add additional information to a given graph, structure or drawing.
Calculate	Find a numerical answer showing the relevant stages in your work (unless instructed not to do so). Showing all stages of work, identify a numerical answer to a given problem.
Describe	Give a detailed account. Include details for given term, process or phenomena.
Distinguish	Give the differences between two or more different items. Describe how two or more items vary.
Estimate	Find an approximate value for an unknown quantity. For an unknown quantity, identify the approximate value or values.
Identify	Find an answer from a given number of possibilities. When given multiple possible answers, select the most appropriate outcome or choice.
Outline	Give a brief account or summary. Include a brief summary or list with details.

Objective 3: Command Terms and Their Meanings

Term	Meaning
Analyse	Interpret data to reach conclusions. Use available data to reach an appropriate conclusion.
Comment	Give a judgment based on a given statement or result of a calculation. Make judgments based on results of calculations or responses.
Compare	Give an account of similarities and differences between two (or more) items, referring to both (all) of them throughout. Refer to all given items with details, including similarities and differences.
Compare and contrast	Give an account of similarities and differences between two or more items or situations, referring to both throughout. Refer to all given items with details, including similarities and differences.
Construct	Represent or develop in graphical or logical form. Use graphs or another appropriate form to represent given data.
Deduce	Reach a conclusion from the information given. Use given information to reach a logical conclusion.
Derive	Manipulate a mathematical relationship(s) to give a new equation or relationship. Using a given mathematical relationship, develop a new equation or appropriate relationship.
Design	Produce a plan, simulation or model. Develop a method, design, model or appropriate simulation.
Determine	Find the only possible answer. Identify the correct answer.
Discuss	Give an account including, where possible, a range of arguments for and against the relative importance of various factors, or comparisons of alternative hypotheses. Argue both for and against the importance of given factors, hypotheses, alternative hypotheses or phenomena with appropriate details.
Evaluate	Assess the implications (strengths and limitations). Identify strengths and limitations with explanations.
Explain	Give a detailed account of causes, reasons or mechanisms. Include specific details for the reasons, causes, methods or processes involved.
Predict	Give an expected result. Using given information, identify an expected outcome or result.
Sketch	Represent by means of a line graph. The axes must be labelled, but the scale is not required. Important features (for example, intercept) must be clearly indicated. Represent given data by means of a line graph with labelled axes and reference points clearly labelled (ex: intercept).
Solve	Obtain an answer using algebraic and/or numerical methods. Using numerical methods, calculate an answer.
Suggest	Propose a hypothesis or other possible answer. Use information to hypothesise or propose a possible answer.

Always underline or highlight the command term when presented with IB questions. Doing so will help you become familiar with the command terms. It will also help you respond with appropriate depth of understanding. For example, in the question presented below, you would highlight or underline the word "explain" and make sure that your response follows the IB expectations outlined for this command term.

> **Explain** how the process of independent assortment leads to genetic variation. (4)

Remember, the number presented in parentheses immediately after each question is the number of correct statements you must include in order to gain full marks. In the example above, you would need to include at least 4 correct statements in your explanation to receive full credit for this question.

STUDY TIPS

In order to study effectively for the IB exam, you should set aside time to review material each day as the exam approaches. Your brain will retain the information better if it is exposed to the material in smaller doses over time than if you try to cram in all your studying in just a few weeks. Make sure to take notes on the material as you review each chapter, and attempt all practice questions. Go back over any material you had trouble comprehending each day as you begin a new review session. This will help to seal the information in your long-term memory.

When responding to multiple-choice questions (Paper 1), make sure to choose the right answer but also understand why the other choices are wrong. Read each question twice, and make sure you understand what the question is asking. In addition, read all the answer choices. One might appear to be the correct response, but a better choice may be available. **Never leave an answer choice blank!** You can only gain points on the IB exam for correct responses. You cannot lose any points for incorrect responses. SL students will have to answer 30 multiple-choice questions in 45 minutes. HL students will have to answer 40 multiple-choice questions in 1 hour. So you will have plenty of time to read each question carefully. Highlighting key words in the question may help you keep focused on finding the best response. Make sure you mark your selection clearly on the answer sheet!

The data-based questions (DBQs) and short-answer sections (Paper 2) will require you to make some predictions or show your knowledge on the subject matter from the SL and additional HL material. Pay attention to all command terms, and keep your answers focused on what IB expects in your response. In addition, pay attention to the number of points assigned to each section. If you are asked to interpret a graph or visual display of data, make sure your answer focuses on only what the graph is displaying. Always discuss the highest and lowest values presented, any outliers in the data presented, any plateaus in the graph(s), and the basic trend(s) shown. Although the question may not directly ask for values, you can often earn points for stating values as you discuss the trends of various figures.

Sometimes you may be asked to make a prediction from the data displayed. Look at the data, and see what can be predicted from it based solely on what is shown. When manipulating data or responding with numerical responses, always include the units for the data shown and show your work for any calculations made. Correct calculation(s) without including the given units can result in failure to earn the point(s).

The last section of Paper 2 will include two (SL) or three (HL) essays. You will choose to respond to one (SL) or two (HL). Pick the essay(s) that you feel you can earn the most points from. The point values will be displayed next to each of the three parts of each essay question. Respond to what is being asked, and give the detail that is required by the command term. Define any terms you use in your essay. Many times, points are awarded for giving definitions of key terms. In addition, make it a habit always to include examples for any biological concept you are explaining. You can earn quality points in the essay section if you can make your essay flow from each part to the next. In other words, part (a) flows to part (b) and part (b) flows to part (c). You can also earn quality points if your essay has good construction. Quality points are extra points awarded beyond the number allotted for each essay. After you finish writing your responses, read the questions again and make sure your responses are focused on the question. Bonus points for quality of construction are awarded only if the examiner does not have to reread your essay for clarification. You can use diagrams to enhance your responses, but they must be fully explained in your response. Always label and, when appropriate, annotate your diagrams. Use complete sentences in your essays.

CORE
(SL AND HL)
MATERIAL

Mathematical Concepts and Statistical Analyses

CENTRAL IDEAS
- Statistical analysis of data
- Variability of data
- Correlation and causation

TERMS
- Causation
- Chi-squared analysis
- Correlation
- Error bars
- Mean
- Normal distribution
- Standard deviation
- *t*-test

INTRODUCTION

The IB Biology Exam includes data analysis sections that require you to understand and carry out mathematical calculations and statistical analyses of data. This section is included as an overview of mathematical concepts that should be reviewed before taking the IB Biology Exam. The mathematical concepts covered will also benefit you as you carry out the laboratory investigations required for your internal assessment (IA).

NOS Connections

- **Repeated measurements and multiple readings of data improve the reliability of the data.**
- **Statistical analysis allows scientists to evaluate the accuracy and precision of data.**
- **"Levels of certainty" can be obtained from data analysis and allow scientists to evaluate the validity of the data collected.**

STATISTICAL ANALYSIS OF DATA

The manipulation of data collected from research allows scientists to analyse the impact of their interventions fully. Statistics allow scientists to support their findings quantitatively (mathematically). Although data may appear to be similar or different in terms of the exact numbers, the two data sets (populations) may or may not be significantly different. Statistics allows scientists to support their conclusions based on statistical analysis of the numerical data collected.

The most basic statistical analysis is calculating the mean ($\overline{x}$) and standard deviation (s) for a set of data. These values give more information about the populations (sets of numbers) than just the data values. In addition, they can be graphed to show a visual interpretation. At the minimum, when conducting experiments for your IA (internal assessment/laboratory investigations), you should use these statistical tests. More advanced statistical analysis involves the use of a *t*-test to compare the means of two populations.

Variability of Data

The variability of data can be shown graphically with the use of **error bars**. Error bars can be used to show the standard deviation or the range of the data. It is important to understand that the word *error* does not refer to mistakes made in data collection. It just gives more information about the variability of the data collected. The graph below uses error bars to show the standard deviation for each set of data displayed. You can easily see that group number 5 had the largest standard deviation. The error bars reach above and below the mean to indicate both the positive and negative spread of the data. Error bars, as shown in Figure 1, can be added to any graph by hand or with the use of spreadsheet software. Displaying the standard deviations graphically with error bars allows for ease of comparison of two or more sets of data.

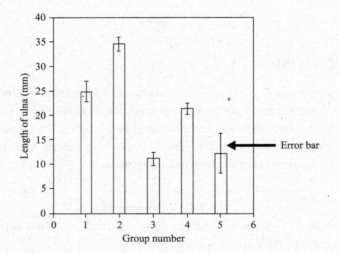

Figure 1. Graph with error bars showing standard deviation

Mean and Standard Deviation

At a minimum, when manipulating data you should calculate the mean and standard deviation. The **mean** ($\overline{x}$) simply gives you the average of all your data combined. To calculate the mean, you add up all your data points and divide the sum of these numbers by the total number of data points.

The mean for the following set of data would be calculated as follows:

$$4 + 6 + 9 + 3 + 8 + 5 + 6 = 41 / 7 = 5.857 \ldots = 5.9 \text{ (rounded up)}$$

The **standard deviation** (*s*) indicates how far the data are spread from the mean—how far the data deviate from the average. A large standard deviation indicates that the data points show a wide range of variation from the mean. A small standard deviation indicates that the data points show little variation from the mean. Biological data often show great variability. Data that show high variability (such as large standard deviations) can still be good data. In all cases, 68% of the data points will fall within 1 standard deviation above or below (+ or −) the mean and 95% will fall within 2 standard deviations. Figure 2 represents a standard bell curve with standard deviations identified. Figure 3 shows sizes of bell curves in relation to standard deviation sizes.

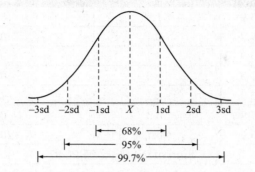

Figure 2. Standard bell curve showing standard deviations

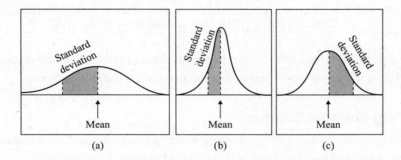

Figure 3. Bell curves displaying (a) large standard deviation, (b) small standard deviation and (c) normal standard deviation

Although you are not required to know the formula for calculating the standard deviation (s), you are expected to understand the formula as shown in Figure 4.

$$s = \sqrt{\frac{\sum(x - \bar{x})^2}{(n-1)}}$$

where:
x = each score
$\bar{x}$ = the mean or average
n = the number of values
Σ means we sum across the values

Figure 4. Standard deviation formula

Table 1 shows how to calculate the standard deviation for the set of numbers 4, 6, 9, 3, 8, 5, 6. Note that n is the number of data points in the set. In this case, $n = 7$.

Table 1: Calculated Standard Deviation

Number (x)	Mean ($\bar{x}$)	Deviation (Difference) from Mean ($\bar{x} - x$)	Squared Deviation (Gets Rid of Negative Values) ($\bar{x} - x)^2$
4.0	5.9	1.9	3.61
6.0	5.9	−0.1	0.01
9.0	5.9	−3.1	9.61
3.0	5.9	2.9	8.41
8.0	5.9	−2.1	4.41
5.0	5.9	0.9	0.81
6.0	5.9	−0.1	0.01
Sum of (Σ) $(\bar{x} - x)^2 = 26.87$			
$n - 1 = 7 - 1 = 6$			
$\sum \dfrac{(\bar{x} - x)^2}{(n-1)} = \dfrac{26.87}{6} = 4.48$			
Square root of 4.48 = (s) standard deviation = 2.11			

You can practice this example using a graphing calculator or a spreadsheet program to help you understand the process and make sure you can obtain the same answer.

Comparing Multiple Samples

When comparing two or more samples, using the mean and standard deviation of each population allows easy comparison. Look at the two populations shown below:

Population A: 1, 2, 8, 10, 12, 3

Mean ($\bar{x}$): 6
Standard deviation (s): 4.60

Population B: 3, 3, 8, 7, 10, 5

Mean ($\bar{x}$): 6
Standard deviation (s): 2.83

If only the means of the two populations were graphed, these two populations would look the same. Calculating the standard deviations (with the use of error bars) allows us to see easily that data set B does not deviate as much from the mean as does data set A. (See Figure 5.)

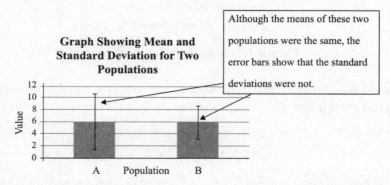

Although the means of these two populations were the same, the error bars show that the standard deviations were not.

Figure 5. Graph showing error bars representing standard deviations

CORE MATERIAL

The *t*-Test

The means of two populations can be further statistically compared with the use of a **t-test**. The *t*-test can be used when two conditions are met. First, the data must show **normal distribution**. In other words, most of the values must fall close the mean with only a few outliers. When graphed, the data form a bell-shaped distribution. Second, the sample size for each population must be at least 10. When performing a *t*-test, a null hypothesis must be stated. The **null hypothesis** states that there will be no statistically significant difference between the means of the two populations. The results of the *t*-test allow you to accept or reject the null hypothesis based on the probability of the two populations being similar. A two-tailed *t*-test can be calculated with a graphing calculator or with the use of spreadsheet software.

IB does not require you to calculate a *t*-test, but you must know when it is appropriate to use one. In addition, you may be required to use a *t*-table. To use a *t*-table, you need to understand how to determine the degrees of freedom for your population set. In statistics, the degrees of freedom represent the number of values in a set of data that can vary. A very simplistic way to try to understand the complex concept of degrees of freedom is to think of the mean as a known entity with the rest of the values in a set as varying in relation to the mean. Since there are 2 means in a *t*-test comparison of 2 populations, the degrees of freedom are calculated as the total number of values minus 2 (the stated means from which all the rest of the values will vary). The **degrees of freedom** (df) for a *t*-test are determined by adding the number of data points (n) in your sample sets and subtracting 2: ($n - 2$). Once you know the degrees of freedom, you find the value on the *t*-table and cross over the table to the 0.05 (5%) range of acceptance (or probability (p) value). This indicates that there is only a 5% (0.05) or lower probability that the difference in the means of the populations is simply due to random variation. In other words, you are 95 percent confident that the difference between the means of the two populations is due to the treatment you manipulated (the independent variable) and not due to random chance.

> The *p*-value refers to how likely differences between data sets are simply due to chance.

Once you determine the degrees of freedom, you need to find where this value surpasses the 0.05 probability column. (See Table 2) The number that intersects the degrees of freedom row and the 0.05 probability column is the critical value.

- If the critical value is smaller than your *t*-value or if the probability is greater than 0.05, you accept the null hypothesis.
- If the *t*-value is larger than your critical value or the probability is less than 0.05, you reject the null hypothesis. You are 95% confident your results are directly related to the manipulation of your independent variable.

Some statistical software programs provide only the *p*-value (p = probability of the event being significant) for a *t*-test. When given only the *p*-value to interpret the results, you reject the null hypothesis if the *p*-value is less than 0.05 (5%) and accept the null hypothesis if the *p*-value is greater than 0.05 (5%). By rejecting the null hypothesis, you are accepting your alternative hypothesis. Accepting the alternative hypothesis means that you are 95% confident that the data you collected are being influenced by the independent variable in your investigation.

> **Null hypothesis**: States that there will be no significant difference between the two sets of data collected (any differences are simply due to chance).
>
> **Alternative hypothesis**: States that there will be a significant difference between the two sets of data collected (any differences are not simply due to chance).

The sample sets below illustrate how a *t*-test is carried out. Figure 6 shows the *t*-test formula.

REMEMBER

In order to carry out a *t*-test, the data must:

- Exhibit a normal distribution (exhibit a bell-shaped spread; most values fall near the mean).
- Have a sample size of at least 10 (for each population).

$$t = \frac{\bar{X}_1 - \bar{X}_2}{S_{\bar{X}_1 - \bar{X}_2}}$$

Figure 6. *t*-test formula.

➥ Sample Worked *t*-Test

Data for sample *t*-test:
Population A: 5, 7, 6, 10, 2, 5, 3, 9, 10, 8

Mean ($\bar{x}$): 6.5
Standard Deviation: 2.3
Population B: 2, 5, 7, 8, 6, 4, 5, 3, 2, 1

Mean: ($\bar{x}$): 4.3
Standard Deviation: 2.8
Calculate the *t*-value with use of a calculator or spreadsheet program.

Calculated *t*-value: 1.92

Determine the degrees of freedom (total data points in both populations: $n-2$).

Degrees of freedom: 18 (20 − 2)

Determine the critical value (use degrees of freedom for the sample and 0.05 probability).

Critical value: 2.101 (See Table 2)

Accept or reject the null hypothesis: Accept the null if your *t*-value is less than the critical value, and reject the null hypothesis if your *t*-value is more than the critical value. A sample *t*-table is shown in Table 2.

For the sample set presented, the *t*-value is 1.92 and the critical value is 2.101. The null hypothesis is accepted. This means that although there is a difference between the means of the two populations (sets of data), that difference is not statistically significant.

The *p*-value for this set of data calculates as $p = 0.07$. Since this value is greater than 0.05, the null hypothesis is accepted.

Table 2: Sample *t*-Table

Degrees of Freedom	Probability			
	0.2	0.1	0.05	0.02
1	3.078	6.314	12.706	31.821
2	1.886	2.920	4.303	6.985
3	1.638	2.353	3.182	4.541
4	1.533	2.132	2.776	3.747
5	1.476	2.015	2.571	3.365
6	1.440	1.943	2.447	3.143
7	1.415	1.895	2.385	2.998
8	1.397	1.860	2.308	2.896
9	1.383	1.833	2.262	2.821
10	1.372	1.812	2.228	2.764
11	1.363	1.796	2.201	2.718
12	1.356	1.782	2.179	2.681
13	1.350	1.771	2.180	2.650
14	1.345	1.761	2.145	2.624
15	1.341	1.753	2.131	2.602
16	1.337	1.746	2.120	2.583
17	1.333	1.740	2.110	2.567
18	1.330	1.734	2.101	2.552
19	1.328	1.729	2.093	2.539
20	1.325	1.725	2.086	2.528

$(n-2) = (20-2) = 18 \rightarrow$ (row 18)

Tables vary in length (number of degrees of freedom) and in the number of probability levels that they include. Remember that you will always use a 0.05 acceptance level (95% confidence level) that the probability of your results are due to the condition investigated (independent variable). The *t*-test compares the means of two populations and is looking for the degree of overlap between the two sets of data (populations). (Refer to Figure 7.)

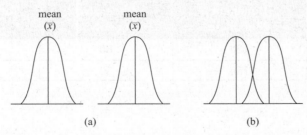

mean $(\bar{x})$ mean $(\bar{x})$

(a)   (b)

Figure 7. Data showing normal distribution (a) with no overlap between the means of the two populations (statistically significant difference) and (b) with a large overlap between the means of the two populations (no significant statistical difference)

Chi-Squared (χ^2) Test

Chi-squared analysis is a statistical tool used to determine how far data you observe (collect) deviates from what you expect to observe (collect). In other words, how far are the observed values from the expected values? A null hypothesis is stated in order to assume that there will be no difference between the observed and expected.

➡ Sample Chi-Squared Problem

Is there a significant difference between the number of expected dominant and recessive phenotypes observed in the results of a monohybrid cross and the number of observed dominant and recessive phenotypes observed in the results of a monohybrid cross?

Null hypothesis: There will be no significant difference between the expected phenotypes and the observed phenotypes resulting from a monohybrid cross.

1. Observed collected data:

Dominant Phenotype	Recessive Phenotype	Total Offspring
120	80	200

2. Calculate the expected value. Monohybrid crosses have an expected phenotypic ratio of 3:1 (3 dominant for every 1 recessive phenotype). For this problem, the expected value would be 75% dominant phenotype and 25% recessive phenotype (3:1 phenotypic ratio).

   - 200 (total offspring) $\times$ 0.75 (75% expected dominant) = 150 expected dominant phenotypes
   - 200 (total offspring) $\times$ 0.25 (25% expected recessive) = 50 expected recessive phenotypes

3. Compare the observed values with the expected values with use of the chi-squared analysis. This can be done on a calculator or with computer software. You will not be required to carry out a chi-squared analysis by hand. However, you may be required to understand how to interpret the results of a chi-squared analysis.

	Dominant Phenotype	Recessive Phenotype	Total Offspring
Observed	120	80	200
Expected	150	50	200

4. After the chi-squared value is calculated, a chi-squared table is used to determine the critical value. For a chi-squared analysis, the degrees of freedom is calculated as $n - 1$, where n represents the total number of levels (categories).

$$n = 2 \text{ (dominant or recessive)} - 1 = 1 \text{ degree of freedom}$$

$$\chi^2 = \sum \frac{(o-e)^2}{e}$$

$$\text{Chi-squared} = \frac{\text{(Sum of (Observed – Expected) Squared)}}{\text{Expected}}$$

5. The calculated chi-squared value for this set of data is 24. Use the chi-squared table and the degrees of freedom (df) to determine the critical value for your set of data. Just like with the t-test, a 5% (0.05) level of probability is always used as our level of acceptance.

Critical value: 3.84 (1 df and 0.05 probability level)

6. Determine if there is a significant difference between the observed and expected values (accept or reject the null hypothesis).

- If the critical value for your set of data is smaller than the chi-squared value, the null hypothesis is rejected. This means that a statistically significant difference between the observed and expected values exists (alternate hypothesis is accepted).
- If the critical value for your set of data is larger than the chi-squared value, the null hypothesis is accepted. This means that a statistically significant difference between the observed and expected values does not exist (alternate hypothesis is rejected).

Sample Chi-Squared Table

Df	0.20	0.10	0.05	0.02
1	1.64	2.71	3.84	5.41
2	3.22	4.61	5.99	7.82
3	4.64	6.25	7.81	9.84
4	5.99	7.78	9.49	11.67

7. Since the critical value is less than the chi-squared value for this set of data, the null hypothesis is rejected (3.84 is much less than 24). There is a significant difference between the expected and observed values of the data collected.

Correlation and Causation

The existence of a **correlation** between two variables does not imply **causation** (causal relationship). An excellent example of this is the fact that we can see a correlation between ice cream sales at the boardwalk and the number of shark attacks. Although we see a correlation between shark attacks and ice cream sales, ice cream sales do not cause (causation) sharks to attack. The correlation exists simply due to the fact that more ice cream is sold in the summer months when more people are in the ocean subjected to shark attack (see Figure 8). In order to exhibit causation, a variable (such as ice cream sales) would have to have a direct impact on a correlating event (such as shark attacks). This reminds us that we need to be careful when interpreting data!

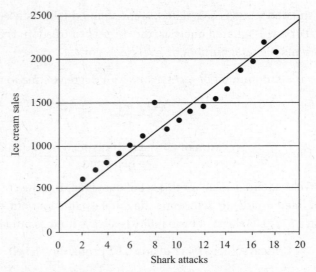

Figure 8. Graph showing correlation between shark attacks and ice cream sales

Without a controlled experiment, causation cannot be determined simply based on correlation.

Percent Change

Data analysis questions often involve calculating percent change. Percent change is calculated as follows:

$$\text{Percent change} = \frac{\text{New value} - \text{Original value}}{\text{Original value}} \times 100$$

STEP 1 The original value is subtracted from the new value in order to calculate the change.

STEP 2 Calculated change is divided by the original value. This will result in a decimal value.

STEP 3 Multiply the result by 100 to display the percent change.

➥ Example

You have 10 amoebas in a petri dish. Three days later, you have 25. What is the percent change in the number of amoebas in the petri dish?

STEP 1 Subtract 10 (original value) from 25 (new value).

$$25 - 10 = 15$$

STEP 2 Divide 15 (change in value) by the original value (10).

$$15/10 = 1.5$$

STEP 3 Multiply 1.5 by 100 to find the percent change.

$$1.5 \times 100 = 150\% \text{ change}$$

1. Which of the following must data exhibit in order to carry out a *t*-test?

   I. A normal distribution
   II. A minimal sample size of 10
   III. At least three populations of data

   A. I only
   B. I and II only
   C. I and III only
   D. I, II and III

2. For data that show normal distribution, what percentage of data will fall outside of 2 standard deviations?

   A. 5%
   B. 68%
   C. 95%
   D. 100%

3. What can be used to show the range of data when constructing a graph?

   A. Means
   B. Legends
   C. Error bars
   D. Titles

4. Data for a sample population that exhibits a large standard deviation:

   A. Must have few samples in the population
   B. Has sample data points that are all close to the mean value
   C. Has sample data points that are spread out far from the mean
   D. Does not exhibit variation

5. The level of significance (probability value, *p*) used for the *t*-test is always:

   A. 0.01
   B. 0.05
   C. 0.10
   D. 1.00

6. Finding the sum of a set of values and dividing it by the number of values is used to calculate the:

   A. Mean
   B. Standard deviation
   C. Range
   D. Mode

7. A scientist measures the average growth of a plant species and finds the average to be 10 mm. Ten days later, the scientist again measures the average growth of a plant species and finds the average to be 16 mm. The calculated percent change in growth of the plant species between the first and second reading is:

A. 0.6%

B. 6%

C. 60%

D. 160%

8. The graph below shows the average number of various bird species seen at seven different locations in Delaware.

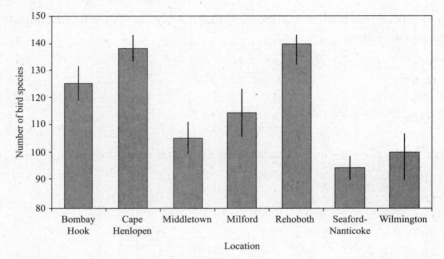

a. Deduce the location where the number of each species seen showed the least variability. (2)
b. Suggest reasons for Cape Henlopen having the greatest number of bird species. (2)
c. Calculate the difference between the number of bird species seen in Wilmington and the number of bird species seen in Rehoboth. (2)

9. The table below shows the heights of plants collected at two different locations.

a. Calculate the mean plant height for the plants sampled at location B. (1)

Location A (cm)	Location B (cm)
10	5
12	7
14	6
11	4
9	10
12	4

b. State the percentage of values for the sample sets that would fall within one standard deviation. (1)
c. Explain why a *t*-test would or would not be appropriate to analyse the data. (3)

ANSWERS EXPLAINED

1. **(B) (I and II only)** In order to perform a *t*-test, the data must exhibit normal distribution and have a sample size of at least 10. A *t*-test is used to compare the means of two populations, which rules out III (at least three populations of data).

2. **(A) (5%)** As a rule, 68% of data always falls within 1 standard deviation and 95% always falls within 2 standard deviations. This leaves 5% outside of the 2 standard deviations.

3. **(C) (Error bars)** Error bars are used to show the standard deviation or range of data for a set of numbers.

4. **(C) (Has sample data points that are spread out far from the mean.)** Remember that the standard deviation indicates how far the data in a sample population are spread from the mean. A small standard deviation indicates a population of values close to the mean. A large standard deviation indicates a population of values spread out farther from the mean.

5. **(B) (0.05)** As a rule, 5% is always the value used for the probability of an occurrence being due to random chance. (This indicates a 95% confidence level that our findings are not due to random chance)

6. **(A) (Mean)** Finding the sum of a set of values and dividing it by the number of values is how to calculate the mean.

7. **(C) (60%)** 16 (new value) – 10 (old value)/10 (new value) = 0.6 × 100 = 60%

8. (a)

   - Seaford-Nanticoke shows the least variability; note that Seaford-Nanticoke and Cape Henlopen are similar.
   - Error bars show variability in data.
   - The error bar is the shortest, indicating the least variability in the data.
   - All other error bars are larger than the error bar for Rehoboth.

   (b)

   - More diverse habitats are found in Cape Henlopen than in other regions.
   - Cape Henlopen has a greater source of food, allowing for more species diversity.
   - Random differences showing Cape Henlopen with the greatest diversity may have occurred only at the time of sampling.
   - There may be fewer birds in each population of each species at Cape Henlopen, allowing more species to exist.
   - Tourists may encourage more species at Cape Henlopen by supplying food.

   (c) 40 (+/– 10) species more at Rehoboth versus Wilmington: 140 – 100 = 40 (+/– 10) species more
   Work must be shown to earn both points.

9. (a) 5 + 7 + 6 + 4 + 10 + 4 = 36/6 = 6 cm
   Work must be shown and units must be included in order to earn the point.

   (b) 68%

   (c) A point is earned for any of the following:

   - A *t*-test is used to compare the means of two populations.
   - The data exhibit normal distribution/a requirement for the *t*-test.
   - There must be at least 10 data points for each population in order to perform a *t*-test and there are only 6/there are not enough data points.
   - A *t*-test would not be appropriate/would be inappropriate.

Cell Biology

CENTRAL IDEAS
- Cell division must be controlled
- Cellular and subcellular structures and functions
- Cell size and magnification calculations
- Membrane fluidity and dynamic processes
- Methods of transport (active and passive)
- Stem cells and therapeutic use
- Cellular evolution

TERMS

- Amphipathic
- Asexual Reproduction
- Binary fission
- Cell
- Cyclins
- Cytokinesis
- Diffusion

- Eukaryotic cell
- Homeostasis
- Magnification
- Metabolism
- Mitosis
- Multicellular
- Organelle

- Osmosis
- Prokaryotic cell
- Sexual Reproduction
- Stem Cell
- Unicellular

INTRODUCTION

Understanding how cells are organized in order to perform cellular functions is vital in comprehending how all cells carry out all the processes necessary for life. This understanding applies to both unicellular organisms (prokaryotes and unicellular eukaryotes) and multicellular organisms. Cellular structure is related to the function of the cell and the tissue it supports. Each organelle carries out a vital function for the cell just as each organ carries out vital functions for the organism. Hence, the organs of an organism are analogous to the organelles of a cell. Cells must be able to produce energy, store genetic information, build and break down molecules and reproduce. In addition, cells must be able to maintain an internal environment that is separate and different from the external environment they live in. The subcellular structures (organelles) and the cell membrane allow the cell to do this. The process of mitosis allows the cell to divide and produce new cellular organisms (unicellular) or new tissues (multicellular).

> **REMEMBER**
>
> Structure always relates to function. If the structure is altered too much, the function cannot be carried out.

NOS Connections

- Observation and experimentation are important processes that provide scientific evidence.
- Technology has allowed scientists to make new discoveries and test existing theories (microscope, electron microscope).
- Models of scientific understanding help scientists explain unobservable processes (atomic structure, cell structure, cellular processes).
- Ethical issues arise with scientific discoveries (stem cell research).
- Scientists play a major role in educating the public about scientific progress (stem cell research, cancer and the environment).
- Scepticism and collaboration among scientists leads to new discoveries and scientific theories (cell membrane structure).
- Causation and correlation must be scientifically supported (cancer and the environment).

1.1 INTRODUCTION TO CELLS

- **All living organisms are composed of cells**. Every living organism identified has been observed to be composed of one or more cells.
- **Cells are the basic and smallest unit of life for all living organisms**. Although cells organize into tissues and organs in order to provide structure to organisms, each cell has its own independent functions as well.
- **Cells can come only from preexisting cells**. The process of mitosis or binary fission (in prokaryotes) is required in order to produce new body cells. Meiosis is required for the creation of reproductive cells. Both processes require the division of previously existing cells.

Evidence for the Cell Theory

- **All living organisms are composed of cells**. Robert Hooke coined the term *cell* when he observed cork under a microscope. Since the cell wall structures of the cork looked like the rooms in a monastery that are called cells, the monk decided to use the term to describe what he was viewing with his rudimentary microscope. Since Hooke's discovery, every living organism observed has been composed of one or more cells.
- **Cells are the basic and smallest unit of life for all living organisms.** This is evident from the observation of living organisms. Both unicellular and multicellular organisms are composed of cells. Although cells come in many different shapes and can exist as single units (unicellular) or as multicellular structures, the smallest unit of an organism that can carry out life functions is the individual cell.
- **Cells can come only from preexisting cells.** The creation of new cells from preexisting cells can easily be observed with the use of a microscope. New cells have never been observed being produced without the process of existing cells dividing into new cells.

Unicellular and Multicellular Organisms

Cells can exist as single living organisms (unicellular) or as multicellular structures. Each individual unit of an organism is a single cell—the building block for all living organisms. Unicellular organisms carry out all the life functions that multicellular organisms carry out. Life processes essential to all living organisms include the ability to reproduce, respond to stimuli (the environment), maintain homeostasis, carry out metabolic processes and grow. Table 1.1 outlines the importance of life processes.

REMEMBER

You cannot go any smaller than an individual cell and maintain the emergent properties of life.

Table 1.1: The Importance of Life Processes

Life Process	Importance
Reproduction	In order for a population to continue to exist, it must be able to produce more fertile offspring (reproduce).
Response	Organisms must be able to respond to the environment in order to make needed internal adjustments and to survive changing external conditions.
Homeostasis	Cells must be able to maintain internal conditions within a narrow range in order to maintain the constant stable conditions required for a cell to continue to survive (e.g., temperature, pH, solute concentration).
Nutrition	Cells must take in energy in the form of nutrients obtained from the environment in order to carry out cellular functions.
Metabolism	Metabolic processes (the sum of all chemical reactions within a cell), such as building (anabolic) and breaking down molecules (catabolic), are necessary for organisms to gain energy and assemble the molecules needed to maintain their life functions.
Growth	Organisms must be able to increase in size and differentiate tissues by creating new cells in order to reach a level of growth and cellular organization optimal for survival and reproduction.

Relative Sizes Using SI Units

The use of technology has allowed us to view cellular and subcellular structures that are far too small to be seen with the naked eye. It is important to understand the relative sizes of various structures as we explore their design. Magnification is simply an apparent increase in size; the object remains the same size, and the image viewed enlarges. When you look and draw what you see under the microscope, you must be able to label your drawing with the appropriate scale bar. Scale bars allow one to determine quickly the actual size of the object, regardless of how large it is drawn. Since these objects are all three-dimensional and may vary in their exact structure, all the sizes listed in Table 1.2 are relative averages. You must be able to compare the relative sizes of the objects listed below. You must also be able to use scale bars to calculate magnification and to use magnification to calculate actual size. Metric units are the appropriate SI (International System) units for use in all scientific calculations.

Table 1.2: Relative Metric Sizes

Object	Relative Size	Example
Molecule	1 nm	Glucose
Cell membrane thickness	10 nm	Cell membrane
Virus	100 nm	HIV
Bacteria (prokaryotic cell)	1 μm	*E. coli*
Organelle	Up to 10 μm	Mitochondrion
Cell (eukaryotic)	Up to 100 μm	Epithelial cell (skin cell)

MEMORY TIP

MCV BOC 1, 10, 100

MCV BOC indicates the first letter of each object from smallest to largest.

1, 10, 100 shows how each object increases in size from nanometres (nm) to micrometres (μm). Refer to Table 1.3.

Metric Reminder

1 metre (m) = 100 cm

1 centimetre (cm) = 10 mm

1 millimetre (mm) = 1000 μm

1 micrometre (μm) = 1000 nm

*You must remember these conversions in order to carry out magnification and size calculations.

Table 1.3: MCV BOC Mnemonic

M molecule	C cell membrane	V virus	B bacterium	O organelle	C cell (eukaryotic)
1 nm	10 nm	100 nm	1 μm	10 μm	100 μm

On the IB Biology Exam, you may be asked to put the objects listed in Tables 1.2 and 1.3 in order of relative sizes, identify their actual size, or compare their relative sizes to each other. The mnemonic shown in the memory tip and in Table 1.3 will help you to memorize these for size calculations as well. For example, if you are asked to calculate the size of a virus from a drawn diagram, you will know you are correct if you get close to the value of 100 nm.

CALCULATING MAGNIFICATION AND THE ACTUAL SIZE OF SPECIMENS

Calculating linear **magnification** and the actual size of a drawing or magnified image of an object can be obtained using the information presented in a scale bar. Scale bars are added to diagrams or images in order to show the actual size of the magnified structures. An example of a scale bar is shown in Figure 1.1.

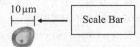

Figure 1.1. Example of a scale bar

Magnification will be stated (e.g., 100X) or a scale bar will be included to indicate the actual size of the object. The methods used to calculate the size of a structure based on the magnification or the magnification based on the scale bar are outlined in Table 1.4.

Table 1.4: Size Calculations

Magnification to Actual Size	
Actual size = Size of object ÷ Magnification	When given the magnification, simply measure the object or given dimension in mm and convert that measurement to the unit requested. (The requested unit is usually μm. So multiply the measured dimension by 1000 for this conversion). Then divide that number by the magnification.
Given Size (Scale Bar) to Magnification	
Magnification = Actual measurement of image ÷ Size of scale bar	When given a scale bar, simply measure the scale bar in real measurements (mm) and then divide the real measurement by the measurement shown on the scale bar. (Make sure to convert the scale bar to units you measured in; units must be the same before dividing.) If the measurement is given in μm, simply multiply by 1000.

FEATURED QUESTION

November 2004, Question #1

If a red blood cell has a diameter of 8 micrometers and a student shows it with a diameter of 40 millimeters in a drawing, what is the magnification of the drawing?

A. X 0.0002
B. X 0.2
C. X 5
D. X 5,000

(Answer is on page 553.)

Sample Calculations—Scale Bar to Magnification

1. Determine the actual size of the scale bar—measure the size bar.
2. Convert the measurement to same units shown on the scale bar.
3. Divide the actual size of the bar by the size shown on the scale bar.

> Scale bar = 10 μm
> Actual size measures 5mm
> 5mm × 1,000 = 5,000 μm (actual size)
> 5,000 μm/10 μm = 500X

Sample Calculations—Magnification to Actual Size

1. Measure the cell.
2. Convert the measurement to the desired units
3. Divide the measurement by the given magnification. (See Figure 1.2.)

500X

Figure 1.2. Image with given magnification

> Magnification = 500X
> Size of cell = 5 mm × 1000 = 5000 μm
> 5000 μm/500X = 10 μm (actual size)

Surface Area to Volume Ratio

Cells are limited in size in order to maintain a high surface area to volume ratio. Although both volume and surface area increase when cells become larger, the surface area increases at a much slower rate. The surface area, or cell membrane, in cells is the site of exchange of material between the intercellular and extracellular environment. If a cell becomes too large, it will not have enough surface area as compared with its volume to get rid of waste and take in nutrients at rates necessary to maintain **homeostasis** in order to survive. If the cell were too large, cellular waste would take too long to reach the cell membrane and be expelled. Likewise, nutrients entering into the cell would have to travel (diffuse) too far to reach the centre of the cell at a rate necessary for all parts of the cell to receive nutrition and survive. This greatly limits the size of **unicellular** organisms. **Multicellular** organisms maintain a large volume while increasing their surface area. They can do this because they consist of many small cells, each with a large surface area to volume ratio. This decreases the overall volume relative to an increased surface area as shown in Figure 1.3.

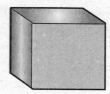

High Surface Area to
Low Volume
(Surface-to-volume
ratio = small)

High Volume to
Low Surface Area
(Surface-to-volume
ratio = large)

	1 mm cube	5 mm cube
Surface area:	6 sides $\times$ 1^2 = 6 mm^2	6 sides $\times$ 5^2 = 150 mm^2
Volume:	1^3 = 1 mm^3	5^3 = 125 mm^3
Surface area to volume:	6:1	1.2 : 1

Figure 1.3. Surface area to volume ratios show (a) high volume to low surface area and (b) high surface area to low volume.

Emergent Properties

Multicellular organisms exhibit emergent properties. In multicellular organisms, cells work together and interact to carry out complex life functions that individual cells would be unable to perform. Cells can combine to form tissues, and tissues can combine to form organs. Organs form organ systems that function together to perform all the required life functions for body systems. These parts working together as a whole are far more productive than they would be as individual components.

BE AWARE

In Figure 1.3, note the distance material has to travel to enter and exit the larger cube. If cells were this large, they could not get rid of or take in materials at a rate fast enough to support cellular processes.

The whole is greater than the sum of its parts. The individual cells that make up an organism each exist independently. However, together they accomplish work that could not be carried out without the cooperation of many cells working together.

The Organization of an Organism

An organism can be described based on the increasing complexity of its organization.

Cells (basic unit of life) $\Rightarrow$

Tissue (group of cells working together to perform a specific function) $\Rightarrow$

Organ (group of tissues working together to perform a specific function) $\Rightarrow$

Organ system (group of organs working together to perform a specific function) $\Rightarrow$

Organism

Multicellular Organisms and Differentiation

Although all cells arise from existing cells, various cells within a multicellular organism can look different from each of the cells from which they came. After beginning life as a single-celled zygote (the cell resulting from the union of sperm and egg during fertilisation), the zygote undergoes cell division, resulting in a multicellular ball of cells that are all identical. Once enough cells are produced from the zygote, the existing cells begin to differentiate—express different genes from the entire collection of genes they possess (genome). By suppressing or expressing different genes (expressing some genes and not others), the resulting cells take on different structures and perform different functions.

The cells in your skin are expressing different genes than the cells in your pancreas even though both cells possess all the genes needed to create all the components that produce the entire organism. For example, pancreatic cells express (turn on) the gene for insulin production while skin cells suppress (turn off) the same gene since skin cells do not produce insulin. This selectivity in which genes are expressed allows cells to carry out specific functions and have specific structures and overall shapes. Figure 1.4 shows how cells differentiate to perform specific functions.

Differential Gene Expression = Different Structure

Genes expressed = Structure = Function

- **Example:** Pancreatic cells express the gene for insulin production.

    Structure: Pancreatic cells possess high level of rough endoplasmic reticulum.

    Function: Pancreatic cells produce and export high amounts of insulin.

- **Example:** Muscle cells express genes for the production of many mitochondria.

    Structure: Muscle cells contain large numbers of mitochondria.

    Function: Muscle cells produce high levels of ATP for use in muscle contraction.

Both cell types possess the same genes. However, they each express different genes to achieve the structure that will allow them to carry out their specific functions.

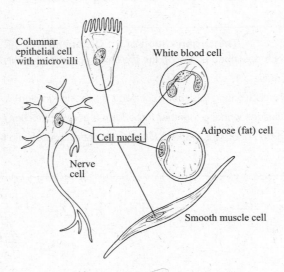

Figure 1.4. Cell specialization

Stem Cells

Undifferentiated cells are referred to as stem cells. **Stem cells** maintain their capacity to divide along different pathways, while differentiated cells can divide along only one pathway. This means that stem cells can become one of many different cell types, while cells that have differentiated have reached their final structure. Various stem cells exhibit different capacities for differentiation. Some stem cells have the capacity to differentiate into an entire organism (express every cell type needed to create a complete organism), while others are limited in their capacity to differentiate. For example, embryonic stem cells have the capacity to produce an entire organism. In contrast, bone marrow stem cells can only differentiate into several blood cell types (red blood cells, white blood cells or platelets). Since embryonic stem cells have the capacity to become complete embryos, they are the subject of great debate. Is it ethical to destroy an early embryo (embryonic stem cell) in order to help discover ways to cure diseases and explore processes that can improve the lives of others? Arguments exist for both sides of this highly debated topic. Figure 1.5 shows stem cell differentiation.

Stem Cells

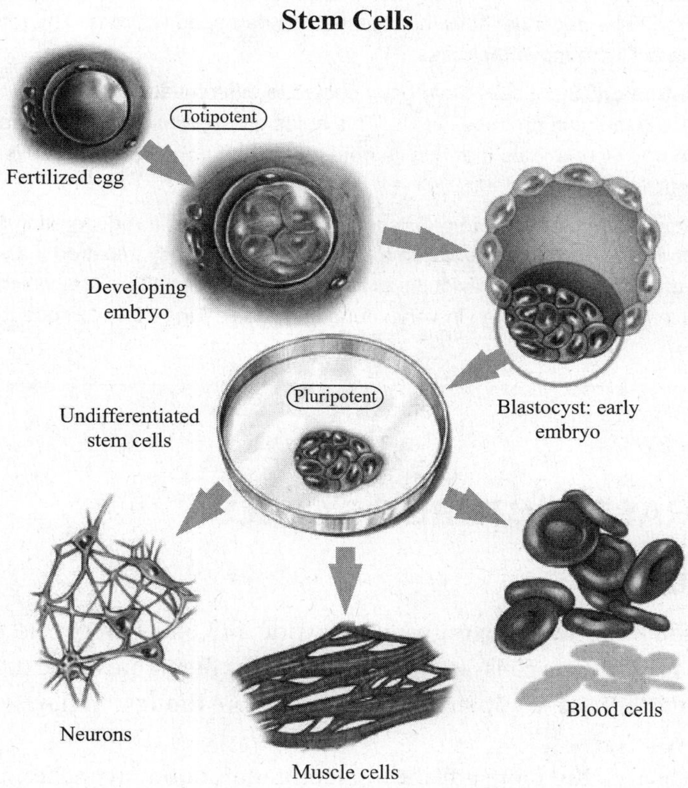

Figure 1.5. Stem cell differentiation

THERAPEUTIC USE OF STEM CELLS

Embryonic stem cells hold all the information on the process of cell differentiation and organism development. Scientists work with stem cells in order to make new discoveries about how cells undergo differentiation. Their discoveries could lead to important information about how cells differentiate and regulate growth processes. Discoveries related to these processes could help in the diagnosis and treatment of many diseases. For example, if scientists gain new information concerning how cells regulate cellular division, they could possibly gain insight into new methods to regulate uncontrolled cellular division (cancer). Currently, stem cells hold great promise for therapeutic uses.

Examples of Therapeutic Uses of Stem Cells

Neuron regeneration: Stem cells have been shown to differentiate into neural tissue and replace damaged neurons. The use of stem cells has shown promise in reversing the damage caused by spinal cord injuries. This holds great promise for individuals affected by brain and spinal cord injuries. Laboratory rats that have had their spinal cords severed have had stem cells injected into the damaged neurons. The rats then regained the ability to move their legs.

Insulin production: Stem cells have been shown to differentiate successfully into pancreatic cells that can produce insulin. This holds great promise for individuals suffering from type I diabetes that results from the loss of pancreatic beta cells that produce insulin.

Treating Stargardt disease: Stem cells have shown promise in regenerating photoreceptors in the eyes and restoring vision to some blind or visually impaired individuals. This holds great promise for individuals suffering from Stargardt disease, which causes a progressive total loss of vision in very young individuals (ages 6 to 12) due to macular degeneration.

1.2 ULTRASTRUCTURE OF CELLS

Prokaryotic Cells

Cells can be divided into two subgroups: prokaryotic and eukaryotic. Due to the less complex structure of **prokaryotic cells**, scientists believe that they were the first type of cells that existed on early Earth. More complex **eukaryotic cells** are thought to have evolved from the primitive prokaryotic cells.

Prokaryotic cells lack the more advanced membrane-bound organelles found in eukaryotic cells. One main distinguishing characteristic is that prokaryotic cells do not have a true nucleus (DNA surrounded by a nuclear membrane), while eukaryotic cells do. The region where the DNA is located in the cytoplasm of the prokaryotic cell is referred to as the nucleoid region. Prokaryotic DNA is not associated with proteins (histones). Since prokaryotic DNA lacks an association with proteins, it is referred to as having naked DNA. Prokaryotic cells are always found as unicellular organisms, while eukaryotic cells can be either unicellular or multicellular. The diagram in Figure 1.6 shows the basic structure of a prokaryotic cell.

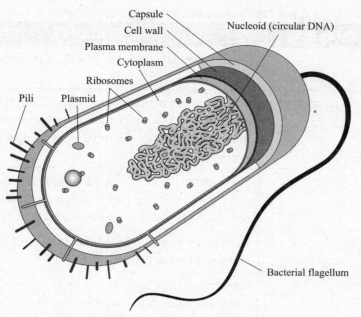

Capsule
Cell wall
Plasma membrane
Cytoplasm
Ribosomes
Pili
Plasmid
Nucleoid (circular DNA)
Bacterial flagellum

Figure 1.6. Diagram of basic prokaryotic cell
(adapted from a work by Mariana Ruiz Villarreal)

PROKARYOTIC CELLULAR STRUCTURE

You may be asked to draw, label, identify and annotate a diagram of the basic structure of a prokaryotic cell. In addition, you may also be asked to identify structures of a prokaryotic cell from an electron micrograph as shown in Figure 1.7. (This is an image taken through an electron microscope. It shows great magnification.)

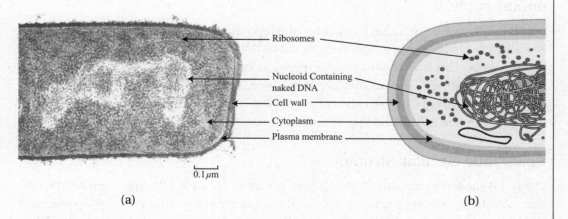

Ribosomes
Nucleoid Containing naked DNA
Cell wall
Cytoplasm
Plasma membrane
0.1 µm
(a)
(b)

Figure 1.7. (a) Electron micrograph of *E. coli* bacterium
with (b) drawn representative structures

Each subcellular organelle of a prokaryotic cell carries out a function that will allow the cell to survive. Table 1.5 outlines the main subcellular structures found in prokaryotic cells.

Table 1.5: Prokaryotic Subcellular Structures

Subcellular Structure	Function
Cell wall	Outer layer of the cell that functions for support and protection
Plasma (cell) membrane	Allows the cell to maintain a separate internal environment than the external environment it lives in (selectively allows material in and out (selectively permeable)
Cytoplasm	Fluid within the cell that contains all the enzymes, ribosomes, nucleoid (DNA) region and molecules necessary for the cell to carry out all its life processes
Pili	Hairlike structures that function for attachment to surfaces or other cells
Flagellum	Protein structure that extends from the cell wall and rotates in order to provide movement for the cell
Ribosomes	Produces proteins for the cell (protein synthesis); referred to as 70S ribosomes; smaller than ribosomes in eukaryotic cells
Nucleoid region	Location of the naked DNA (DNA not associated with histone proteins)
Capsule	A protective structure that surrounds the cell wall of some bacteria.

BINARY FISSION

Prokaryotic cells carry out a very simple form of reproduction that is known as **binary fission**. A unicellular organism can simply copy its naked DNA and split the existing parent cell into two new cells. This is a form of asexual reproduction that results in two new cells that are genetically identical to the cell they came from and genetically identical to each other. This process of asexual reproduction is very fast and allows bacteria to reproduce at a rate of more than 1 million new cells in less than 8 hours!

Eukaryotic Cellular Structure

Figure 1.8 shows the ultrastructure of a liver cell as an example of a typical eukaryotic cell. You should immediately notice the increased complexity of the eukaryotic cell as compared with the prokaryotic cell. Eukaryotic cells have a nuclear membrane that encloses the DNA. Eukaryotic cells possess many membrane-bound **organelles**. Structures within eukaryotic cells are often referred to as being compartmentalized since most organelles possess membranes that separate their interior material from the cytoplasm.

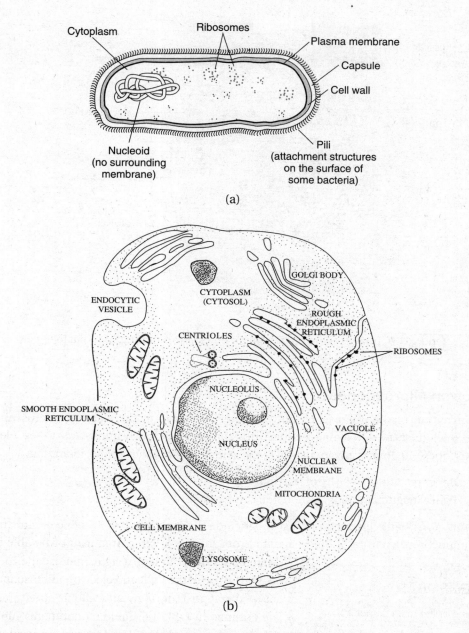

Figure 1.8. Prokaryotic (a) and eukaryotic (b) cells

You may be asked to draw, label, annotate a diagram of and identify the structures of a eukaryotic cell. In addition, you may be asked to identify eukaryotic structures of a liver cell from an electron micrograph. (See Figure 1.9.)

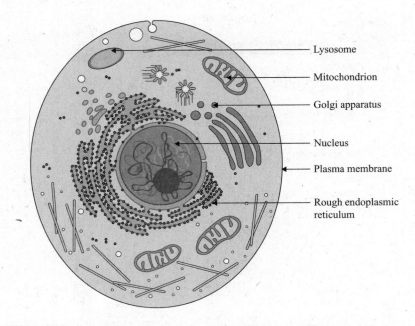

Figure 1.9. Images that can be seen in an electron micrograph of a liver cell

Electron Microscopes

The invention of the microscope and the later invention of the electron microscope has allowed scientists to view a microscopic world that was previously impossible to see. Electron microscopes offer many advantages over the standard light microscope, including:

- Higher resolution (clarity of the image)
- Higher magnification (greater detail can be seen)

Limitations to the electron microscope include the inability to view living organisms and the monochrome nature of the image. In order to view a living organism or be able to see the colours of an image, a light microscope must be used. The use of the light and electron microscopes has allowed scientists to view subcellular structures that are outlined in Table 1.6. Table 1.7 compares subcellular structures found in prokaryotic and eukaryotic cells.

> **REMEMBER**
>
> The functions of some of the main organelles:
>
> - Nucleus: The *brain* of the cell
> - Mitochondria: The *powerhouse* of the cell
> - Golgi apparatus: The *post office* of the cell

Table 1.6: Eukaryotic Subcellular Structures

Eukaryotic Organelle	Function
Free ribosomes (not attached to the rough endoplasmic reticulum)	Produce proteins that will be used by the cell; referred to as 80S ribosomes; larger than ribosomes in prokaryotic cells
Rough endoplasmic reticulum (ER)	Produces and modifies proteins that will be shipped out of the cell; produces the membrane-bound organelle known as the lysosome
Lysosome	Membrane-bound organelle produced by the rough ER that contains digestive enzymes; digests material no longer needed by the cell and also foreign microbes that may threaten the cell
Golgi apparatus	Receives proteins from the rough ER; packages and modifies those proteins for transport out of the cell
Mitochondrion	Produces energy in the form of ATP
Nucleus	Contains the cell's DNA; is enclosed in the nuclear membrane
Nuclear membrane	A double membrane (two layers of phospholipids) that surrounds the nucleus (DNA) and controls what material will be allowed to enter or exit the nucleus
Cell membrane	A single membrane (one layer of phospholipids) that surrounds the cell and controls what enters and exits the cell; is selectively permeable
Nucleolus	Site of the production of ribosomal subunits since each ribosome consists of two subunits (parts)
Smooth endoplasmic reticulum (ER)	Functions for production of lipids and detoxification (destroys toxins)
Cytoplasm	The fluid inside of the cell that contains all the organelles from the cell membrane to the nuclear membrane
Vesicles	Stores material and transports material within and out of the cell; moves material between the rough ER, the Golgi and the cell membrane; can store digestive enzymes or other compounds to keep them separate from the cytosol (intracellular fluid)

Table 1.7: Comparisons Between Prokaryotic and Eukaryotic Cells

Feature	Prokaryotic Cells	Eukaryotic Cells
Location of DNA	Not enclosed in a nuclear membrane; found in the nucleoid region	Enclosed in the nuclear membrane
Packaging of DNA	No proteins (histones) associated with the DNA (not packaged)	Proteins (histones) associated with the DNA for packaging
Shape of DNA	Circular (one continuous loop of DNA)	Linear (contain separate pieces of linear DNA—humans have 46)
Membrane-bound organelles	Do not contain membrane-bound organelles; e.g., do not contain mitochondria	Contain membrane-bound organelles; e.g., do contain mitochondria
Internal membranes	No internal membranes; do not compartmentalize their functions	Contain internal membranes to compartmentalize their functions
Ribosomes	Smaller than eukaryotic ribosomes; called 70S* ribosomes	Larger than prokaryotic ribosomes; called 80S* ribosomes

*70S/80S—The S signifies the sediment rate of the organelle. Larger structures will settle out of solution more rapidly, hence the higher S rate of 80S. Theodor Svedberg discovered this, and the "S" is often thought to refer to his name as well.

Similarities Between Prokaryotic and Eukaryotic Cells

Both prokaryotic and eukaryotic cells possess:

- Cytoplasm
- DNA—although not packaged (not associated with proteins) in prokaryotic cells, the DNA is structurally the same
- Ribosomes—although they are larger in eukaryotic cells, they are of the same composition
- Cell membrane

Refer to Figure 1.8 for a visual comparison of prokaryotic and eukaryotic cells.

All cells must have a way to separate the internal environment from the external environment (cell membrane), store genetic information (DNA), produce proteins (ribosomes) and have a fluid component in which many molecules can be located and easily transported (cytoplasm).

Make sure you can identify both similarities and differences between the two cell types when asked to compare them. If you are asked to distinguish between them, you will gain points only by stating the differences, not the similarities.

Plant and Animal Cells

Many different types of eukaryotic cells exist, and they are grouped according to their similarities. One major division between eukaryotic cell types involves the distinction between the animal and the plant cell. (See Figure 1.10 and Table 1.8.) Plant cells have several unique organelles that make them structurally and functionally different from animal cells.

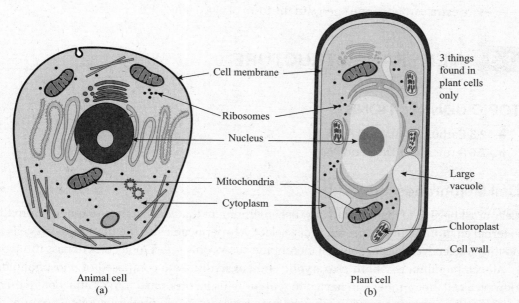

Figure 1.10. Diagrams of (a) a typical animal cell and (b) a typical plant cell

Table 1.8: Major Distinctions (Differences) Between Plant and Animal Cells

Feature	Plant Cell	Animal Cell
Cell wall	Yes	No
Chloroplasts	Yes	No
Vacuole	One large central vacuole	Many small vacuoles
Centrioles	No	Yes
Carbohydrate storage compound	Starch (amylose)	Glycogen
Shape	Generally square	Generally round

DISTINCTIONS EXPLAINED

- Plant cells have cell walls made of cellulose located outside the cell membrane. Cell walls protect and support the plant cell. Animal cells do not have cell walls outside of their cell membranes.
- Plant cells have photosynthetic cells that contain chloroplasts. These photosynthetic cells carry out photosynthesis and make their own carbohydrates. Animal cells do not have chloroplasts and cannot carry out photosynthesis to make their own carbohydrates.
- Plant cells have one large central vacuole that can fill up with water and help support the cell (create turgor pressure). Animal cells have many small vacuoles that can store

water and other materials. However, the vacuoles in animal cells are too small to provide support for the cell.

- Animal cells have centrioles that aid in cell division, while plant cells lack centrioles.
- Plant cells are generally square in shape. Animal cells are generally round but can exhibit many different shapes depending on their function.
- Plant cells store extra carbohydrate energy in the form of starch (amylose). Animal cells store extra carbohydrate energy in the form of glycogen.

1.3 MEMBRANE STRUCTURE

TOPIC CONNECTIONS

- 2.3 Carbohydrates and lipids
- 2.6 Structure of DNA and RNA

Cell Membranes

Cells must be able to maintain an internal environment that is separate and unique from the external environment. The presence of a selectively permeable cell membrane allows cells to do this. A selectively permeable cell membrane allows only select molecules to pass through.

All cell membranes—both prokaryotic and eukaryotic—are composed of phospholipids. However, cell membranes can have different proteins incorporated into or attached to them to give each membrane structure the ability to carry out its required functions. The proteins add permeability to the cell membrane by altering what molecules are allowed to pass through the membrane. Some proteins have gates (doors that can open and close) that control when the cell membrane will allow specific molecules to pass through.

PLASMA MEMBRANE STRUCTURE

The IB exam refers to the cell membrane as the plasma membrane. Other references to the cell membrane may include the plasma membrane, the phospholipid bilayer or the fluid mosaic model. Figure 1.11 shows the basic structure of the plasma membrane.

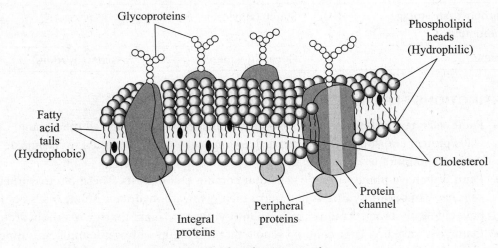

Figure 1.11. The plasma membrane.

The phospholipids that make up the cell membrane contain a *hydrophilic* (water-loving) head (phosphate and glycerol group) and two *hydrophobic* (water-fearing) tails (fatty acids). This means that the head will naturally be attracted to water (hydrophilic), and the tails will naturally repel water (hydrophobic). The combination of both the hydrophilic and hydrophobic properties of phospholipids promotes the rapid formation of phospholipid bilayers when they are placed into aqueous solutions. The tails automatically face away from the water and the heads towards the water. The attraction of the tails to each other and the heads to the surrounding aqueous solution allows the cell membranes to maintain a stable structure.

> Molecules that contain both a hydrophilic and a hydrophobic region are referred to as *amphipathic*. Phospholipid molecules are amphipathic.

The term *polar* is also used to describe a molecule that is attracted to water due to the presence of a partial charge. Hydrophilic molecules include polar molecules and ions. Nonpolar molecules do not have a charge and are not attracted to water. The tails of the phospholipid molecules are nonpolar and exhibit hydrophobic properties. The heads of the phospholipid molecules are polar and exhibit hydrophilic properties. (See Figure 1.12.)

REMEMBER

Hydro: water

Phobic: fear of

Philic: love of

Hydrophobic: fear of water (repel)

   Nonpolar molecules are hydrophobic.

Hydrophilic: love of water (attract)

   Polar molecules are hydrophilic.

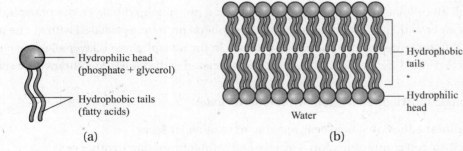

Hydrophilic head (phosphate + glycerol)

Hydrophobic tails (fatty acids)

(a)

Water

Hydrophobic tails

Hydrophilic head

Water

(b)

Figure 1.12. (a) A phospholipid molecule and (b) a section of the phospholipid bilayer

Cholesterol

Mammalian cell membranes contain cholesterol that helps buffer the fluidity of the membrane. This makes the membrane more stable while, at the same time, allowing it to pinch in and out during endocytosis and exocytosis. In addition, since cholesterol is hydrophobic (nonpolar), it limits the permeability of the membrane to small polar molecules.

Membrane Proteins

Membrane proteins are vital in order for cells to regulate how they respond to and interact with the surrounding environment. One of the main functions of membrane proteins is to regulate selectively what is allowed to enter and exit the interior (cytoplasm) of the cell.

In order to allow molecules to move through the cell membrane, the protein must span the cell membrane. Proteins that span the membrane are referred to as *integral* proteins. Integral proteins must exhibit both hydrophilic and hydrophobic properties (amphipathic molecules) in order to span both the hydrophilic and the hydrophobic regions of the cell membrane. Any part of the protein in contact with the hydrophobic tails must have hydrophobic properties. The rest of the protein must be hydrophilic to exist in the aqueous (high water content) environment. Proteins that are found on the surface of the cell membrane (inside or outside the cell) are referred to as *peripheral* proteins. Peripheral proteins do not possess hydrophobic regions since they do not span the nonpolar region of the cell membrane. Although only integral proteins can allow the passage of material that cannot freely pass through the phospholipid bilayer, both integral and peripheral proteins allow for many reactions and cellular processes to occur.

Only very small, nonpolar molecules can pass freely through the cell membrane. Any partially charged (polar) or large molecule must use a protein channel or pump to enter or exit the interior of the cell. Protein channels can have *gates*. These are structures that open to allow the passage of material. These gates open only when membrane receptors are activated. The gates allow the cell to control when material will be allowed to pass through the cell membrane.

MEMBRANE PROTEIN FUNCTIONS

Although all cellular membranes are composed of a phospholipid bilayer, the phospholipid bilayer can have many different integral and peripheral proteins associated with it. The presence of different integral and peripheral proteins in the phospholipid bilayer allows each cell type to carry out different functions. Proteins associated with cellular membranes are shown in Figure 1.13.

The main functions of membrane proteins include:

- **Cellular adhesion**—hold cells together or to other surfaces
- **Cell-to-cell communication**—receive and send information to other cells
- **Hormone binding sites**—receive chemical messages
- **Immobilized enzymes**—catalyse reactions on or in the cell membrane
- **Channels for passive transport**—allow a passageway for passive movement of material that cannot diffuse through the cell membrane
- **Pumps for active transport**—use ATP (energy) to transport material against its concentration gradient

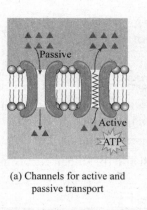

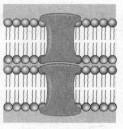

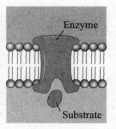

(a) Channels for active and passive transport

(b) Cellular adhesion (cell-to-cell)

(c) Immobilized enzyme

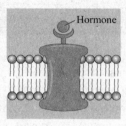

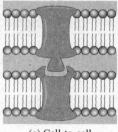

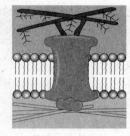

(d) Hormone binding site

(e) Cell-to-cell communication

(f) Cellular adhesion (cell-to-extracellular matrix)

Figure 1.13. Functions of the plasma membrane proteins

Models of Membrane Structure

The discovery of the structure of the cell membrane is a perfect example of the nature of science. Collaboration among scientists, the discovery of new microscopic technology and scientific scepticism all contributed to our current understandings of the cell membrane structure.

In the early 1930s, the Davson–Danielli model of the cell membrane was proposed. This model described the cell membrane as a phospholipid bilayer consisting of two layers of globular proteins surrounding each side of the phospholipid bilayer. The discovery of the electron microscope in the 1950s allowed scientists to view and analyse the structure of the cell membrane more accurately. New information based on analysis of electron micrographs led to scrutiny of the Davson–Danielli model of the cell membrane by the scientific community. Singer and Nicolson proposed the idea of the cell membrane existing as a fluid mosaic design with proteins embedded in and on the periphery of the phospholipid bilayer. The Singer-Nicolson model (fluid mosaic model) of the cell membrane replaced the Davson–Danielli model as the accepted model and remains the currently accepted model for cell membrane structure. Both models of membrane structure are shown in Figure 1.14 and outlined in Table 1.9.

REMEMBER

Passive: No energy required
- Diffusion/facilitated diffusion

Active: Energy required—ATP
- Protein pumps

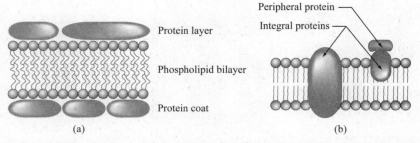

Figure 1.14. Membrane models (a) Davson–Danielli and (b) Singer–Nicolson

Table 1.9: Cell Membrane Models

	Davson–Danielli Model	Singer–Nicolson Model (Fluid Mosaic Model)
Date proposed	1935	1972
Proposal	Solid structure of phospholipids in a bilayer completely surrounded by globular proteins	Fluid structure with freely moving proteins attached to and spanning the membrane that function for transport and other cellular processes
Acceptance	No longer accepted by the scientific community due mostly to new evidence available from electron micrographs	The currently accepted model of cell membrane structure

1.4 MEMBRANE TRANSPORT

TOPIC CONNECTIONS

- 6.5 Neurons and synapses

Diffusion and Osmosis

Diffusion is defined as the passive movement of molecules from a high to a low concentration. Molecules are always moving, even when they are equally distributed (have reached equilibrium). Molecules will always move from an area of high concentration to an area of low concentration. When molecules move from an area of high to low concentration, they are referred to as moving with their concentration gradient. Molecules always diffuse with their concentration gradient and do not need energy to do so.

Osmosis describes the passive movement of water through a selectively permeable membrane. A selectively permeable membrane allows only select materials to pass through. In osmosis, water moves from an area of high concentration to an area of low concentration. No energy in the form of ATP is required for either diffusion or osmosis. (Refer to Figure 1.15.)

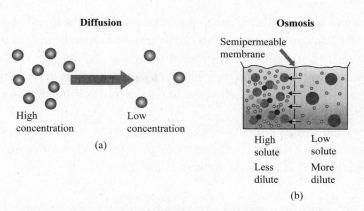

Figure 1.15. Illustrated processes of (a) diffusion and (b) osmosis

Passive Versus Active Transport

Molecules always diffuse from an area of high concentration to an area of low concentration. When molecules are moving from an area of high concentration to an area of low concentration, they are moving *with the concentration gradient*. This process is passive, meaning it does not require energy in order to proceed.

When molecules move from an area of low concentration to an area of high concentration, they are moving *against their concentration gradient*. This process is active, meaning it does require energy (ATP) in order to proceed.

SIMPLE AND FACILITATED DIFFUSION

Passive transport (movement with the concentration gradient) across a cell membrane can occur by two different methods.

1. **Simple diffusion:** No energy or membrane channel protein is required. Material simply diffuses across the membrane from a high to a low concentration. The molecule must be small enough so it can cross the membrane. The molecule must also be nonpolar (not charged) so it can span the hydrophobic (nonpolar) region of the membrane.
2. **Facilitated diffusion:** No energy is required, but the molecule(s) must use a protein channel to cross the membrane. The molecule is either too large to undergo simple diffusion or is polar and cannot cross through the hydrophobic region of the cell membrane. A small, polar molecule, such as an ion, must cross the membrane via facilitated diffusion.

Large or charged molecules cannot simply diffuse through the cell membrane. Large molecules cannot fit through the space between the phospholipids that make up the cell membrane. Charged (polar) molecules cannot pass through the hydrophobic (nonpolar) region containing the hydrophobic tails that make up the centre of the cell membrane. Protein channels can be incorporated into the cell membrane in order to allow large or partially charged molecules to pass through selectively.

> The movement of sodium and potassium through protein channels during the generation of action potentials is an excellent example of facilitated diffusion.

Channel proteins selectively allow molecules to pass through the membrane based on their size and structure. Each protein channel or carrier protein is specific in structure for the molecule it is designed to transport. Carrier proteins recognize the molecule they are designed to transport based on the size and structure. However, carrier proteins differ from protein channels because carrier proteins alter their shape to help in transporting the molecule across the membrane.

Both channel and carrier proteins facilitate the movement of molecules from an area of high to low concentration. Hence, both processes are referred to as forms of *facilitated diffusion*, and no energy (ATP) is required. (Refer to Figure 1.16.)

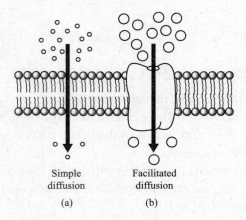

Simple diffusion (a)

Facilitated diffusion (b)

Figure 1.16. Types of diffusion: (a) simple diffusion and (b) facilitated diffusion

PROTEIN PUMPS AND ACTIVE TRANSPORT

If molecules need to be transported against their concentration gradient or if the cell needs to speed up the passive movement of molecules, the cell must use energy to do so. Adenosine triphosphate (ATP) is the most efficient and frequently used energy transport molecule of the cell. Adenosine triphosphate consists of an adenine base bonded to a 5-carbon sugar (ribose) and 3 phosphate groups. The adenine and ribose are together referred to as adenosine. Figure 1.17 shows the structure of ATP and energy release from its phosphate bonds.

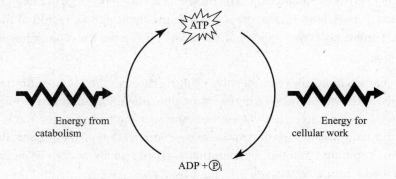

ATP

Energy from catabolism

Energy for cellular work

ADP + P_i

Figure 1.17. (a) Adenosine triphosphate

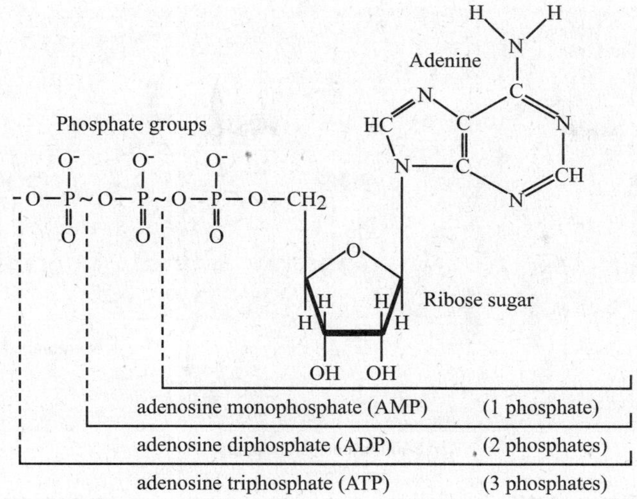

Figure 1.17

Phosphate groups

Adenine

$$-O-P\sim O-P\sim O-P-O-CH_2$$

Ribose sugar

OH   OH

adenosine monophosphate (AMP)   (1 phosphate)

adenosine diphosphate (ADP)   (2 phosphates)

adenosine triphosphate (ATP)   (3 phosphates)

Figure 1.17. (b) releasing energy by losing a phosphate bond

Potential energy that can be used by cells is stored in each of the phosphate bonds. By breaking the phosphate bonds, the cell can release the stored energy and use the free energy to drive cellular reactions. Cells constantly break the phosphate bonds to release the energy and use energy to form new phosphate bonds. This creates an energy use and energy storage cycle involving ATP becoming ADP and vice versa. The energy released by the breaking of phosphate bonds in ATP drives many cellular reactions. The cycles of adding and removing phosphate bonds allow energy to be stored and released throughout the cell.

Adenosine triphosphate is the energy source for protein pumps to move material across the cell membrane. Protein pumps function like protein channels and protein carriers. However, protein pumps must use energy to move molecules generally against their concentration gradient. Rarely, protein pumps can be used to speed up the movement of molecules with the concentration gradient. Figure 1.18(a) shows active transport, and Figure 1.18(b) shows co-transport with the sodium-potassium pump.

ACTIVE TRANSPORT

Energy and a protein carrier are both required. (The molecule is too large, is charged and is moving *against its concentration gradient*.)

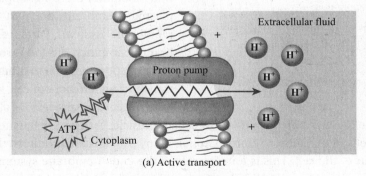

(a) Active transport

Figure 1.18. (a) Active transport.

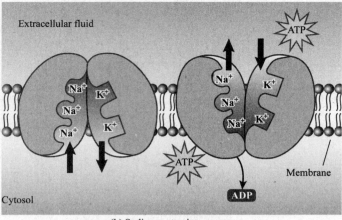

(b) Sodium-potassium pumps

Figure 1.18. (b) sodium-potassium pump (co-transport)

The sodium-potassium pump is an excellent example of active transport. The integral protein uses ATP to pump 3 sodium ions at a time out of the cell while, at the same time, pump 2 potassium ions into the cell. The movement of two different substances at the same time is referred to as co-transport.

Endomembrane System (Vesicle Transport)

The cell membrane is often referred to as a fluid mosaic model because it is flexible and contains proteins that can move around in the membrane. The phospholipids give the membrane a highly fluid nature. In order to keep the cell membrane from being too fluid, cholesterol is incorporated into the cell membrane. Cholesterol *buffers the fluidity* of the cell membrane by associating with both the heads and tails of the phospholipids, making them more stable.

All membrane-bound organelles in the cell consist of membranes that are of the same phospholipid structure as the cell membrane. Since the membranes are all of the same structure, they can effectively pinch on and off of each other. The flexible fluid nature of these membranes allows them to form vesicles and transport material.

The membranes of the rough endoplasmic reticulum, the Golgi apparatus and the plasma membrane are part of the endomembrane system. Parts of these membranes can pinch off and form vesicles that can transport material to other organelles or transport material out of cell. When material exits the cell, the process is called *exocytosis.*

> ### REMEMBER
>
> When referring to the Golgi apparatus:
> *cis:* The side (face) of the Golgi that faces the rough ER
> *trans:* The side (face) of the Golgi that faces the cell membrane

The membrane that has pinched off is now called a vesicle. Vesicles are the method of transport used by cells to transport or store material. Many proteins are produced that need to be transported out of the cell. This is a major role of the endomembrane system. The steps of membrane transport are outlined below and correspond to each numbered step identified on Figure 1.19.

STEPS OF MEMBRANE TRANSPORT

1. Proteins for export are produced on the ribosomes attached to the rough endoplasmic reticulum.

2. The proteins enter the endoplasmic reticulum and are modified (e.g., folded or associated with other molecules).

3. The protein exits the endoplasmic reticulum in a vesicle and is transported to the Golgi apparatus.

4. The vesicle fuses with the Golgi apparatus, and the protein is released into the Golgi apparatus.

5. The protein is further modified and packaged for export from the Golgi apparatus to the cell membrane.

6. The vesicle containing the protein fuses with the cell membrane, and the protein is released from the cell (exocytosis).

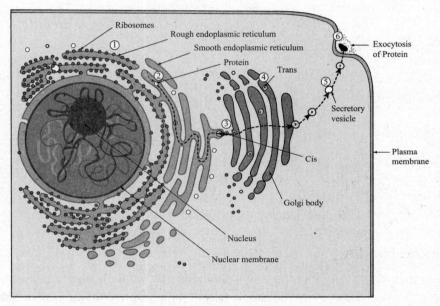

Figure 1.19. The endomembrane system (vesicle transport) showing the steps involved in exocytosis

Lysosomes are also produced by the endomembrane system. Remember that lysosomes contain digestive enzymes. If these enzymes were not stored in vesicles, they would be free to digest cellular components and damage the cell. Lysosomes are produced in the same manner as export vesicles with the exception that lysosomic enzymes packaged in vesicles do not undergo exocytosis.

In addition to material exiting the cell, material can also be brought into the cell by the pinching in of the cell membrane. When material exits the cell, the process is referred to as exocytosis. When material enters the cell, the process is called endocytosis. Both of these processes require the use of energy and allow cells to take in and send out large quantities of material. There are two main types of endocytosis, phagocytosis and pinocytosis. (See Figure 1.20.)

REMEMBER

Functions of the Endomembrane System

- Exocytosis of material
- Transport vesicles
- Production of lysosomes

> **Phagocytosis:** Taking in of solid material (cell eating)
>
> **Pinocytosis:** Taking in of liquid material (cell drinking)

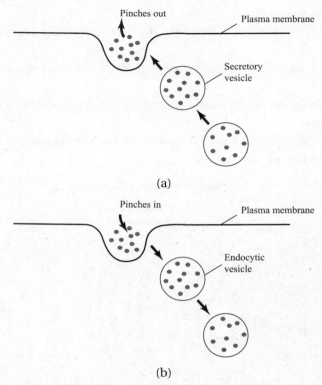

(a)

(b)

Figure 1.20. (a) Exocytosis and (b) endocytosis

Membrane Fluidity

All cellular membranes are built from phospholipids that naturally form a bilayer based on the hydrophilic heads and the hydrophobic tails of the phospholipid molecules. The tails of the phospholipids contain both saturated and unsaturated fatty acids that contribute to the fluid nature of membranes. The presence of double bonds in the unsaturated fatty acid tails of phospholipids gives the membrane a more fluid nature since unsaturated fatty acids form kinks in the tails, making it more difficult to pack the tails tightly together. (See Figure 1.21.)

REMEMBER

Endo: enter   **Cyto:** cytoplasm

Exo: exit   **Sis:** state of

Endocytosis: The state of entering the cytoplasm of the cell

Exocytosis: The state of exiting the cytoplasm of the cell

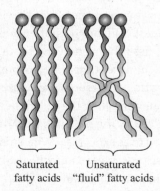

Figure 1.21. Fluidity of unsaturated fatty acids

The fluid nature of the membrane allows it to change shape, break and reform during exocytosis and endocytosis. The flexible nature of the membrane allows it to fold inwards and outwards. When a break forms in the membrane during exocytosis and endocytosis, the self-sealing nature of the membrane quickly fuses the two open ends of the exposed membrane. The individual phospholipid molecules are attracted to each other based on their hydrophilic and hydrophobic components. Collections of phospholipids form attractions between the hydrophilic heads and the hydrophobic tails that allow the heads to be exposed to the aqueous solution and the tails to face away from the aqueous solution. The weak association between individual phospholipid molecules allows the membrane to open and close at specific sites between the phospholipids. The hydrophobic tails quickly re-associate between the sites left open after vesicle formation to avoid exposure to the polar aqueous solution. Likewise, the heads are attracted back together due to their shared hydrophilic (polar) nature.

1.5 ORIGIN OF CELLS

TOPIC CONNECTIONS

- 5.1 Evidence for evolution

Abiogenesis (Spontaneous Generation)

The belief that life could arise from nonliving structures (abiogenesis) was dispelled due to investigations carried out by Francesco Redi (1600s) and Louis Pasteur (1880s). Redi provided evidence that maggots could not spontaneously generate from rotting meat as previously believed. His famous experiment involved the use of jars containing rotting meat. Redi wanted to prove that maggots were produced from eggs laid by flies that landed on the meat and not by spontaneous generation. He placed meat into jars. Some jars were uncovered, leaving the meat fully exposed. Some jars were covered with mesh to keep out flies. The rest of the jars were completely covered. Only those jars that allowed flies access to the meat became infested with maggots. This experiment showed that multicellular life could not arise from nonliving material. Redi's experiment is shown in Figure 1.22.

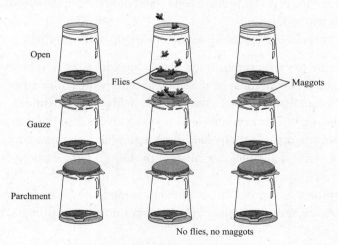

Figure 1.22. Redi's experiment

The question as to whether microscopic life could arise through spontaneous generation remained until Louis Pasteur carried out his famous experiment. Pasteur used specially shaped flasks to show that once microorganisms were killed in broth, they could not spontaneously regenerate. He introduced the idea that microorganisms are carried on dust particles that settle in solutions that allow the microorganisms to reproduce. The S-shape on the opening to the flask allowed air to enter but kept the dust particles carrying the microorganisms from having access to the broth. No microorganisms appeared in the broth that had no access to dust particles. The experiment provided evidence against the spontaneous generation of life (abiogenesis). Pasteur's experiment is shown in Figure 1.23.

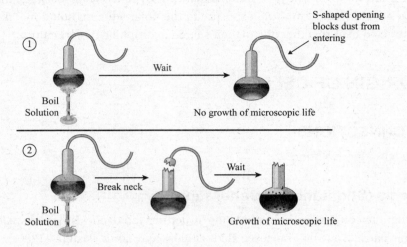

Figure 1.23. Pasteur's experiment

Origin of Organic Compounds

In order for life to appear on Earth, organic compounds would need to be synthesized from inorganic material. These organic compounds would need to form polymers that would eventually allow the formation of cellular components. Cellular components could assemble into primitive prokaryotic cells. Prokaryotic cells could then undergo processes that allowed them to evolve into eukaryotic cells. Early eukaryotic cells could eventually evolve into the many life forms we see on Earth today.

Four processes are needed for the spontaneous origin of life on Earth.

- In order for life to exist on Earth, the ability to synthesize simple organic compounds from inorganic material must be possible. In other words, monomers must be created: amino acids, monosaccharides, glycerol, fatty acids and nucleotides.
- These simple organic compounds are necessary in order to build more complex structures. The simple organic molecules must assemble into more complex organic molecules. As a result, polymers are created: proteins, lipids, carbohydrates and nucleic acids.
- As more complex molecules are assembled, some must be molecules of heredity that can self-replicate in order to pass on hereditary information (traits). This is the creation of DNA and RNA.
- The complex molecules must be packaged in a way that allows them to exist in an internal environment that is different and separate from the external environment. This involves the formation of cell membranes.

Formation of Organic Compounds

In an effort to discover how organic material could have originated, Stanley Miller and Harold Urey put together a model of what they believed simulated the primitive conditions of early Earth. The model, shown in Figure 1.24, included gases that were thought to be present when Earth formed. These gases included methane (CH_4), ammonia (NH_3), hydrogen (H_2) and water (H_2O). Water simulated Earth's early ocean, and sparks of energy were added to simulate the turbulent atmosphere thought to exist at the time. Sparks of energy were used to simulate lightning from the atmosphere that scientists believe may have been the source of energy for organic material formation. Miller and Urey's experiment showed that simple organic molecules (including 15 of the 20 amino acids) could be formed from inorganic material under the modelled conditions. (See Figure 1.24.)

> And WHAM!, organic compounds were created. Use this as a mnemonic to help you remember the gases used to simulate early Earth: **W**ater, **H**ydrogen, **A**mmonia and **M**ethane.

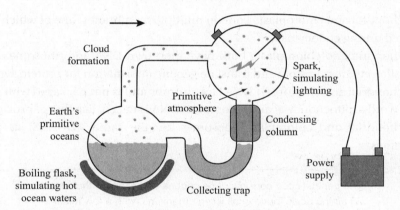

Figure 1.24. Miller and Urey's experiment (1953)

Some scientists argue that Miller and Urey's experiment showed only that organic material could be produced from inorganic material in the laboratory environment. Whether these compounds could be created outside controlled experimental conditions, such as those that existed during early Earths' formation, remains a question yet to be answered. Questions also remain about whether the gases used in the experiment were actually present on early Earth and whether the energy from the atmosphere could have been as intense as it was represented as in Miller and Urey's experiment.

> **Biogenesis:** Life arising from life
>
> **Abiogenesis (spontaneous generation):** Life arising from non-living structures

Origin of Eukaryotes—Endosymbiosis

The first cells to appear on Earth were prokaryotic cells that arose from nonliving material. The eukaryotic cells evolved over time from their ancestral prokaryotes. The evolution of mitochondria and chloroplasts within eukaryotic cells has been best explained by the theory of endosymbiosis. Eukaryotic cells possess subcellular structures such as mitochondria and chloroplasts. According to the theory of endosymbiosis, these organelles were originally independent prokaryotic entities themselves. These independently living structures were taken in by other prokaryotes through the process of endocytosis and incorporated into the cytoplasm. The aerobic prokaryotic bacterium that was taken in through endocytosis provided the "host" cell with energy. The engulfed photosynthetic cyanobacterium produced carbohydrates for the "host" cell. The benefits of harbouring the newly engulfed prokaryotes kept the host from rejecting or digesting them. Other organelles, such as the nucleus, may have formed from infoldings in the plasma membrane. Eventually, the process of endosymbiosis produced the first ancestral eukaryotic cell.

Evidence for the theory of endosymbiosis and the theory that mitochondria and chloroplasts were once prokaryotic independently living cells include the following:

- Mitochondria and chloroplasts contain multiple membranes, one of which could have resulted during endocytosis.
- Mitochondria and chloroplasts both have their own DNA and ribosomes that would have allowed them to store and translate genetic information for protein synthesis.
- Mitochondrial and chloroplast DNA lacks introns and is not packaged with proteins. In other words, mitochondrial and chloroplast DNA is naked, like that of prokaryotic cells.
- Mitochondrial and chloroplast ribosomes are the same size (70S) as prokaryotic ribosomes.

> The genetic code consists of 64 codons that code for the same 20 amino acids found in all living organisms (with a few minor variations). This supports the belief that all life originated from a common ancestor.

1.6 CELL DIVISION

In order to carry out **asexual reproduction**, grow (increase in size), undergo embryonic development and replace or repair damaged tissue, a cell must be capable of dividing in order to form new cells. The cell must first replicate its genetic material (DNA) so each new cell will have all the information it needs to carry out all cellular processes. **Mitosis** is the process of nuclear division in which the cell creates two exact genetic copies of its DNA. Most cells follow mitosis with a process called **cytokinesis** in which the cell completely separates the existing cytoplasm and organelles into two new cells. The cells resulting from mitotic divisions are all genetically identical to the cell from which they originated. The dividing cell is referred to as the parent cell, and the resulting new cells are referred to as daughter cells. (See Figure 1.25.)

REMEMBER

Cytokinesis results in 2 new, genetically identical daughter cells.

Some cells are multinucleated. In other words, they have more than one nucleus in the same cell. They achieve this by carrying out mitosis without undergoing cytokinesis. An example of this is human muscle cells. They are multinucleated due to the process of mitosis and lack of cytokinesis.

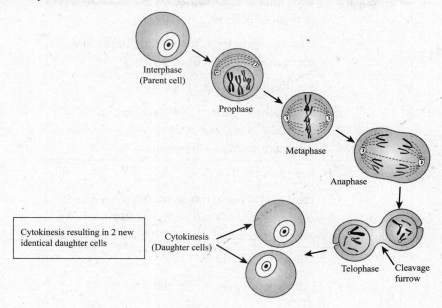

Figure 1.25. The stages of cellular division (mitosis and cytokinesis)

The Cell Cycle

Each cell goes through a cycle that regulates when it will divide. This cycle is known as the cell cycle, and each stage consists of unique activities the cell is carrying out. Some cells spend more time in an active, nondividing state known as interphase, while other cells divide more frequently. The cell cycle stages and the main events of each stage are shown in Figure 1.26 and Table 1.10.

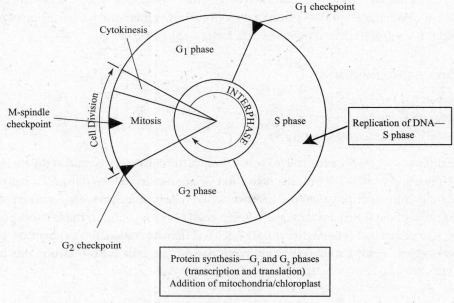

Figure 1.26. Stages of the cell cycle

Table 1.10: Cell Cycle Stages and Main Events

Stage	Main Events
G_1	**G**ap/**g**rowth of cell Protein synthesis (transcription/translation); many metabolic events occurring
S	**S**ynthesis (replication) of DNA The existing (parent) DNA is copied so each new (daughter) nucleus will receive a complete set of chromosomes
G_2	**G**ap/**g**rowth of cell Protein synthesis (transcription/translation); production of microtubules; many metabolic events occurring
M	**M**itosis Creation of 2 genetically identical nuclei Complete separation of replicated DNA (chromosomes) Consists of 4 parts: prophase/metaphase/anaphase/telophase
Cytokinesis	Separation of the cytoplasm and the creation of 2 new identical cells

Cancer and Cell Division

Multiple checkpoints exist within the cell cycle in order to regulate when the cell divides and the speed at which the cell moves through the cell cycle. **Cyclins**, which are a group of regulatory proteins, control the movement of a cell through the cell cycle.

CELL CYCLE CHECKPOINTS

- Between G_1 and the S-phase
- Between G_2 and mitosis
- At the metaphase stage of mitosis

Checkpoints occur before the cell leaves G_1, before the cell leaves G_2 and at the metaphase stage of mitosis. If a cell is not fully ready to enter G_1 or mitosis, cellular division is halted. The checkpoint in mitosis during metaphase ensures that the chromosomes are lined up correctly and that the spindle fibres are in place. When a cell fails to respond to these checkpoints, it may begin uncontrolled cellular division. The cell will then be considered a cancer, or tumour, cell. Any organ or tissue can become cancerous if any of its cells start to exhibit this uncontrolled cell division.

CANCER

Cancers can be the result of oncogenes (genes with the potential to cause cancer), mutagens (an agent that can change the genetic code) or metastasis (a cancer originating in one tissue and travelling to another). Several oncogenes have been discovered to exist and can be detected by genetic testing. Individuals found to carry a specific oncogene may be counseled as to how to help prevent the formation of the potential cancer. Agents such as those found in cigarettes have been correlated with high incidences of cancer and should be avoided.

Unicellular organisms do not develop cancer because they do not consist of tissues. Since mitosis is a form of asexual reproduction for them, if a unicellular organism starts to speed up the cell cycle rate, the organism is simply reproducing more rapidly!

Interphase of the Cell Cycle

The cell cycle stages are further divided into three different phases, including interphase, mitosis and cytokinesis. Stages G_1, S and G_2 all occur during what is termed interphase. Interphase is an active period in the life of a cell that includes processes such as protein synthesis (G_1 and G_2), DNA replication (S) and an increase in the number of mitochondria and/or chloroplasts (G_1/G_2). During interphase, the chromosomes are collectively referred to as chromatin. The chromosomes cannot be distinguished from each other.

Mitosis

Mitosis is the creation of two new, genetically identical nuclei. This part of the cell cycle is divided into four phases—prophase, metaphase, anaphase and telophase. Each phase is associated with unique events that occur at each stage as the nucleus divides. Table 1.11 outlines the stages and the main events that occur at each stage.

> **REMEMBER**
>
> Use the mnemonic *PMAT* to help you remember the stages of mitosis.

Table 1.11: Stages and Main Events of Mitosis

Stage	Main Events
Prophase	Supercoiling of chromosomes, causing them to now be seen as individual units. Breakage (breakdown) of nuclear membrane.
Metaphase "chromosomes have MET in the middle" (meta means "middle")	Chromosomes are lined up at the middle of the cell (equator). Spindle fibres attach to the centromeres of the chromosomes.
Anaphase "ANA bANAna split"	Centromeres split, separating the sister chromatids.
Telophase	Movement of the chromosomes to opposite ends of the cell (poles). Reformation of nuclear membranes.

Prophase

During prophase, the nuclear membrane breaks down and the individual chromosomes become visible. This is due to supercoiling of the DNA that makes up the chromosomes. Before mitosis, the DNA is not supercoiled. Individual chromosomes cannot be seen. Before mitosis, the DNA is referred to as chromatin.

The two replicated copies of each chromosome (chromatids) that were produced in the S-phase during DNA replication are held together by a structure called the centromere. The matching chromatids of a chromosome are referred to as sister chromatids. They are genetically identical. The region around the centromere where the spindle fibres will eventually attach to pull the chromatids apart is referred to as the kinetochore. Once the spindle fibres shorten and pull the two chromatids apart, the two separate chromatids are referred to as individual chromosomes. The only time you have chromatids is before the replicated chromosomes separate. (See Figure 1.27.)

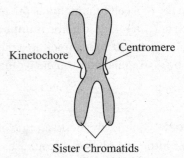

Figure 1.27. Replicated chromosome structure

Replicated chromosomes are evident by the presence of a centromere that is holding together the two exact copies of the chromosome (chromatids). Unreplicated chromosomes exist as singular units with individual centromeres that are not holding together the two copies of the chromosome. Prophase ends when the replicated chromosomes line up at the centre of the cell, signalling the beginning of metaphase.

Metaphase

During metaphase, the chromosomes line up at the centre of the cell. The centre of the cell is referred to as the equator (sometimes called the metaphase plate) of the cell. This is analogous to the imaginary equator that divides Earth into two hemispheres. The opposite ends of the cell are referred to as poles of the cell. This is analogous to Earth's two poles (opposite ends). Microtubules attach to the centromeres of each replicated chromosome. These specialized microtubules are referred to as spindle fibres and are vital for separation of the sister chromatids.

REMEMBER

Metaphase: The replicated chromosomes have "*MET*aphase" in the middle.

Anaphase

During anaphase, the sister chromatids separate and are pulled to opposite *poles* of the cell. Each centromere connecting the sister chromatids splits in half so that each chromosome has its own centromere. The separation of the chromatids is accomplished by the shortening of the spindle fibres that *pulls* apart each chromatid. Each new chromosome moves to the opposite pole of the cell.

Telophase

During telophase, the nuclear membranes reform around the sets of chromosomes at the opposite poles of the cell. The spindle fibres are broken down, and the cell prepares to enter interphase. If the cell is not going to remain multinucleated (having more than one nucleus in the cell), which is usually the case, the cell membrane begins pinching in to begin the process of cytokinesis. (See Figure 1.28.) The region in which the animal cell membrane pinches in is referred to as the cleavage furrow.

REMEMBER

Anaphase: ANA b**ANA**na split!

Cytokinesis

Complete separation of the cytoplasm of the cell into two new identical daughter cells is referred to as cytokinesis. Cytokinesis involves the pinching in of the cell membrane (cleavage furrow) in animal cells. However, plant cells have to undergo the process of forming a cell plate (a cell wall that separates the new cells) because the existing cell wall is too rigid to pinch inwards. (See Figure 1.28.)

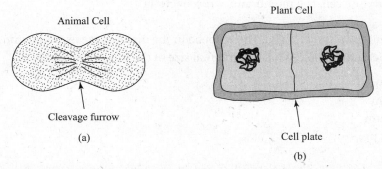

Figure 1.28. Cytokinesis showing (a) cleavage furrow and (b) cell plate

Mitosis produces two genetically identical nuclei by separating the replicated chromosomes into two new unreplicated chromosome sets. Each set contains an exact copy of all the original chromosomes. After the two identical sets of chromosomes have moved to opposite poles of the cell, the nuclear membrane reforms around the chromosomes. Unless the cell is going to remain multinucleated, the process of cytokinesis fully separates the cytoplasm of the parent cell into two new identical daughter cells.

Mitotic Processes

Mitosis is involved in the processes of embryonic development, growth, tissue repair and asexual reproduction of unicellular organisms.

HINT

Use the mnemonic **GATE** (first letter of each process) to help you remember the mitotic processes!

Growth: The process of organisms increasing in size and producing new tissues.

Asexual reproduction: The reproduction of genetically identical cells (unicellular organisms).

Tissue repair: Replacing cells that have been lost due to damaged tissue (injury or disease).

Embryonic development: The process of the zygote (resulting cell from union of the sperm and egg) becoming a multicellular organism.

CELL BIOLOGY

1. Which of the following are found in both prokaryotic and eukaryotic cells?

    I.   Cell membrane
    II. Ribosomes
    III. Mitochondria
    A. I only
    B. I and II only
    C. II and III only
    D. I, II and III

2. Which of the following are in the correct order of INCREASING size?

    A. Virus, bacteria, eukaryotic cell, cell membrane
    B. Bacteria, virus, eukaryotic cell, cell membrane
    C. Cell membrane, virus, bacteria, eukaryotic cell
    D. Eukaryotic cell, cell membrane, virus, bacteria

3. A student draws an amoeba. The amoeba in the drawing measures 20 mm. The magnification is 200X. What is the actual size of the drawn amoeba?

    A. 10 μm
    B. 50 μm
    C. 100 μm
    D. 1000 μm

4. During which phase of the cell cycle do cells experience both replication and protein synthesis?

    A. G_1
    B. S
    C. Interphase
    D. Mitosis

5. Plant and animal cells both have:

    A. Cell walls and ribosomes
    B. Cell membranes and mitochondria
    C. Large vacuoles and mitochondria
    D. Chloroplasts and mitochondria

6. In which of the following structures can cancerous tumours form?

    I. Human skin
    II. Plant stems
    III. Amoebas

    A. I only
    B. I and II only
    C. I and III only
    D. I, II and III

7. Which of the following are true statements concerning prokaryotic DNA?

    I. It is found in the nucleoid region.
    II. It is associated with proteins.
    III. It is linear in structure.

    A. I only
    B. I and II only
    C. I and III only
    D. I, II and III

8. Which statement about the phospholipid membrane is true?

    A. The membrane consists of only hydrophilic regions.
    B. The membrane consists of hydrophilic regions in the interior of the membrane.
    C. The membrane consists of hydrophobic regions on the exterior of the membrane.
    D. The membrane consists of both hydrophilic and hydrophobic regions.

9. Diffusion is defined as:
    A. The passive movement of water through a membrane.
    B. The active movement of molecules from a low to high concentration.
    C. The passive movement of molecules with their concentration gradient.
    D. The passive movement of molecules against their concentration gradient.

10. Which of the following is correctly matched?

	Type of Movement	Energy Needed	Protein Channel
A.	Simple diffusion	No	No
B.	Facilitated diffusion	Yes	No
C.	Active transport	No	Yes
D.	Osmosis	Yes	No

11. Which of the following are functions of membrane proteins?

   I. Transport
   II. Cell Recognition
   III. Energy storage

   A. I only
   B. I and II only
   C. I and III only
   D. I, II and III

12. Which of the following is paired correctly?

	Passive Transport	Active Transport
A.	Molecules moving against their concentration gradient	Molecules moving with their concentration gradient
B.	Molecules moving with their concentration gradient	Osmosis
C.	Osmosis	Simple diffusion
D.	Partially charged molecules moving with their concentration gradient	Partially charged molecules moving against their concentration gradient

13. What is a stem cell?

   A. An undifferentiated cell that has the potential to become many cell types.
   B. A cell that has differentiated into a specific pathway of development.
   C. A cell that has become cancerous and divides uncontrollably.
   D. A cell that can no longer divide and has reached its final life stage.

14. Which of the following molecules were included in the solution Miller and Urey used to simulate conditions on early Earth in order to show how the production of simple organic compounds could have occurred?

   I. Methane
   II. Water
   III. Ammonia

   A. I only
   B. I and II only
   C. I and III only
   D. I, II and III

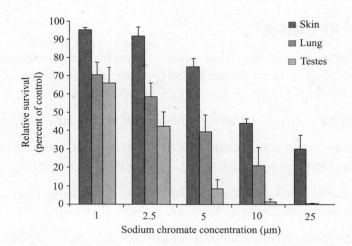

15. The effect of sodium chromate concentration on the survival of skin, lung and testes tissue in sea lions is shown in the figure above.
    a. Deduce which tissue has the greatest sensitivity to sodium chromate. (2)
    b. Calculate the percent change between the effect of 1 μm of sodium chromate and 25 μm of sodium chromate on the skin cells of sea lions. (2)
    c. Predict the effect of sodium chromate levels on reproductive abilities in sea lions. (2)

16. State four functions of membrane proteins. (4)

17. Draw a prokaryotic cell. (4)

18. Outline the cell theory. (3)

19. Define osmosis. (2)

20. a. Draw the plasma membrane. (5)
    b. Explain how molecules can travel through the cell membrane. (8)
    c. Compare prokaryotic and eukaryotic cells. (4)

ANSWERS EXPLAINED

1. **(B)** (I and II only) All cells have a cell (plasma) membrane and ribosomes in order to regulate what enters and exits the cell and to produce proteins. Prokaryotic cells do not have membrane-bound organelles such as the mitochondria.

2. **(C)** Cell membrane (10 nm), virus (100 nm), bacteria (1 μm), eukaryotic cell (100 μm).

3. **(C)** (100 μm) 20 mm (measured size × 1000 = 20,000 μm); 20,000 μm/200 (magnification) = 100 μm. Units can be in mm as well (20 mm/200X = 0.10 mm).

4. **(C)** (Interphase) Only in interphase do BOTH protein synthesis (G_1 and G_2) and replication (S-phase) occur. (Remember: Interphase includes G_1, S and G_2).

5. **(B)** (cell membranes and mitochondria) Animal cells do not have cell walls, a large vacuole or chloroplasts.

6. **(B)** (I and II only) Any tissue can form tumours (plant, animal, fungi, protist). Unicellular organisms (amoebas) cannot form tumours because they are not made of tissues.

7. **(A)** The DNA of a prokaryotic cell is found in the nucleoid region. It is not associated with proteins (naked) and is circular (eukaryotic DNA is linear).

8. **(D)** The phospholipid membrane consists of hydrophilic regions (heads) on the exterior of the membrane and hydrophobic regions (tails) in the interior of the membrane.

9. **(C)** Diffusion is the passive movement of molecules with their concentration gradient.

10. **(A)** Simple diffusion requires no energy and does not need a protein channel. Active transport requires energy. Facilitated diffusion requires a channel protein. Osmosis is the passive movement of water through a membrane.

11. **(B)** Membrane proteins can function for transport or for cell recognition. Energy storage is not a function of membrane proteins.

12. **(D)** Partially charged molecules can move passively with their concentration gradient but require active transport (energy) to move against their concentration gradient.

13. **(A)** Stem cells are undifferentiated cells that have the potential to become many cell types.

14. **(D)** Methane, water and ammonia were all used in Miller and Urey's experiment to simulate conditions that may have been present on early Earth.

15a.
- The testes tissue was most sensitive to the sodium chromate.
- At 25 μm, most of the testes tissue was destroyed by the sodium chromate.
- All levels of sodium chromate affected the survival of testes cells.
- The skin was the least sensitive to the sodium chromate.
- The lungs exhibited intermediate sensitivity to the sodium chromate.

b. At 1 μm, sodium chromate skin cell survival was 95%. At 25 μm, survival was 30%. $\frac{(30-95)}{95} \times 100 = -68\%$ (±4).*Percent change is equal to the (new value minus the old value)/old value × 100.

c.
- High levels of sodium chromate can cause sterility in sea lions.
- 25 μm and higher would result in virtually no surviving testicular cells.
- Testicular cells are important for spermatogenesis/sperm production.
- Levels above 2.5 μm of sodium chromate have the greatest effect on testicular cells and would affect the survival of sperm the greatest.
- Sodium chromate must affect rapidly dividing cells more than cells that are not dividing as often, and testicular cells are rapidly dividing cells.
- All levels of sodium chromate affect the survival of testicular cells and would have an impact on fertility in the sea lions.

16. One point is awarded for each correctly identified function:
- Hormone binding sites
- Enzymes/immobilized enzymes
- Cell adhesion/holding cells together
- Cell-to-cell communication
- Passive transport
- Active transport/pumps for transport

17. One point is awarded for each correctly drawn and labelled structure:
- Cell membrane
- Cell wall
- Pili
- Flagella
- Capsule
- DNA/nucleoid region
- Ribosomes
- Cytoplasm
- Plasmid

18. One point is awarded for each correctly outlined part of the cell theory:
- All organisms are made of cells/the basic building block of life is the cell.
- All organisms (both unicellular and multicellular) have been observed to be made of cells.
- The microscope has allowed us to see that all organisms are made of cells.
- Cells are the smallest unit of life.
- Cells cannot survive if they are broken down into their parts.
- Each cell carries out life functions.
- Cells come from preexisting cells.
- Cells divide by mitosis or binary fission (prokaryotes) to produce new cells.

19. One point is awarded for the correct definition, and one point is awarded for including added detail:
- Definition: Osmosis is the passive movement of water through a semipermeable membrane/cell membrane.

Details:

- Osmosis does not require energy.
- Osmosis is a special type of diffusion.
- Osmosis always involves water moving from an area of high concentration to an area of low concentration.

20a. Points are awarded only for each correctly labelled structure:

- Phospholipid/phospholipid bilayer
- Cholesterol
- Hydrophilic head/polar head
- Hydrophobic tails/nonpolar tail
- Integral protein/channel protein
- Peripheral protein
- Glycoprotein

b. One point is awarded for each correct statement:

- The cell membrane is selectively permeable/has different proteins imbedded in it to regulate what it allows to pass.
- All cells have a cell membrane consisting of phospholipids.
- Small/nonpolar molecules (O_2/CO_2/N_2) can move through by simple diffusion/do not need a transport protein/travel through the phospholipid bilayer.
- Large or charged (polar/ions) molecules must use a channel protein/facilitated diffusion.
- No energy is required for simple or facilitated diffusion.
- Simple diffusion occurs when molecules move with the concentration gradient/molecules move from high to low concentration.
- When a molecule moves through a protein in the cell membrane, the process is called facilitated diffusion.
- Active transport is required to move molecules against the concentration gradient/from low to high concentration.
- ATP is the energy molecule for active transport.
- Membrane proteins must have a hydrophobic and a hydrophilic region to span the membrane/their structure allows them to span the membrane and aid in transport through channels.
- Channels can be gated and regulate what moves through/open based on cellular signals.

c. One point is awarded for each correct comparison:

- Prokaryotic cells have smaller (70S) ribosomes, while eukaryotic cells have larger (80S) ribosomes.
- Prokaryotic cells lack a true nucleus/have a nucleoid region where the DNA is located, while eukaryotic cells have a nuclear membrane/DNA is found in the nucleus.
- Prokaryotic cells lack membrane-bound organelles/mitochondria, while eukaryotic cells have membrane-bound organelles/mitochondria.
- Prokaryotic cells are always unicellular, while eukaryotic cells can be unicellular or multicellular.
- Prokaryotic cells have naked DNA/their DNA not associated with proteins, while eukaryotic cells have DNA associated with proteins (histones).

Molecular Biology

CENTRAL IDEAS

- Chemical elements and reactions
- Properties of water
- Organic and inorganic compounds
- The storage of genetic information
- Protein synthesis
- Enzymatic control of cellular processes
- Organic compounds and energy transfer

TERMS

- Active site
- Aerobic
- Anaerobic
- Cellular respiration
- Condensation
- Denaturation

- Enzyme
- Gene
- Hydrolysis
- Inorganic
- Ion
- Organic

- Photosynthesis
- Proteome
- Specific heat
- Transcription
- Translation

INTRODUCTION

The basic unit of all matter is the atom. Atoms join together (bond) in many different ways in order to make up all the structures that are found in both organic and inorganic materials. If a substance is composed of only one type of atom, it is referred to an element. Atoms bond together to make molecules, and molecules can join together to make larger structures. The properties of atoms determine how they will react with other atoms and what type of bonds they will form.

Atoms can be further broken down into subatomic structures. The three main subatomic structures are the proton, neutron and electron. Atoms form bonds with other atoms in order to reach a stable state. A stable state is reached when an atom's outer valence shell is full. Atoms can obtain stable outer shells by sharing electrons (covalent bonds). They can also obtain stable outer shells by either giving away or taking on elections (ionic bonds). When atoms give away electrons, they become positively charged. When atoms take on electrons, they become negatively charged. In both cases, the charged atoms are now called ions. Positively charged ions and negatively charged ions attract each other, forming ionic bonds. Ionic bonds form from the opposite charges attracting each other and not from the sharing of electrons. Once atoms have achieved a stable state, they resist interacting with other atoms.

The electrons shared in a covalent bond are not always shared equally between the atoms. If electrons are shared equally in a bonded molecule, the molecule is nonpolar. If the electrons are shared unequally, the molecule is polar. The sharing of electrons is what makes covalent molecules either hydrophilic (polar) or hydrophobic (nonpolar). Recall that the cell membrane has both hydrophilic and hydrophobic regions.

Atoms join together in unique ways to make both living and nonliving structures. Carbon bonds with other atoms to make up the four organic compounds found in living organisms. The four organic compounds and their features are outlined in Table 2.1. Refer to this table as you review this unit.

All organic compounds contain carbon bonded with oxygen and hydrogen in unique ways. (Proteins and nucleic acids also bond with other types of atoms.) Some inorganic molecules exhibit the same atomic bonding pattern as organic molecules. Two examples of this are carbon dioxide (CO_2) and bicarbonate ions (HCO_3^-). For this reason, an added distinction for organic molecules is that living organisms produce them.

NOS Connections

- Theories support scientific concepts and can be modified as new discoveries are made (nucleic acid structure—DNA, RNA).
- Collaboration and cooperation between scientists lead to new theories and scientific discoveries (nucleic acid structure—DNA, RNA).
- Patterns, trends and discrepancies exist in the natural world. (Some organic molecules can be synthesized in the laboratory.)
- Models can be used to show scientific structures (molecular models).
- Ethical issues exist within scientific research (animal experimentation).

2.1 MOLECULES TO METABOLISM

TOPIC CONNECTIONS

- Option B—Biotechnology and bioinformatics

Chemical Elements

Carbon, hydrogen, oxygen and nitrogen are the most frequently occurring chemical elements in living organisms. These elements form the organic compounds that make up all living organisms. Carbon is most abundant of all elements in living organisms and is found in all organic molecules. Since carbon has only 4 electrons on its outer valence shell (second shell), it needs to obtain 4 more to become stable. To achieve a stable state, carbon can form 4 covalent bonds with other atoms (including other carbon atoms), can form linear or ring structures, and can form single, double or triple bonds. The ability of carbon to form so many different structures makes it the perfect element to use to build the many and varied organic compounds. Since carbon loves to bond with nitrogen, oxygen and hydrogen, these 4 elements are the key elements found in all organic compounds.

Organic Molecule	Monomer	Bond	Uses
Carbohydrates (hydrate of water) contain C, H, O bonded in a 1:2:1 ratio of CH_2O	Monosaccharide	Glycosidic linkages	**Instant energy** Monosaccharides **Stored energy** Polysaccharides: glyco[gen (ani]mals)/amylose (starch i[n...]) **Structure** Polysaccharides: cellulose (p[...] cell wall)/chitin (fungi cell wall) **Component of nucleic acids** Monosaccharides: ribose/ deoxyribose
Proteins contain C, H, O, N and sometimes S	Amino acids	Peptide bonds	**Transport** Haemoglobin (oxygen) **Communication** Hormones (insulin) **Protection** Antibodies **Catalyst** Enzymes **Recognition** Glycoproteins (cell membrane) **Movement** Muscles (actin and myosin) **Structure** Collagen and keratin
Lipids contain C, H, O	Glycerol and fatty acids	Ester bonds	**Long-term energy storage** Fats, waxes and oils: Insulation and buoyancy (fat and whale blubber) **Protection:** Fat surrounds all organs **Cell membrane**: Component of phospholipid bilayer
Nucleic acids contain C, H, O, N, P	Nucleotide	Phospho-diester bonds	Information storage: DNA Information transfer: RNA Energy transfer: ATP/ADP Coenzymes (hydrogen carriers): $NADP/FADH_2/NADPH$

MOLECULAR BIOLOGY

...iety of other elements are important to living organisms. Among these elements are
...ur, calcium, phosphorus, iron and sodium. The importance of each element to living
...nisms is presented in Table 2.2.

Table 2.2: Elements and Their Roles in Living Organisms

Element	Role in Animals	Role in Plants	Role in Prokaryotic Cells
Sulphur	Component of amino acids	Component of amino acids	Component of amino acids and used by some prokaryotes (chemoautotrophs) as an energy source
Calcium	• Bone and teeth formation • Muscle contraction • Nerve impulse • Cell messenger • Blood clotting	• Component of cell wall for added strength • Used for geotropism (locating gravity)	Signalling and cell structure
Phosphorus	Component of nucleic acids (including DNA, RNA and ATP)	Component of nucleic acids (including DNA, RNA and ATP)	Component of nucleic acids (including DNA, RNA and ATP)
Iron	Component of enzymes and haemoglobin (oxygen-carrying molecule in red blood cells)	Component of enzymes (proteins that catalyse reactions)	Energy production
Sodium	• Active transport • Water balance (osmoregulation) • Neural impulses	• Active transport • Water potential (water balance in cells)	Flagellum movement

Organic Compounds

Organic compounds contain carbon and are made by living organisms. Inorganic compounds
may contain carbon, but living organisms do not usually make them. Some molecules, such
as CO_2 and HCO_3^-, are exceptions to this rule. Although living organisms usually do not produce inorganic compounds, many inorganic compounds are vital for life functions.

Amino acids, glucose, ribose and fatty acids are four organic molecules that are important
in cellular processes and structures. Be prepared to identify as well as sketch the basic structure of these molecules on the IB Biology Exam. (Refer to Table 2.3.)

Table 2.3: Basic Structures of Amino Acids, Glucose, Ribose and Fatty Acids

Amino Acid

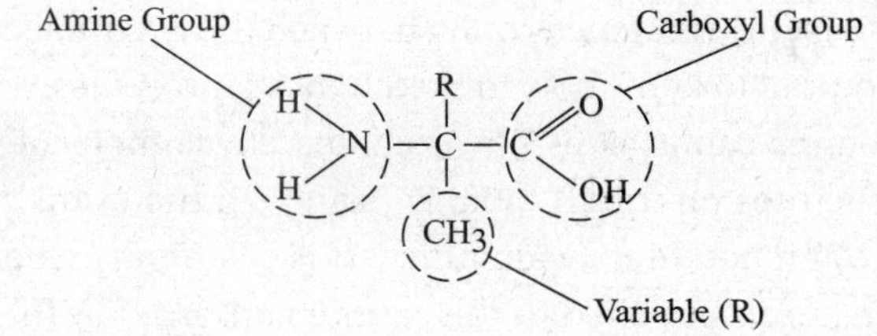

Amine Group

Carboxyl Group

R

H — N — C — C = O / OH

CH₃

Variable (R)

Glucose

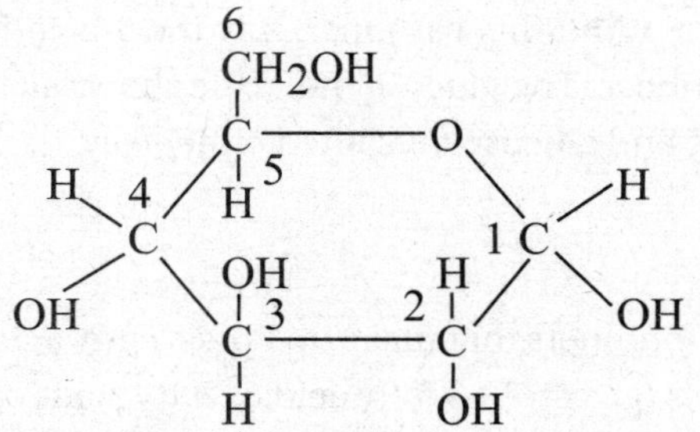

6 CH_2OH

C 5

H 4 C / OH

OH

C 3

H

O

1 C / H

H 2 C / OH

OH

H

Ribose

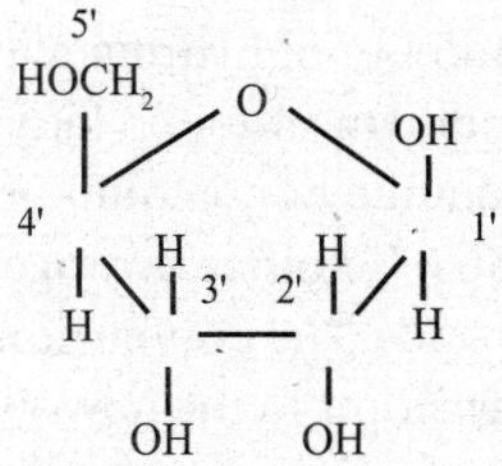

5' $HOCH_2$

4'

H
H

O

3' 2'

H
H

OH   OH

OH

1'

H
H

Fatty Acids

Saturated

H   H   H   H   O
| | | | //
H — C — C — C — C ··· C
| | | | \
H   H   H   H   OH

Length of chain varies

Unsaturated

H   H   H   H   O
| | | | //
H — C — C — C = C ··· C
| | \
H   H   OH

Length of chain varies

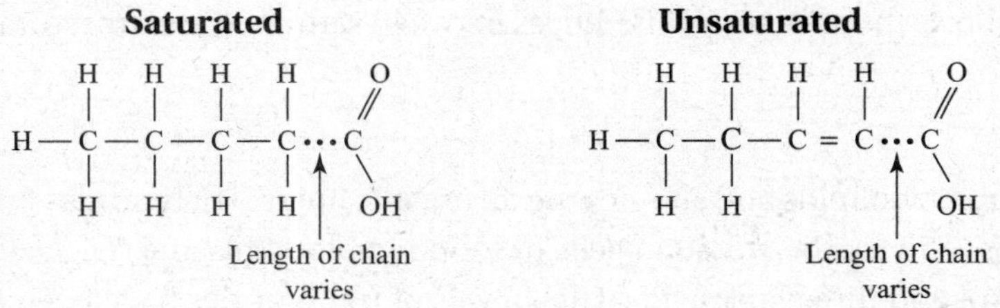

NOTE
Some organic molecules, such as urea, can be artificially created in the laboratory.

AMINO ACIDS

There are a total of 20 different amino acids each with a different R-group. The R-group is the only part of the molecule that differs among amino acids. The amine group consists of the $-NH_2$ portion of the molecule, and the carboxyl group consists of the –COOH portion of the molecule. The carboxyl group is acidic due to its ability to donate (lose) a hydrogen atom (proton) to solution. This is where the name *amino* (the amine group) *acid* (the carboxyl group) comes from. Amino acids are the building blocks (monomers) for proteins. Refer to the amino acid depicted in Table 2.3.

GLUCOSE AND RIBOSE

Glucose and ribose are both simple carbohydrates that have slightly different structures. Glucose contains 6 carbon atoms, and ribose contains 5 carbon atoms. If you remember this, it will help you distinguish (state differences only) between the two molecules simply by counting carbons. You are required to know how to sketch these molecules.

Although both structures can exist as linear chains, these molecules will always be presented in their ring structures on the IB Biology Exam. On the exam, all the carbons found in these molecules are often not directly identified. It is customary when drawing these molecules to show only the atoms attached to the central carbons. For this reason, you need to understand that at each bend in the ring structure is a central carbon between the atoms shown. As shown in the monosaccharide ribose in Table 2.3, the numbers 1 through 4 represent the location of the four additional carbons. Only the 5th carbon is directly shown since it is part of a side chain of ribose. The glucose molecule shows all 6 of its carbons labelled. Be prepared to see both ribose and glucose presented either way.

FATTY ACIDS

Fatty acids are part of the monomers that make up lipids. Fatty acids are often associated with glycerol to form triglycerides (glycerol + 3 fatty acids). Fatty acids consist of long chains of carbon and hydrogen atoms covalently bonded (hydrocarbon chain) with a carboxyl group on one end of the molecule. Saturated fatty acids have no double bonds. In other words, saturated fatty acids contain the maximum number of hydrogen atoms. In contrast, unsaturated fatty acids have at least some double bonds. Unsaturated fatty acids do not contain the maximum number of hydrogen atoms. Each double bond could potentially be broken into two single bonds, thereby allowing the addition of another hydrogen atom. Unsaturated fatty acids are easier to digest than saturated fatty acids. The double bonds in unsaturated fatty acids can be broken, allowing the addition of new atoms to the now-available single bonds. Double bonds are also more flexible than single bonds. As a result, unsaturated fatty acids are liquid at room temperature while saturated fatty acids are solid at the same temperature. The dots in the diagram of the fatty acids shown in Table 2.3 illustrate that the chain can be of great length. Table 2.3 also shows the structural differences between saturated and unsaturated fatty acids.

LIPIDS

Triglycerides, phospholipids and steroids are all organic lipids. Triglycerides are an important source of energy and can be stored in cells or found in blood plasma. Triglycerides are composed of glycerol and three fatty acids. High levels of triglycerides have been linked to coronary heart disease and therefore should be limited in the diet. Phospholipids are an important component of the cell membrane. Steroids have many functions in the body, including communication (hormones), mammalian cell membrane structure (cholesterol), aiding in immune function and regulating reproductive processes. Steroids can be quickly identified based on the presence of 4 carbon rings. On the exam, you need to recognize these molecules from basic structural drawings. (See Figure 2.1.)

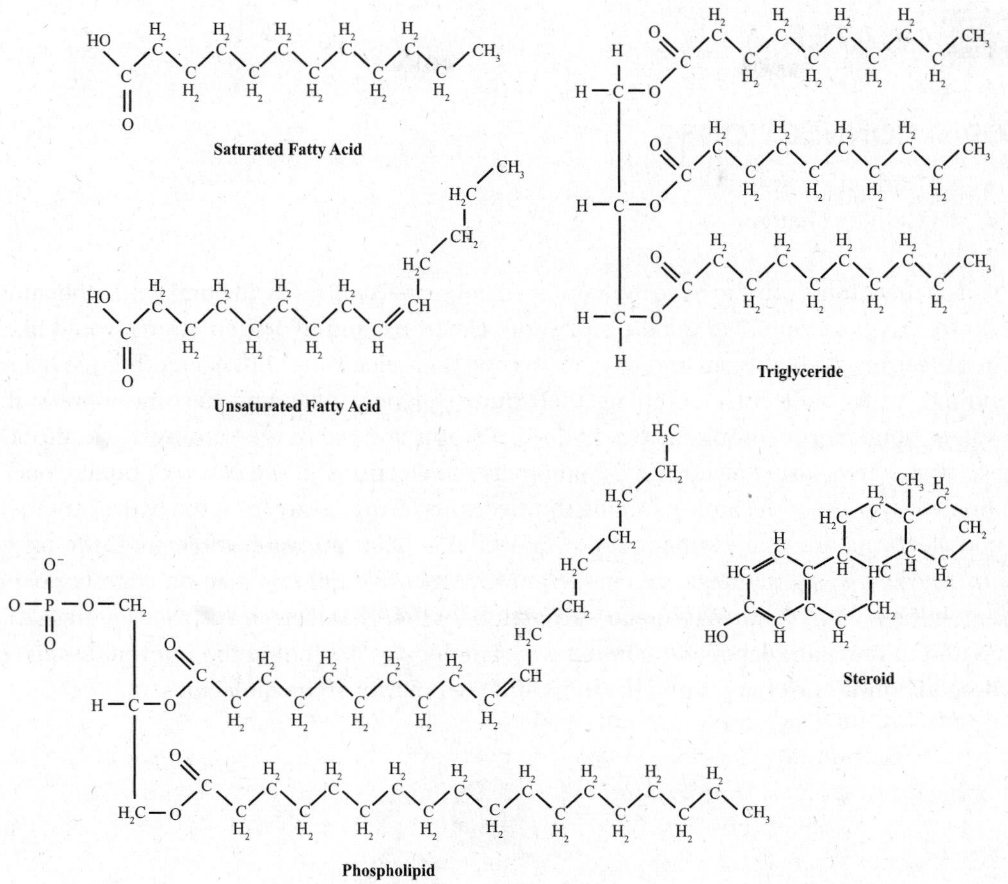

Figure 2.1. Lipid molecules

Try to limit saturated fatty acids in your diet!

- **Saturated fatty acids** increase the levels of bad cholesterol (LDL) in your body. High levels of LDL are linked to coronary heart disease. Sources of saturated fatty acids include butter, cheese, fatty meats and cream. Limit these in your diet!
- **Unsaturated fatty acids** increase the levels of good cholesterol (HDL) in your body. Sources of unsaturated fatty acids include olive oil, nuts and avocado. Include these in your diet!

Metabolism

The sum of all reactions occurring inside of cells is referred to as metabolism. Metabolic processes involve the building up and breaking down of substances.

- **Anabolic reactions**: These reactions are the building of complex molecules from simple molecules. The formation of a polymer from monomers through the process of condensation is an anabolic reaction.
- **Catabolic reactions**: These reactions are the breaking down of complex molecules into simple molecules. The formation of monomers from a polymer through the process of hydrolysis is a catabolic reaction.

REMEMBER

Anabolic reactions build up.

Catabolic reactions break down.

2.2 WATER

TOPIC CONNECTIONS

- 4.3 Carbon cycling
- 4.4 Climate change

Recall that atoms want to have their outer valence shells full and will bond with other atoms to do so. Oxygen would like to take on 2 more electrons, and hydrogen atoms would like to gain 1 electron. By hydrogen and oxygen sharing their electrons with each other (covalently bonding), water molecules complete their outer valence shells and become more stable. Covalent bonds form among the two hydrogen atoms and the oxygen atom (H_2O). Although these atoms are sharing electrons, oxygen pulls the electrons in the covalent bond closer to its nucleus (centre of the atom), forcing the electrons farther away from the hydrogen nuclei. When electrons are shared unequally, a molecule is polar. In water molecules, the oxygen atom takes on a slightly negative charge and the hydrogen atoms take on slightly positive charges due to oxygen having a greater attraction for the shared electrons. (See Figure 2.2.) Be prepared to draw and label two bonded water molecules, including the internal bonds, the hydrogen bonds, the charges on the atoms and the polarity of the molecules.

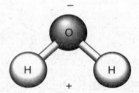

Figure 2.2. Water molecule

The polarity of water leads to the formation of hydrogen bonds between water molecules. Hydrogen bonds form due to an attraction between a slightly positive hydrogen atom in one polar molecule and a slightly negative oxygen atom in another polar molecule. Hydrogen bonding is an important feature of water molecules that gives water many unique properties that are important to living organisms. (See Figure 2.3.)

> ### REMEMBER
>
> Water is not the only molecule that has hydrogen bonding. The formation of the DNA double helix involves hydrogen bonding as well.

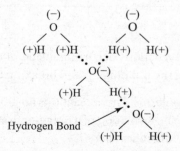

Figure 2.3. Cohesive forces in water molecules

MOLECULAR BIOLOGY

Properties of Water

Water exhibits thermal, cohesive and solvent properties that are due to the hydrogen bonds that form among water molecules. Hydrogen bonds are individually weak. Collectively, though, they can be quite strong. Although they are not as strong as covalent bonds, hydrogen bonds resist breaking. The strong attraction among water molecules gives water many unique properties.

The attraction of water molecules to each other is called *cohesion*. The attraction of water molecules to other surfaces is called *adhesion*. Both adhesion and cohesion are due to the formation of hydrogen bonds. Adhesion and cohesion give water the properties that are important to living organisms. (Refer to Table 2.4.)

Table 2.4: Characteristics of Water as They Relate to Living Organisms

Characteristic	Property	Importance to Living Organisms
High specific heat: Water resists temperature changes.	Thermal	Organisms can maintain stable body temperatures in varied climates for a period of time. Aquatic environments can maintain fairly stable temperatures to support multiple life-forms.
Cohesion: Water attracts other water molecules.	Cohesive	**Transpirational pull:** Water passively moves from the roots to the leaves of plants, supplying the entire organism with water. As one molecule evaporates from the leaves, another is pulled into the roots, thereby forcing the entire water column upwards. **Surface tension**: The attraction of the water molecules on the surface of water creates a thin "film" of tightly held water molecules. This provides a surface for small insects to walk on.
Universal solvent: Water can dissolve many substances. The polarity of water allows it to dissolve both positively and negatively charged substances (ions, polar molecules).	Solvent	Water dissolves polar substances and ions in order for them to be easily transported into the tissues of living organisms. Since water can dissolve many substances, it is the site of many metabolic processes.
High heat of vaporization: Water absorbs a lot of heat when it changes from a liquid state of matter to a gaseous state of matter.	Thermal	**Evaporative cooling**: Since heat is lost from the body when water evaporates, sweating is an efficient method of cooling. Evaporation of water also helps cool plant tissues.

FEATURED QUESTION

May 2004, Question #5

Which feature of water determines its solvent properties?

A. Peptide bonds
B. Hydrophobic interactions
C. Ionic bonds
D. Polarity

(Answer is on page 553.)

(Answer is on page 553.)

REMEMBER

Animals are made up of about 70% water, making the characteristics of water vital to life functions!

How a compound is transported throughout the body depends on the compound's solubility in water.

- **Glucose, polar amino acids, oxygen and sodium chloride are soluble in blood plasma due to their hydrophilic nature.**
- **Fats and cholesterol are insoluble in water due to their hydrophobic nature and must be transported within protein capsules (lipoproteins). High levels of lipoproteins in the blood are dangerous since they are insoluble and can block blood flow.**

2.3 CARBOHYDRATES AND LIPIDS

TOPIC CONNECTIONS

- Option B—Biotechnology and bioinformatics

Monomers are the most basic (smallest) unit of structure used to build the four major organic compounds: carbohydrates, proteins, lipids and nucleic acids. When two monomers bond together, a dimer is formed. When more than two monomers bond together, polymers are formed. One polymer can include thousands of monomers. Monomers of carbohydrates are called monosaccharides (single sugar). Dimers of carbohydrates are called disaccharides (two monosaccharides bonded). Polymers of carbohydrates are called polysaccharides (many monosaccharides bonded). Creating different levels of carbohydrate structure allows the molecules to perform specific functions or create specific structures. Table 2.5 provides examples of carbohydrates, and Table 2.6 outlines the uses of carbohydrates in living organisms.

Mono: single (one subunit)

Di- double (two subunits)

Poly- multiple (many subunits)

MOLECULAR BIOLOGY

Table 2.5: Examples of Monosaccharides, Disaccharides and Polysaccharides

Monosaccharides	Glucose
	Fructose
	Galactose
Disaccharides	Sucrose (glucose + fructose)
	Maltose (glucose + glucose)
	Lactose (glucose + galactose)
Polysaccharides	Cellulose
	Starch
	Glycogen

Table 2.6: Uses of Monosaccharides, Disaccharides and Polysaccharides in Animals and Plants

Animal Carbohydrate	Structural Level	Function (Use)
Glucose	Monosaccharide	Cellular respiration: used for producing ATP
Lactose	Disaccharide	Found in milk to nourish infants
Glycogen	Polysaccharide	Energy storage
Plant Carbohydrate	**Structural Level**	**Function (Use)**
Fructose	Monosaccharide	Makes fruit sweet
Sucrose	Disaccharide	Main sugar transported in plants (table sugar)
Cellulose	Polysaccharide	Component of cell wall (provides strength and rigidity)
Amylose (starch)	Polysaccharide	Energy storage

Cellulose, amylose (starch) and glycogen are examples of polysaccharides. Cellulose is found in the cell wall of plants. It provides support and protection. Starch is the storage molecule for carbohydrates in plants and can exist as amylose or as amylopectin. Amylose consists of unbranched chains of glucose, while amylopectin consists of branching side chains of glucose molecules. Glycogen is the storage molecule for carbohydrates in animals. The structures of these molecules are outlined in Table 2.7.

Table 2.7: The Structure of Starch, Glycogen, and Cellulose

Starch: stored energy in plants	Storage organ: potato	■ Amylose: straight chains of glucose ■ Amylopectin: straight chains of glucose with side branches of glucose chains
Glycogen: stored energy in animals	Storage organ: muscle cells	Straight chains of glucose with branching side chains
Cellulose: structural component of cell wall	Component of plant cell walls	Layered straight chains of glucose

Condensation and Hydrolysis

In order to utilize the monomers that are bonded in large polymers, your body must be able to digest polymers into their individual monomers. In addition, your body must be able to build new polymers from the monomers that result from digestion of polymers.

The process of joining monomers together to create polymers is referred to as condensation. Condensation describes the process of creating dimers (the joining of two monomers) and polymers (the joining of many monomers) with a resulting loss of water. Whenever monomers are joined together to produce organic macromolecules, the removal of water is involved regardless of the type of organic compound involved. The formation of a dipeptide results from the bonding together of two amino acids through the process of condensation. Triglycerides form from the condensation of three fatty acids and one glycerol. The formation of a triglyceride releases three water molecules. Figure 2.4 shows the formation of a dipeptide and a triglyceride by the process of condensation.

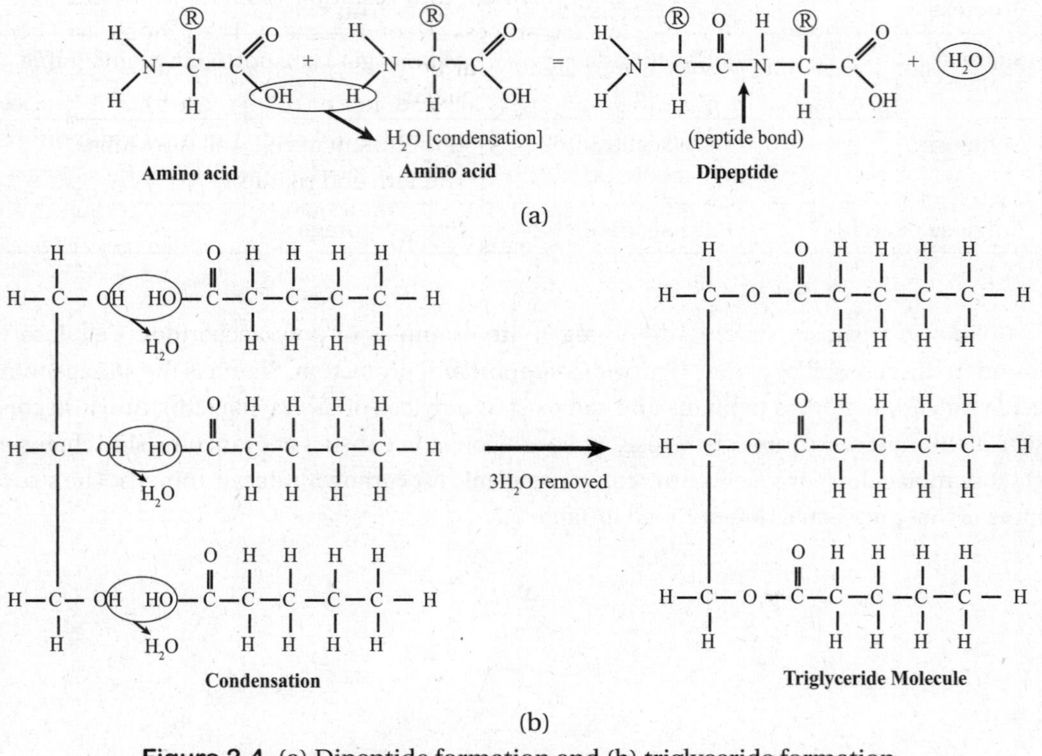

Figure 2.4. (a) Dipeptide formation and (b) triglyceride formation

Fatty acids can be saturated, monounsaturated or polyunsaturated. Fatty acids that are unsaturated are either found as *cis* or *trans* isomers. The position of the hydrogen atoms on the same side of the double bond in the *cis* isomer allows the molecule to form a bent structure. The position of the hydrogen atoms on the opposite sides of the double bond in the *trans* isomer allows the molecule to form a straight chain structure. Fatty acid structures are shown in Figure 2.5.

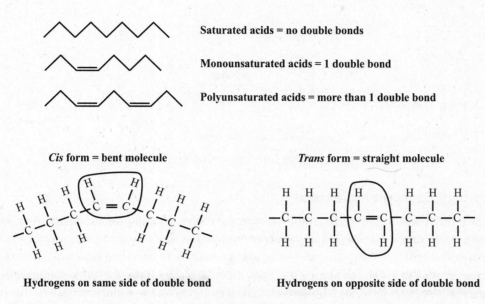

Figure 2.5. (a) Types of fatty acids and (b) *cis* and *trans* unsaturated fatty acids

Hydrolysis is the process of breaking polymers into resulting monomers. This reaction involves the addition of water—the reverse process of condensation. The processes of condensation and hydrolysis occur continually in your body as you break down polymers into monomers and then use the resulting monomers to build new polymers. The phrase "you are what you eat" really is true. You are ingesting large polymers, digesting them into monomers, and using the resulting monomers to build your own polymers. The protein in the steak you ate last night might now be a part of the protein found in your own muscles today! Figure 2.6 shows the formation of polymers from monomers by condensation and the breaking of polymers into monomers by hydrolysis.

Hydrolysis

Hydro: water

Lysis: to break open

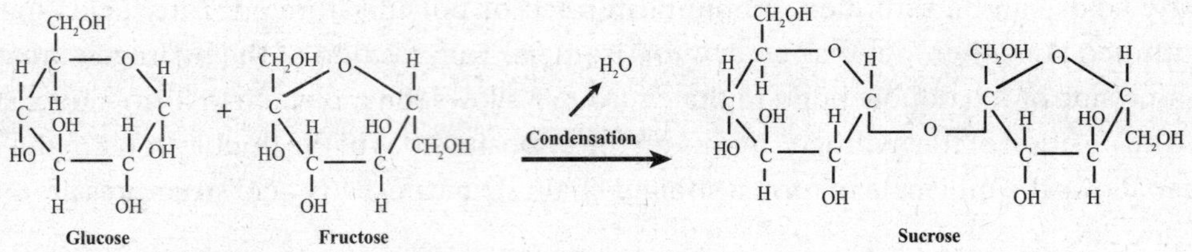

(a) The formation of sucrose from the condensation of glucose and fructose

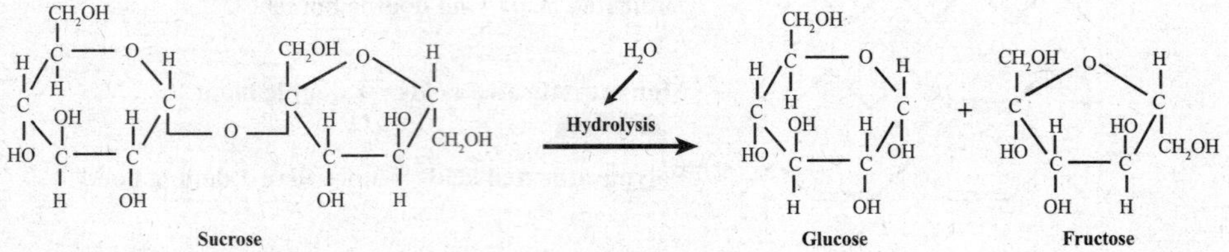

(b) The formation of glucose and fructose from the hydrolysis of sucrose

Figure 2.6. (a) Condensation and (b) hydrolysis

Energy Storage Molecules

Lipids function for long-term energy storage, for thermal insulation, for cushioning, and as a component of the cell membrane. Both carbohydrates and lipids can be used for long-term energy storage. However, the amount of energy they each store and how quickly these molecules can be digested for energy use does differ. Carbohydrates are broken down and used for energy more rapidly than lipids can be broken down and used for energy. For this reason, the energy stored in carbohydrates will be used before any stored lipids. This is the reason that marathon racers eat a big pasta meal before competing in a race. The extra carbohydrates in the pasta will be stored in their body for quick access during the marathon.

When compared to carbohydrates, lipids store twice as much energy gram for gram as carbohydrates do. Although lipids take more time to digest, twice the amount of energy will be released from the digestion of lipids than from the digestion carbohydrates. Since lipids can store more energy per gram, they are lighter to store than carbohydrates. This makes lipids more suitable for long-term energy storage. For this reason, humans store some excess carbohydrates as glycogen but convert most of the excess glucose into lipids. Many birds use lipids to store most of their energy to reduce the effect of the weight of stored energy on flight ability.

2.4 PROTEINS

Protein Structure

All proteins are formed from different combinations of the same 20 amino acids. These amino acids are linked together by condensation, forming peptide bonds between adjacent amino acids. Organisms synthesize different proteins to carry out the many different functions. The entire set of proteins that can be expressed by an individual is referred to as their proteome. Each individual possess a unique proteome.

Following protein synthesis at the ribosome, the polypeptide must undergo structural changes in order for it to carry out its intended function. There are four levels of protein structure, each involving unique bonding patterns. How a protein folds is based on the composition of the R-groups of the amino acids in the primary chain. Each level of protein folding is outlined in Table 2.8.

Table 2.8: Protein Structure

Structural Level	Features	Visual Display
Primary structure	Peptide bonds Linear structure of amino acids (order based on mRNA codons) Produced by translation at the ribosome	gly phe val ala ser cys tyr ala
Secondary structure	Hydrogen bonds increase folding of the polypeptide Folds may exhibit an alpha helix or a beta pleat	Hydrogen bonds show as dashes phe val ser gly ala cys ala tyr Alpha helix Beta pleat
Tertiary structure	Further folding into a 3-D structure involving disulphide bonds Determines the protein's function	
Quaternary structure	2 or more polypeptides noncovalently bonded together Prosthetic groups may be bonded with the polypeptides (for example, iron is a prosthetic group in haemoglobin)	Haemoglobin Polypeptide chain — Polypeptide chain Polypeptide chain — Heme — Polypeptide chain

Globular and Fibrous Proteins

Depending on their function, proteins may exist as either globular or fibrous structures. Globular proteins are spherical in shape with a compact structure. They usually carry out metabolic processes, have complex tertiary or quaternary structure, and are soluble in water. In contrast, fibrous proteins are linear in shape with an extended structure. They rarely have complex tertiary structure and usually perform structural roles in the body. Table 2.9 outlines globular and fibrous protein structures.

Table 2.9: Globular and Fibrous Proteins

Globular Proteins	
Example	**Function**
Haemoglobin	Transports oxygen
Enzymes (amylase, helicase, etc.)	Catalyse reactions
Antibodies	Protection in the immune system by targeting antigens
Fibrous Proteins	
Example	**Function**
Keratin	Gives strength to hair and nails
Collagen	Connective tissue
Thrombin	Involved in blood clotting

Polar and Nonpolar Proteins

The R-groups of amino acids vary in structure in order for proteins to express both polar (hydrophilic) and nonpolar (hydrophobic) properties. Channel proteins must consist of both polar and nonpolar amino acids in order to span the nonpolar interior of the plasma membrane and to exist within the polar cytoplasm. In addition, enzymes must be able to catalyse reactions that involve both polar and nonpolar substrates. Enzymes that catalyse reactions involving polar substrates must have polar amino acids in their active sites in order to attract polar substrates. Enzymes that catalyse reactions involving nonpolar substrates must have nonpolar amino acids in their active sites in order to attract nonpolar substrates. This ensures an attraction between the enzyme and substrate in order to catalyse a specific reaction efficiently. The structural folding of proteins is due to the polar attraction of hydrophilic R-groups to water and the nonpolar repulsion of hydrophobic R-groups away from water.

Protein Function

Proteins have many functions:

- **Catalysing reactions**: Examples: enzymes such as amylase (digests amylose), rubisco (catalyses carbon fixation)
- **Transport**: Example: haemoglobin transporting oxygen
- **Communication** (hormones): Example: insulin
- **Movement**: Examples: actin and myosin in muscle cells

- **Protection**: Example: immunoglobulins (antibodies)
- **Blood clotting**: Examples: thrombin and fibrin
- **Vision**: Example: rhodopsin (pigment that perceives light)
- **Structure**: Examples: keratin (in hair in nails), collagen (in connective tissue) and spider silk (in spider webs)

2.5 ENZYMES

TOPIC CONNECTIONS

- 8 AHL—Metabolism, cellular respiration and photosynthesis

Enzyme Function

> **Enzyme:** A biological catalyst that speeds up a reaction by lowering the activation energy.

Activation energy is the amount of energy that must be invested in a reaction in order for the reaction to proceed forward. Regardless of whether the reaction is exergonic (releases energy) or endergonic (takes in energy), the activation energy is always lowered by the work of enzymes. Enzymes are fairly specific as to which substrates they will act on. The enzyme and its substrate have molecular structures that encourage them to form bonds at the active site.

> **Active site:** The site on an enzyme where the substrate will bond.

The active site is where the reaction is catalysed. The reaction can be anabolic (bringing more than one substrate together—building molecules) or catabolic (breaking substrates into smaller units—breaking down molecules).

Enzyme-Substrate Specificity

The three-dimensional structure of each enzyme determines which substrate it will react with. The name of the enzyme is often, but not always, based on the substrate with which it catalyses a reaction. To determine the name of the enzyme, the suffix *–ase* is added to the name of the substrate. For example, the catabolic enzyme lactase breaks the substrate lactose into glucose and galactose. Without the enzyme, the activation energy required for the conversion would be much higher. As a result, the reaction would take much longer and your body would not be able to use the energy from the resulting glucose and galactose monomers efficiently.

> **REMEMBER**
>
> You can usually identify enzymes because most of their names end in *–ase*.
>
> Remember that enzymes are proteins built from amino acids.

The enzyme shown in Figure 2.7 forms an enzyme-substrate complex with substrate *A* at the active site. Notice that substrate *B* does not fit into the active site. A reaction involving substrate *B* would be catalysed by an enzyme with an active site matching the shape of substrate *B*. This is known as the *lock-and-key* model. The enzyme and substrate fit together just like a particular lock can be opened only with a particular key that has a particular shape.

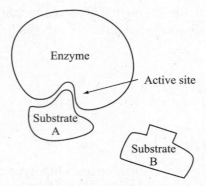

Figure 2.7. Enzyme-substrate specificity

Denaturation of Enzymes

> **Denaturation (denature):** The loss of an enzyme's three-dimensional structure, leading to the loss of its ability to catalyse a reaction.

Enzymes will catalyse reactions most efficiently in specific environments. Collisions between enzymes and substrates allow enzymatic reactions to occur. Each enzyme has an optimal temperature, pH, and substrate concentration at which it functions most effectively. Most enzymes in the human body function effectively at core body temperature (37°C). As the temperature gets higher, the enzyme will start to lose its three-dimensional biological structure (denature) and will no longer bond to its substrate. Because reactions in the body are dependent upon enzymes, a very high fever can be life threatening due to the inability of enzymes to catalyse biological reactions. At lower temperatures, the reactions will slow down due to slowed molecular motion. However, the enzyme will not lose its structure.

Since the optimal pH for most of the human body is close to 7.4, most enzymes function best at this pH level. If an enzyme is exposed to an environment with a pH different from the optimal pH for the enzyme, the reaction the enzyme catalyses will be slowed. If the pH alters too much, the enzyme will denature and not be able to catalyse the reaction at all. Some enzymes, such as gastric enzymes, function best at a very low pH (3–4). In contrast, enzymes in your mouth function best at around a pH of 6. When you swallow and enzymes from your mouth reach the acidic gastric juices, they lose their three-dimensional structure (denature) and can no longer catalyse reactions. The same is true for gastric enzymes that function best at a pH of 3–4. When they reach the intestines, they denature in the more alkaline environment, pH 7–8. (See Figure 2.8.)

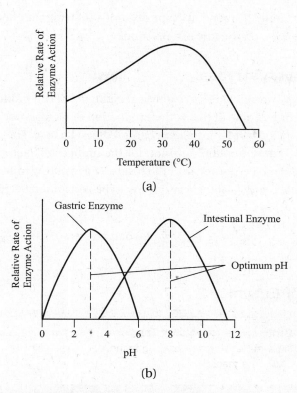

(a)

(b)

Figure 2.8. The effect of (a) temperature and (b) pH on the rate of enzyme-catalysed reactions

Substrate concentration also affects the rate at which enzymes function. Enzyme function increases as substrate concentration increases—but only up to the point of saturation. Once all the available enzymes are working as rapidly as they can on the available substrates, the enzymes cannot work any faster regardless of increasing substrate concentrations. This is analogous to assembly line production. The workers can make more products as more supplies are added. At some point, though, they will be working as fast as they can regardless of the amount of supplies available. (Refer to Figure 2.9.)

REMEMBER

Enzyme structure directly relates to enzyme function!

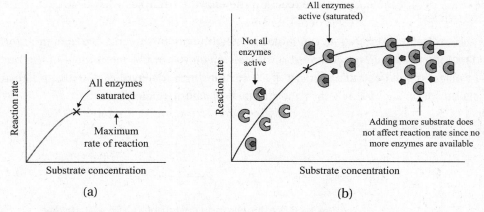

(a)   (b)

Figure 2.9. Graphs of the effect of substrate concentration on the rate of enzyme-catalysed reactions. Graph (b) is a visual representation of (a).

You may be asked about the general shape of graphs showing the effect of temperature, pH, or enzyme concentration on the rate of a reaction.

Industrial Uses of Enzymes

Industry uses immobilized enzymes to catalyse reactions that can produce products useful to humans. An excellent example of this is the production of lactose-free milk. (Refer to Figure 2.10.) Individuals who are lactose intolerant cannot digest lactose. They experience problems with indigestion and cannot obtain nutrition for the undigested lactose. Lactase is attached to alginate beads, and milk is poured over the beads. The resulting milk is lactose free (lactose is digested into glucose and galactose) and ready to be consumed by individuals with lactose intolerance.

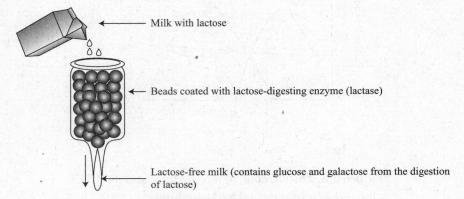

Milk with lactose

Beads coated with lactose-digesting enzyme (lactase)

Lactose-free milk (contains glucose and galactose from the digestion of lactose)

Figure 2.10. Production of lactose-free milk

2.6 STRUCTURE OF DNA AND RNA

TOPIC CONNECTIONS

- 2.2 Water
- 3.5 Genetic modification and biotechnology
- 7 Nucleic acids

Nucleotides

Deoxyribonucleic acid (DNA) is the molecule used by organisms to store genetic information. The information is stored based on the arrangement of the monomers of the nucleic acid—nucleotides. Nucleotides consist of a 5-carbon sugar (deoxyribose), a phosphate and a nitrogenous base. Figure 2.11 shows the structure of a basic nucleotide.

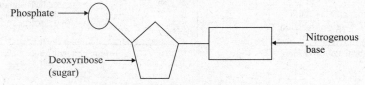

Phosphate

Deoxyribose (sugar)

Nitrogenous base

Figure 2.11. Nucleotide structure

MOLECULAR BIOLOGY

Four different nitrogenous bases differentiate the four nucleotides used to build DNA. These bases are adenine, guanine, cytosine and thymine. A fifth base, uracil, is found only in ribonucleic acid (RNA). Uracil is used in place of thymine when building RNA. In DNA, each nitrogenous base is bonded to a molecule of deoxyribose (a 5-carbon sugar) that is also bonded to a phosphate. Figure 2.12 shows the structures of all five nitrogenous bases.

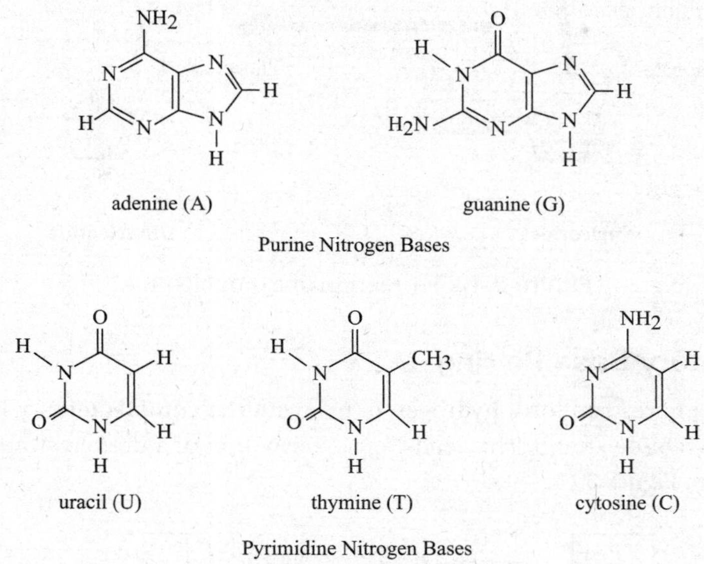

adenine (A)   guanine (G)

Purine Nitrogen Bases

uracil (U)   thymine (T)   cytosine (C)

Pyrimidine Nitrogen Bases

Figure 2.12. Nitrogenous bases

Adenine and guanine are purines. Cytosine, thymine and uracil are pyrimidines. Purines (adenine and guanine) have double-ring structures. Pyrimidines (cytosine, thymine and uracil) have single-ring structures. A purine always bonds with a pyrimidine, and a pyrimidine always bonds with a purine. Adenine will bond with thymine and guanine with cytosine. Remember that when building RNA, uracil is a pyrimidine because it replaces thymine and bonds with adenine.

> **REMEMBER**
>
> You can remember the pyrimidines in DNA because they each have a "y" in their name.
>
> Pyrimidines: thymine (uracil in RNA), cytosine,
>
> Purines: guanine, adenosine

DNA	RNA
A – T	A – U
C – G	C – G

Polynucleotides

Nucleotides are the monomers of nucleic acids. Neocleotides join together to form dinucleotides and, ultimately, polynucleotides such as DNA and RNA. The formation of covalent bonds between the nucleotides leads to the formation of a single-strand of nucleotides. Figure 2.13 shows the formation of a phosphodiester (covalent) bond joining two nucleotides together.

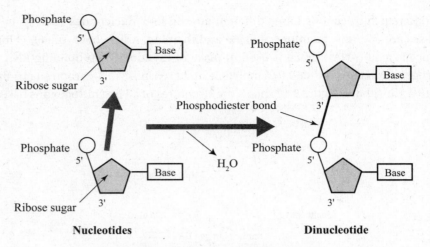

Figure 2.13. Formation of a dinucleotide

Complementary Base Pairing

Polynucleotide strands can form hydrogen bonds with the complementary bases found on another polynucleotide strand. This leads to the formation of a double-stranded DNA molecule as shown in Figure 2.14.

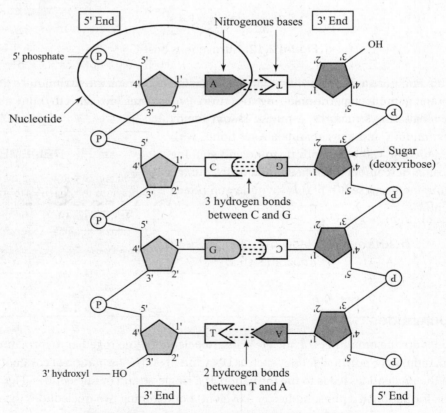

Figure 2.14. Formation of double-stranded DNA

MOLECULAR BIOLOGY

Recall that the monomers of nucleic acids are nucleotides. Nucleotides are joined together by covalent (phosphodiester) bonds to create a single strand of DNA. Two single strands are joined together with hydrogen bonds to create the double-stranded structure of DNA. Two hydrogen bonds form between adenine and thymine, and three hydrogen bonds form between guanine and cytosine. (See Figure 2.14.) Adenine and thymine are complementary bases. This means that their molecular structures attract each other and hydrogen bonds quickly form between them. The same is true for guanine and cytosine. Note that the two strands run opposite of each other. One strand will have the phosphate at the top bonded to the fifth (5') carbon, and the opposite strand will have the phosphate on the top bonded to the third (3') carbon. Knowing this is important if you are going to study HL material.

> **Covalent (phosphodiester) bonds** form between deoxyribose and phosphates. These are much stronger than hydrogen bonds.
>
> **Hydrogen bonds** form between the complementary nitrogenous bases. 2 hydrogen bonds form between adenine and thymine; 3 hydrogen bonds form between guanine and cytosine.

In addition to DNA, another type of nucleic acid is involved in storing and transferring genetic information. Ribonucleic acid (RNA) is made of nucleotides like DNA but has several distinct differences as compared with DNA. First, RNA does not contain the nitrogenous base thymine. Instead, thymine is replaced by uracil. Second, the sugar ribose in used instead of deoxyribose. Third, RNA is single stranded. Finally, RNA is a much shorter molecule than DNA since RNA codes for only one trait at a time. In eukaryotic cells, RNA must be able to leave the nucleus and travel to a ribosome in order to make proteins. The single-stranded structure of RNA allows it to fit through the nuclear pores, which are too small for DNA to travel through. The structure of RNA allows it to carry out its functions. Table 2.10 and Figure 2.15 outline the main differences between DNA and RNA.

Table 2.10: Comparisons between DNA and RNA

Feature	DNA	RNA
Number of strands	Double stranded	Single stranded
Nitrogenous bases	Adenine (A), thymine (T), guanine (G), cytosine (C)	Adenine (A), uracil (U), guanine (G,) cytosine (C)
Sugar present (a 5-carbon pentose)	Deoxyribose	Ribose
Length	Long (codes for all the genes)	Short (codes for only one gene at a time)
Location	In the nucleus (eukaryotes) or nucleoid region (prokaryotes)	In the nucleus (eukaryotes) and the cytoplasm (both eukaryotes and prokaryotes)
Composition	Nucleotides	Nucleotides

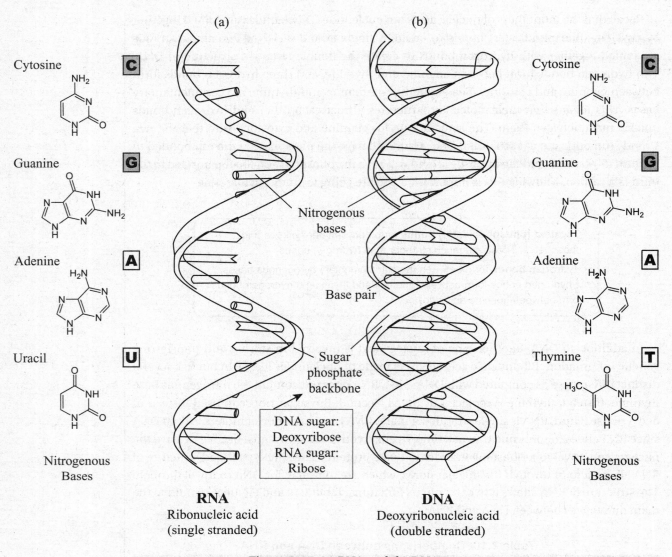

Figure 2.15. (a) RNA and (b) DNA

2.7 DNA REPLICATION, TRANSCRIPTION AND TRANSLATION

TOPIC CONNECTIONS

- 3.5 Genetic modification and biotechnology
- 7.2 Transcription and gene expression
- 7.3 Translation

DNA Replication

Replication of DNA is required for cellular division to create two exact copies of all the genetic material cells need to carry out life processes. In order to replicate (copy) DNA, the two strands of DNA must be separated. Each exposed strand serves as a template to make the two new double-stranded DNA molecules. The enzyme helicase unwinds the DNA helix strands and breaks the hydrogen bonds between the complementary bases. Free nucleotides with complementary bases are added to both exposed DNA strands by the enzyme DNA polymerase. When DNA separates, a *replication bubble* forms. Replication occurs at the two sites where the DNA is separating. These two sites are referred to as replication forks. The replication forks form at the opposite ends of the opened sections of the DNA molecule.

Two single strands of DNA result from the opening of the DNA molecule at each replication fork. The two strands are referred to as the *leading* and *lagging* strand. On the leading strand, replication occurs in a continuous fashion as DNA polymerase works towards the replication fork. On the lagging strand, DNA polymerase works away from the replication fork and adds the nucleotides in a discontinuous fashion. The two polymerase molecules are part of one large molecule, so they must work in close association with each other. This is why DNA replication on the lagging strand is discontinuous. The polymerase working on the lagging strand is constantly pulled back by the polymerase continually moving forward on the leading strand. Since the DNA polymerase working away from the replication fork is continually being pulled back and has to start over adding nucleotides, replication takes a bit longer on the lagging strand. This explains the name *lagging strand*. (See Figure 2.16.)

Polymerase is an enzyme that makes polymers by joining monomers together (*polymer*: many; *-ase*: enzyme).

DNA polymerase joins DNA monomers (nucleotides), creating long chains of DNA.

- **The leading strand** is formed continuously. DNA polymerase works towards the replication fork.
- **The lagging strand** is formed discontinuously. DNA polymerase works away from the replication fork.

RNA polymerase joins RNA monomers (nucleotides), creating a long chain of RNA.

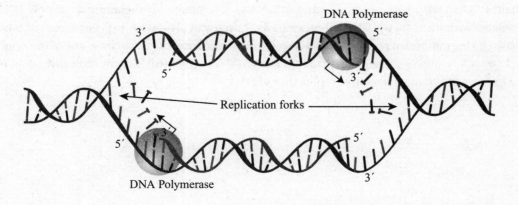

Figure 2.16. (a) DNA replication bubble

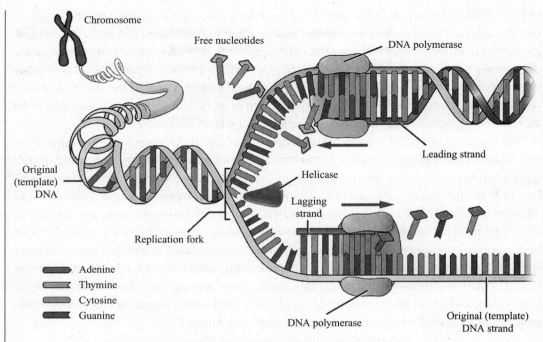

Figure 2.16. (b) processes at replication fork

COMPLEMENTARY BASE PAIRING

The replication of DNA involves the addition of new bases to the growing strands in a complementary manner. Adenine is always added when thymine is present, and guanine is always added when cytosine is present. A purine always bonds to a pyrimidine. By matching complementary base pairs, the DNA code can be maintained exactly as it appears on the original strand. In addition, the complementary base pairing allows for the DNA molecule to maintain an even width of 2 nm.

SEMICONSERVATIVE REPLICATION

Replication of DNA occurs by using the two original strands as templates for the two new strands. Two new DNA molecules are produced that each contain one of the original strands (half) of DNA. Using an original strand of DNA as a template for the addition of new DNA nucleotides conserves the original order of the bases and also helps prevent errors in base pairing. This method of replication is referred to as semiconservative because one of the original strands is always conserved (saved) in each of the newly formed DNA molecules. Figure 2.17 shows the semiconservative replication of DNA.

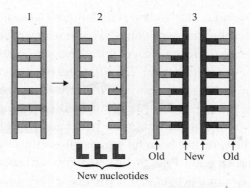

Figure 2.17. Semiconservative replication

Transcription

DNA contains all the information needed for a cell to synthesize all the proteins necessary to carry out life functions. The information stored in the genes of DNA must be copied and carried to a ribosome in order to produce proteins. The molecule that contains a copy of the DNA code for a protein and carries the message to the ribosome is a ribonucleic acid (mRNA). Both DNA and mRNA are nucleic acids that store information in the genetic code based on the order of their bases (nucleotides). The processes of transcription (copying the genetic code from DNA to mRNA) and translation (using the genetic code in RNA to produce proteins at the ribosome) are collectively referred to as *protein synthesis*. Transcription occurs in the nucleus, and translation occurs in the cytosol at the ribosome. (See Figure 2.18.)

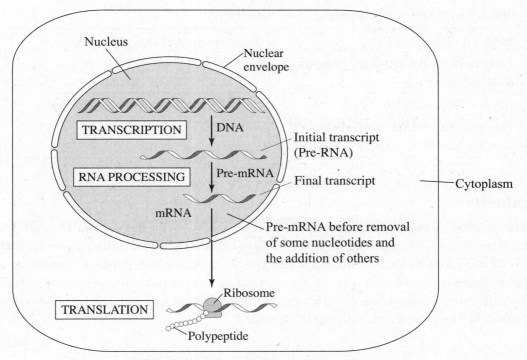

Figure 2.18. Transcription and translation

MOLECULAR BIOLOGY

The process in which sections of DNA that code for traits (genes) are copied into complementary RNA molecules is called transcription. The enzyme that catalyses this process is RNA polymerase. RNA polymerase recognizes where the gene begins on the DNA strand based on a region of DNA called the *promoter* region. The promoter region is a series of bases prior to the gene that is recognized by RNA polymerase. RNA polymerase opens up the DNA strand at this promoter site in order to begin making a copy of the DNA template for that specific gene.

During transcription, complementary bases are matched up just like they are in DNA replication. However, after the gene has been fully transcribed, the DNA molecule re-forms its hydrogen bonds and closes up again. Prior to the RNA molecule leaving the nucleus, it is often referred to as pre-mRNA (pre-messenger RNA). The pre-mRNA undergoes processing in the nucleus in order to become mature mRNA (messenger RNA). This processing involves the removal of some nucleotides and the addition of others. The newly formed messenger RNA (mRNA) molecule containing the complementary base sequences then leaves the nucleus for translation at the ribosome.

Three types of RNA are made: messenger RNA (mRNA), ribosomal RNA (rRNA) and transfer RNA (tRNA). Messenger RNA is the molecule formed from DNA that carries the genetic message to the ribosome to code for the production of a protein. Ribosomal RNA makes up a large portion of the ribosome. Transfer RNA transfers the correct amino acid to the ribosome. Table 2.11 differentiates among the different types of RNA.

Table 2.11: Types of RNA

Type of RNA	Function
mRNA—messenger RNA This is a copy of the DNA template.	Carries the genetic code for a protein to the ribosome
tRNA—transfer RNA Each molecule has a specific amino acid attached to it.	Transfers the amino acid from the cytoplasm to the ribosome
rRNA—ribosomal RNA This, along with associated proteins, form ribosomes.	Catalyses the formation of polypeptides (proteins)

CODONS

The genetic code used to produce a protein is stored in mRNA in the form of triplet bases. Each triplet combination is referred to as a *codon* and consists of three mRNA nucleotides. Each of the 64 codons codes for one amino acid that will be brought to the ribosome to assemble a polypeptide. The first codon is always the start codon, and the last is always the stop codon. The stop codon does not code for an amino acid. Instead, it just signals the end of polypeptide formation. The genetic code is shown in Figure 2.19.

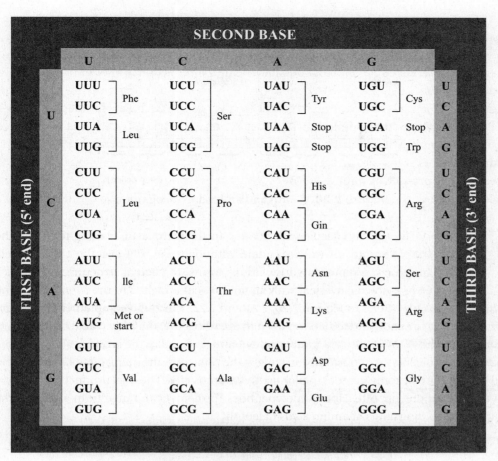

SECOND BASE

		U	C	A	G	
	U	UUU ⎤ Phe UUC ⎦ UUA ⎤ Leu UUG ⎦	UCU ⎤ UCC ⎥ Ser UCA ⎥ UCG ⎦	UAU ⎤ Tyr UAC ⎦ UAA Stop UAG Stop	UGU ⎤ Cys UGC ⎦ UGA Stop UGG Trp	U C A G
	C	CUU ⎤ CUC ⎥ Leu CUA ⎥ CUG ⎦	CCU ⎤ CCC ⎥ Pro CCA ⎥ CCG ⎦	CAU ⎤ His CAC ⎦ CAA ⎤ Gln CAG ⎦	CGU ⎤ CGC ⎥ Arg CGA ⎥ CGG ⎦	U C A G
	A	AUU ⎤ AUC ⎥ Ile AUA ⎦ AUG Met or start	ACU ⎤ ACC ⎥ Thr ACA ⎥ ACG ⎦	AAU ⎤ Asn AAC ⎦ AAA ⎤ Lys AAG ⎦	AGU ⎤ Ser AGC ⎦ AGA ⎤ Arg AGG ⎦	U C A G
	G	GUU ⎤ GUC ⎥ Val GUA ⎥ GUG ⎦	GCU ⎤ GCC ⎥ Ala GCA ⎥ GCG ⎦	GAU ⎤ Asp GAC ⎦ GAA ⎤ Glu GAG ⎦	GGU ⎤ GGC ⎥ Gly GGA ⎥ GGG ⎦	U C A G

FIRST BASE (5' end) — THIRD BASE (3' end)

Figure 2.19. The genetic code

There are far more (64) codons (triple bases on mRNA) than amino acids (20). This makes the genetic code redundant; some codons code for the same amino acid. However, each codon always codes for only one amino acid. Notice that leucine (Leu) is coded for by the codons CUU, CUC, CUA, CUG as well as UUA and UUG. This shows the redundancy of the genetic code. However, each of the individual codons codes for only one amino acid, showing that the genetic code is not ambiguous.

Translation

Translation is the process of reading the mRNA triplet code in order to form a polypeptide (chain of many amino acids held together by peptide bonds). When the mRNA molecule joins the ribosome in the cytoplasm, translation begins. Each transfer RNA (tRNA) molecule includes an anticodon on one end and an amino acid on the other end. An anticodon is a nucleotide triplet that matches up with its complementary codon on the mRNA strand. This matching of anticodon and complementary codon ensures that the correct amino acid is brought to the growing polypeptide chain. Figure 2.20 illustrates the interaction between tRNA and mRNA.

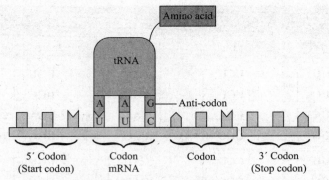

Figure 2.20. Codon and anticodon recognition

REMEMBER

The prefix *poly-* means "many."

The tRNA molecules are floating in the cytoplasm, waiting to match their anticodons to an mRNA codon. Although you will see the tRNA molecular structure represented in various forms, the actual structure resembles an upside-down cloverleaf with an amino acid attached to the 3′ end of the molecule. (Refer to Figures 2.20, 2.21 and 2.22.) Once the complementary bases of the codon and anticodon align at the ribosome, the amino acid brought to the ribosome by the tRNA is released from the tRNA and is added to the growing chain of amino acids. tRNA molecules are constantly arriving at the ribosome, dropping off their amino acids, and leaving the ribosome in order to pick up new amino acids. The final product of translation is a completed polypeptide. Human polypeptides (proteins) can range from a few hundred amino acids to thousands of amino acids in length!

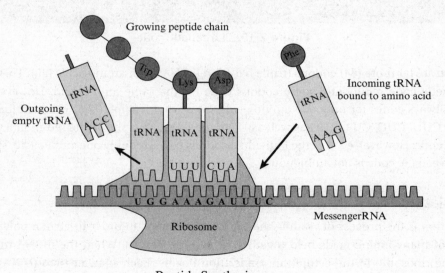

Peptide Synthesis

Figure 2.21. Translation (polypeptide formation)

The terms *cytoplasm* and *cytosol* are both used when referring to the fluid inside of the cell. Technically, the cytosol is the fluid found between the cell membrane and the nuclear membrane, while the cytoplasm includes the fluid plus all of the organelles.

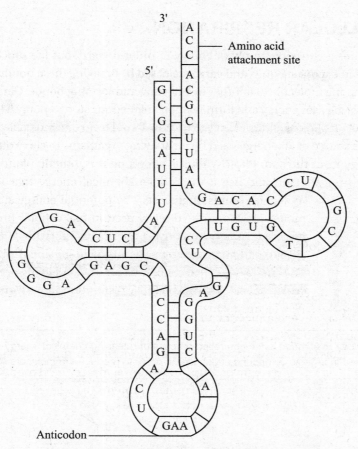

Figure 2.22. Transfer RNA (tRNA)

POLYPEPTIDE FORMATION

The nitrogenous bases in DNA and RNA are used to produce triple-base arrangements of codons. This is analogous to how we rearrange the letters of the alphabet to make different words. For example, the word "bed" can be rearranged to make the name "Deb." The same letters (nitrogenous bases) are used. However, the meaning of each word (codon) is different. Nitrogenous bases can be arranged in different ways to code for different amino acids. There are only 20 different amino acids, but they are coded for by 64 different codons. Obviously, some of the amino acids are coded for by more than one codon. Each chain of amino acids is coded for by one gene. That gene produces one specific polypeptide through the processes of transcription and translation.

The transcription of DNA followed by the translation of mRNA into a single protein (one gene—one polypeptide) is often referred to as the *central dogma* of biology.

Each gene located in DNA codes for one polypeptide. However, some traits are expressed by a combination of multiple polypeptides working together. Therefore, it is inappropriate to state that one gene codes for one trait. For example, haemoglobin is produced from four polypeptide chains. This means that four different genes are involved in forming haemoglobin.

> **REMEMBER**
>
> One gene codes for one polypeptide.
> This is the central dogma of biology.

2.8 CELLULAR RESPIRATION

The body requires a constant supply of energy in order to carry out life processes. Potential energy is stored in chemical bonds and can be released by breaking these bonds. Energy can be transferred from one molecule to another by breaking and forming bonds. Generally, breaking molecular bonds releases energy and forming molecular bonds stores energy. The breaking and forming of molecular bonds allows energy to be transferred from one molecule to another.

The ultimate source of all energy used by most living organisms to carry out life processes is radiant energy from the sun. Photosynthesis allows photosynthetic plant cells to absorb radiant energy from the sun and use it to produce chemical energy that is stored in the molecular bonds of glucose. The potential energy stored in glucose molecules cannot be directly used to fuel cellular processes. All living organisms must convert the energy stored in the bonds of glucose to energy stored in the bonds of adenosine triphosphate (ATP). This process is called cellular respiration. All living organisms use the energy released during cellular respiration to perform life functions. (See Figure 2.23.)

> **REMEMBER**
>
> ATP is the main molecule used by all living organisms to carry and transfer energy throughout living systems.

> *Cellular respiration* is the controlled process of releasing chemical energy stored in organic compounds (glucose) in order to produce the molecule ATP.

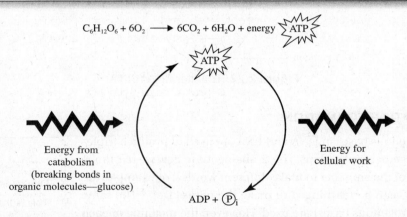

$$C_6H_{12}O_6 + 6O_2 \longrightarrow 6CO_2 + 6H_2O + energy \quad ATP$$

Energy from catabolism (breaking bonds in organic molecules—glucose)

Energy for cellular work

$ADP + P_i$

Figure 2.23. Energy transfer and adenosine triphosphate (ATP)

Glycolysis

> *Glyco:* sweet (sugar)
> *Lysis:* break
> *Glycolysis:* breaking sugar

Cellular respiration consists of three main parts—glycolysis, the Krebs cycle and the electron transport chain. The first step in cellular respiration, glycolysis, occurs in the cytoplasm (cytosol) of the cell. Glycolysis can be summarized as follows:

- The chemical bonds in glucose are broken, releasing stored energy.
- ATP is formed from ADP, thereby transferring the released energy to the ATP.
- Pyruvate is formed.

Specifically, the 6-carbon glucose molecule is split, resulting in the production of two 3-carbon pyruvate molecules. The energy released from this process is used to produce a small yield of ATP (2 ATP per glucose). The first step of glycolysis—splitting the glucose molecule—requires an energy investment of 2 ATP. This energy investment phase is followed by an energy payoff phase, resulting in the production of 4 ATP. Since 2 ATP were invested to begin the process and 4 ATP were gained, the overall net energy gain is only 2 ATP. (See Figure 2.24.)

$$4 \text{ ATP produced} - 2 \text{ ATP invested} = 2 \text{ ATP net gain}$$

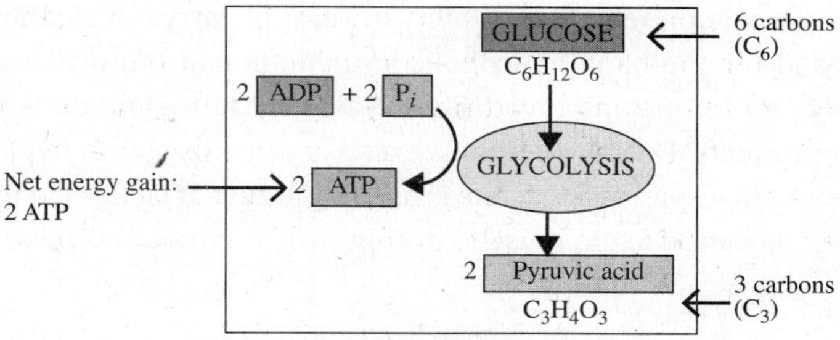

Figure 2.24. Glycolysis in the cytosol

Anaerobic Respiration

Glycolysis is considered to be an anaerobic process since it does not require the presence of oxygen. The Krebs cycle and electron transport chain, though, are aerobic processes. They break down pyruvate and yield significantly higher amounts of ATP (34–36) than does glycolysis. (See Figure 2.25.) In order to complete aerobic cellular respiration, the presence of oxygen is required. If oxygen is present, the pyruvate produced in glycolysis enters the mitochondria, and aerobic respiration proceeds.

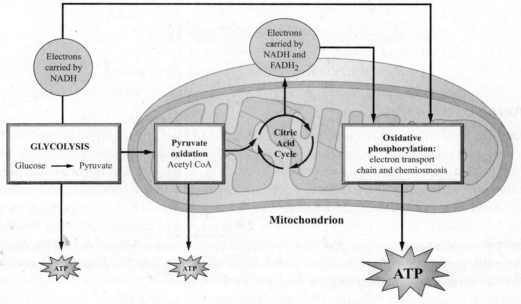

Figure 2.25. Cellular respiration and ATP production
($NADH$ and $FADH_2$ are discussed in HL Topic 8.)

REMEMBER

Lactic acid fermentation occurs in animals.

Alcoholic fermentation occurs in plants and yeast.

If no oxygen is available, cells carry out glycolysis followed by anaerobic fermentation. Lactic acid fermentation occurs in animals, and alcoholic fermentation occurs in plants and yeast. (See Figure 2.26.) The two types of fermentation differ in the final molecules that are produced. In animals, lactic acid fermentation produces lactate. In plants and yeast, alcoholic fermentation produces ethanol with the release of carbon dioxide. Fermentation does not yield any further ATP. However since fermentation must include glycolysis, it is considered to yield a total of 2 ATP (from glycolysis). Note that pyruvate is a reactant for both fermentation in the cytosol and aerobic respiration in the mitochondria.

Beer and wine production rely on the ability of yeast to carry out alcoholic fermentation and produce ethyl alcohol (ethanol). Alcoholic fermentation also provides the carbon dioxide that bakers rely on for making bread rise. Lactic acid fermentation is used to produce cheese and dairy products. During strenuous exercise when oxygen is depleted, the human body turns to anaerobic respiration. Lactate is the by-product of lactic acid fermentation that causes the burning sensation felt in muscles during such strenuous exercise.

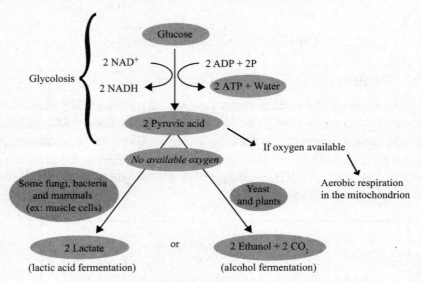

Figure 2.26. Lactic acid and alcoholic fermentation

Aerobic Respiration

The process of glycolysis is common to all living organisms (prokaryote and eukaryote). Since it does not require oxygen or the presence of mitochondria, scientists believe that it was the first method of cellular energy production to evolve. This is supported by the belief that prokaryotes lived on early Earth without any oxygen available and prokaryotes lack mitochondria for use in aerobic respiration. The small yield of ATP from glycolysis is not enough to support the life processes of more complex organisms. The evolution of the mitochondrion allowed eukaryotic cells to carry out aerobic respiration and produce large yields of ATP. In the mitochondria, pyruvate produced in glycolysis is further broken down during aerobic respiration. During aerobic respiration, carbon dioxide and water are released. (See Figure 2.27.) The carbon dioxide diffuses through the cell membranes, into the blood and then out to the lungs. Then it is exhaled from the body.

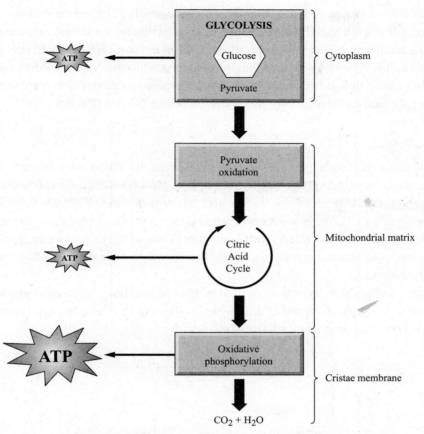

Figure 2.27. The aerobic breakdown of glucose

The only reason you need to breathe in oxygen is so you can carry out aerobic respiration since oxygen functions as the final electron acceptor!

2.9 PHOTOSYNTHESIS

TOPIC CONNECTIONS

■ 2.5 Enzymes

Light-Energy Conversion

According to the laws of thermodynamics, energy is never lost or gained—it is transferred from one form to another. Photosynthesis is the process that converts light (radiant) energy into chemical energy. Radiation (light energy) from the sun arrives to Earth in packets of energy called photons. This radiant energy is converted into chemical energy by the process of photosynthesis. Plants, some protists and some bacteria can carry out the process of photosynthesis.

> **Photo:** light (photons)
> **Synthesis:** to build
> $6CO_2 + 12 H_2O \rightarrow C_6H_{12}O_6 + 6O_2 + 6 H_2O$

Radiant energy arrives to Earth in a range of wavelengths. The colour that you see is the colour that is being reflected by the surface. The colours you do not see are being absorbed by the surface. Each wavelength of light produces a different colour when it bounces off the surface of a structure. The different wavelengths of light produce the visible colours red, orange, yellow, green, blue, indigo and violet (ROY G BIV). The shortest wavelength of the visible spectrum of light is violet (400 nm), and the longest wavelength is red (700 nm).

CHLOROPHYLL

Molecules that can absorb light are referred to as pigments. Different pigments absorb different wavelengths of light. The main photosynthetic pigment is chlorophyll, and it does not absorb green light. Since green is not absorbed, it is reflected back from the surface, making most photosynthetic plants appear green. Some plants have other pigments that function as their main pigment and can take on different colours due to the difference in absorption. All photosynthetic organisms have at least some chlorophyll, making chlorophyll the main photosynthetic pigment.

Chlorophyll absorbs wavelengths of light in the red and blue range most efficiently and reflects green, as shown in Figure 2.28. This means that the wavelengths of red and blue light are the ones that fuel photosynthesis most efficiently.

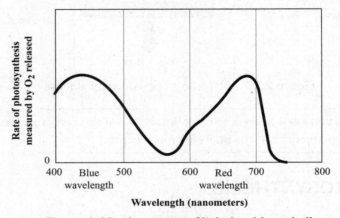

Figure 2.28. Absorption of light by chlorophyll

Light-Dependent Reactions

Photosynthesis is divided into two processes: the light-dependent reactions and the light-independent reactions. The light-dependent reactions occur in the thylakoid membrane, and the light-independent reactions occur in the stroma. (See Figure 2.29.) The light-independent reactions are also called the Calvin cycle.

The light-dependent reactions require sunlight and the sun's radiant energy to split water, releasing oxygen and hydrogen. The process of splitting water into 1 oxygen and 2 hydrogen ions is referred to as *photolysis*.

REMEMBER

All the oxygen given off by plants was originally part of a water molecule!

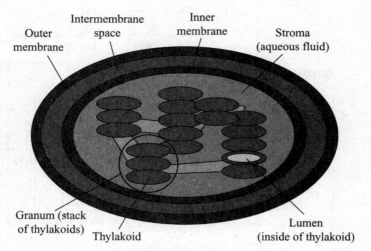

Figure 2.29. Chloroplast

> **Photo:** light
>
> **Lyse:** to break open
>
> **Photolysis:** Breaking open a water molecule into its atomic parts using energy derived from the sun.

In addition to photolysis, the light-dependent reactions also convert the radiant energy from the sun into chemical energy stored in the molecules ATP and NADPH. ATP stores energy in its phosphate bonds. NADPH stores energy in its hydrogen bonds. The hydrogen in NADPH was originally part of a water molecule split by photolysis. Breaking the bonds in water releases its stored energy, thereby making that energy available to be stored in ATP and NADPH.

Light-Independent Reactions

The energy stored in ATP and NADPH during the light-dependent reactions is used in the light-independent reactions to produce carbohydrates such as glucose. Stored energy is released from ATP by breaking a phosphate bond and from NADPH by breaking a hydrogen bond. The released energy is used to form new chemical bonds in carbohydrates (sugars).

This process of breaking bonds and forming new bonds in ATP and NADPH occurs in a cycle between the light-dependent reactions (forming the bonds) and the light-independent reactions (breaking the bonds). This cycle is the process of photosynthesis, as shown in Figure 2.30. Table 2.12 compares and contrasts the light-dependent and light-independent reactions.

Table 2.12: Light-Dependent and Light-Independent Reactions

	Light-Dependent Reactions	Light-Independent Reactions (Calvin Cycle)
Location	Thylakoid membrane	Stroma
Reactants	ADP, NADP, water	ATP, NADPH, CO_2
Products	ATP and NADPH	Carbohydrates (glucose)
By-products	Oxygen	Water

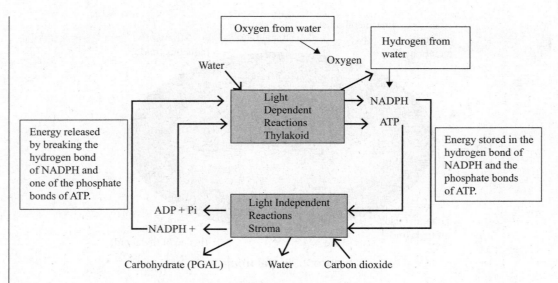

Figure 2.30. Photosynthesis

Measuring Photosynthesis Rates

The rate of photosynthesis can be directly measured by determining the amount of carbon dioxide that is being taken up by the plant or by measuring the rate at which oxygen is being given off. The rate can be measured indirectly by measuring the increase in biomass. Biomass is the dry weight (weight with water removed) of a living organism. Since photosynthetic organisms take in carbon dioxide and water in order to produce carbohydrates, the process adds mass to the organisms. This increase in biomass can be used to determine the rate of photosynthesis indirectly. Something similar happens with people. When they eat a lot of food, their size (biomass) increases! See Table 2.13 for more information about measuring the rate of photosynthesis.

$$\text{Light}$$
$$6CO_2 + 12H_2O \rightarrow C_6H_{12}O_6 + 6H_2O + 6O_2$$

FACTORS AFFECTING PHOTOSYNTHESIS RATES

The rate of photosynthesis changes as the environment the plant lives in fluctuates. Three environmental factors influence the rate of photosynthesis: temperature, light intensity and carbon dioxide concentration. Photosynthesis occurs more rapidly at higher temperatures than it does at lower temperatures. Heat increases the rate of all reactions since it makes molecules speed up and collide more often. However, if the temperature becomes too high, it can cause molecules to lose their structure and ability to function. For example, enzymes needed for the light-dependent reaction will denature and no longer catalyse reactions to produce carbohydrates. Each of these variables is considered to be limiting factors to the rate of photosynthesis. The graphs in Figure 2.31 illustrate the effect of light intensity, temperature and carbon dioxide concentration on the rate of photosynthesis.

Table 2.13: Methods for Measuring the Rate of Photosynthesis

Measurement	Relationship to Photosynthetic Rates	Method
Oxygen production	Oxygen is given off during photosynthesis due to the photolysis of water.	**Direct:** Collect oxygen bubbles given off by aquatic plants and measure the amount collected.
Carbon dioxide uptake	Carbon dioxide is taken up during the process of photosynthesis in order to provide carbon atoms for the productions of organic compounds.	**Indirect:** Measure changes in the pH of water with aquatic plants. As carbon dioxide is removed from water, the pH will increase.
Increase in biomass	Plants take in carbon dioxide and other molecules needed in order to carry out photosynthesis. As a result, their biomass increases.	**Indirect:** Harvest groups of plants at a series of time (from same area and quadrant size) and compare the changes in biomass (dry weight) in order to calculate a rate of increase.

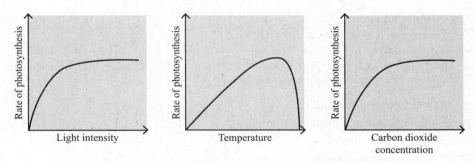

Figure 2.31. Factors that limit the rate of photosynthesis

1. Two water molecules are bonded together by:
   A. Hydrogen bonds
   B. Covalent bonds
   C. Ionic bonds
   D. Peptide bonds

2. An example of a monosaccharide is:

   A. Sucrose
   B. Galactose
   C. Maltose
   D. Cellulose

3. The building blocks of nucleic acids are:

   A. Monosaccharides
   B. Fatty acids
   C. Nucleotides
   D. Glucose and fructose

4. When DNA replicates, each new DNA strand contains half of the original strand. What term describes this?

   A. Universal
   B. Antiparallel
   C. Semiconservative
   D. Double helix

5. Translation is a process that occurs in the

   A. nucleus at the ribosome.
   B. cell membrane at a protein channel.
   C. cytoplasm at a ribosome.
   D. cytoplasm at a lysosome.

6. Glycolysis results in a net gain of:

   A. 2 ATP
   B. 4 ATP
   C. 6 ATP
   D. No ATP

7. Which of the following wavelengths (colours) of light does chlorophyll most efficiently absorb?

   A. Red and green
   B. Yellow and blue
   C. Red and blue
   D. Yellow and red

8. Which of the following are the main functions of lipids?

    I.   Energy storage
    II. Insulation
    III. Information storage

    A. I only
    B. I and II only
    C. I and III only
    D. I, II and III

9. The main function of glycogen in animals is for

    A. structure.
    B. building DNA.
    C. cushioning.
    D. energy storage.

10. The nitrogenous bases in DNA are held together by

    A. phosphodiester bonds.
    B. ionic bonds.
    C. hydrogen bonds.
    D. covalent bonds.

11. Which features of DNA are correctly identified in the table below?

	Number of Bonds Between Adenine and Thymine	Bonds Between Sugars and Phosphates	Type of Bond Between Adenine and Thymine
A.	Three	Phosphodiester	Hydrogen
B.	Two	Phosphodiester	Hydrogen
C.	Two	Hydrogen	Phosphodiester
D.	Three	Hydrogen	Phosphodiester

12. What base is found in both DNA and RNA?

    A. Thymine
    B. Uracil
    C. Thymine and uracil
    D. Guanine

13. What process occurs at the ribosome in eukaryotic cells?

    A. Transcription
    B. Replication
    C. Translation
    D. Transcription and translation

14. If the triplet base arrangement is AAA on DNA, what is the anticodon triple base pair that codes for the same amino acid?

   A. AAA
   B. GGG
   C. UUU
   D. TTT

   [handwritten: AAA → RNA UUU → CODON AAA]

15. Which of the following molecules are produced by both aerobic and anaerobic respiration?

   I. ATP
   II. Pyruvate
   III. Acetyl CoA

   A. I only
   B. I and II only
   C. I and III only
   D. I, II and III

16. Which of the following are products of the light-dependent reactions that are used in the light-independent reactions?

   I. ATP
   II. NADPH
   III Oxygen

   A. I only
   B. I and II only
   C. II and III only
   D. I, II and III

17. The enzymes used by the light-independent reactions are located in the:

   A. Stroma of the chloroplast
   B. Grana of the chloroplast
   C. Inner membrane of the chloroplast
   D. Outer membrane of the chloroplast

18. The site on an enzyme that binds to the substrate is known as the:

   A. Alternate site
   B. Activation site
   C. Active site
   D. Residual site

19. Photosynthesis involves:

    A. The conversion of radiant energy into chemical energy
    B. The conversion of chemical energy into radiant energy
    C. The exchange of radiant energy for more radiant energy
    D. The loss of chemical energy by radiant energy transfer

20. Where do the light-dependent reactions occur?

    A. In the thylakoid membrane
    B. In the stroma
    C. In the cytosol
    D. In the matrix

21. Which bond is responsible for the formation of the tertiary structure of a protein?

    A. Peptide
    B. Hydrogen
    C. Disulphide
    D. Ionic

22. Which of the following are main functions of proteins?

    I.   Catalysing reactions
    II. Transporting of oxygen
    III. Long-term energy storage

    A. I only
    B. I and II only
    C. I and III only
    D. I, II and III

The following information pertains to question 23.

Aerobic respiration uses oxygen and gives off carbon dioxide. Cells that are actively carrying out cellular respiration will release more carbon dioxide and use more oxygen in comparison to cells that are not carrying out cellular respiration or are carrying out lower rates of cellular respiration. In aquatic environments, an increase in carbon dioxide production can be measured indirectly by changes in the pH of the water. As more carbon dioxide is released into the water, the pH will become more acidic. The pH drops because carbon dioxide combines with water and forms carbonic acid.

Elodea canadensis is an aquatic plant found throughout North America. The pH and temperature of two environments in which *Elodea* is found were analysed, and the averages were calculated. The southern environment was found to have an average daily temperature of 30 degrees Celsius, and the northern environment was found to have an average daily temperature of 10 degrees Celsius. The pH of the water in which the *Elodea* live was taken over a period of 12 days. The graphs below show the results.

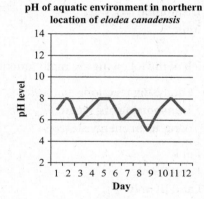

23. (a) Outline the relationship between pH level and location of *Elodea canadensis*. (2)
     (b) Suggest a reason for this relationship. (2)
     (c) State the day in which the pH was the highest for the southern location of *Elodea canadensis*. (1)
     (d) Calculate the average pH for the southern location. (2)

_____

24. Define enzyme. (1)

25. List two factors that affect the rate of photosynthesis. (2)

26. State the term that describes a protein's loss of its three-dimensional structure.

27. Draw the formation of a dipeptide. (4)

28. (a) Draw a chloroplast. (4)
     (b) Outline the light-dependent reactions. (6)
     (c) Explain direct methods that can be used to measure the rate of photosynthesis. (4)

MOLECULAR BIOLOGY

ANSWERS EXPLAINED

1. **(A)** Water molecules are bonded by hydrogen bonds between the oxygen of one water molecule and a hydrogen of the other water molecule.

2. **(B)** Galactose is a monosaccharide.

3. **(B)** The building blocks of nucleic acids are nucleotides.

4. **(C)** The process of DNA replication that results in two new strands with each containing half of the original strand is referred to as semiconservative replication.

5. **(C)** Translation follows transcription and results in the production of a protein. Although transcription occurs in the nucleus, translation occurs on ribosomes that are found in the cytoplasm of the cell.

6. **(A)** Glycolysis results in a net gain of 2 ATP.

7. **(C)** Chlorophyll most efficiently absorbs wavelengths of the colours red and blue. Green is reflected.

8. **(B)** Lipids function for energy storage and insulation.

9. **(D)** Energy from carbohydrates is stored as the polysaccharide glycogen in animals.

10. **(C)** Hydrogen bonds hold together the complementary bases in DNA.

11. **(B)** Adenine and thymine share two hydrogen bonds. Phosphodiester bonds hold together the sugar and phosphate backbone of DNA. The complementary bases are held together by hydrogen bonds.

12. **(D)** Guanine. In RNA, thymine is replaced by uracil.

13. **(C)** Of the processes listed, only translation occurs at the ribosome. Replication and transcription occur in the nucleus of eukaryotic cells.

14. **(A)** The complementary bases are AAA. The DNA codes for mRNA codon UUU. So the anticodon on tRNA is AAA.

15. **(B)** Anaerobic respiration produces ATP and pyruvate only. If oxygen is present, the pyruvate enters the mitochondria and Acetyl CoA is produced.

16. **(B)** ATP and NADPH are both products of the light-dependent reaction that are used to fuel the light-independent reaction. Oxygen is a by-product of the light-dependent reactions.

17. **(A)** The enzymes used by the light-independent reactions are located in the stroma where the light-independent reactions occur.

18. **(C)** The active site is the term used to refer to the location on an enzyme where the substrate bonds.

19. **(A)** Photosynthesis is the conversion of radiant (light) energy into chemical energy (ATP and organic compounds).

20. **(A)** The light-dependent reactions occur in the thylakoid membrane.

21. **(C)** Disulphide bonds are formed in the tertiary structure of proteins. Sulphur atoms are found in the R-group of certain amino acids. The location of the amino acids containing sulphur determines the structure of protein folding.

22. **(B)** I and II only. Enzymes are proteins that catalyse reactions, and haemoglobin is a protein that transports oxygen. Long-term energy storage is not a main function of proteins. Lipids function for long-term energy storage.

23. (a)
    - pH is lower (4–7 range) in southern environment
    - pH is higher in northern environment (5–8 range)
    - Both environments show fluctuation in pH levels
    - Days 6–9 were lowest for both environments

    (b)
    - Slower reactions occur at lower temperatures; less carbon dioxide is taken out of water
    - Differences in rates of gas exchange between plants and environment
    - Carbon dioxide uptake changes (increases) pH
    - More photosynthesis occurring in the south decreases carbon dioxide levels
    - Carbon dioxide combines with water to form carbonic acid

    (c) Day 11

    (d) Accept answers ranging from 5.0–5.5. Calculating the average must be shown to receive both points.

24. An enzyme is a biological catalyst that speeds up reactions/lowers the activation energy.

25. Light intensity, carbon dioxide concentration or temperature

26. Denaturation is the loss of a protein's three-dimensional structure.

27.

Amino Acid + Amino Acid ⟶ Dipeptide

28. a. Award a point for each correctly drawn and labelled structure:
    - Outer membrane
    - Inner membrane
    - Thylakoid membrane
    - Grana
    - Stroma
    - Inner membrane space

b.

- Light is required.
- Light-dependent reactions occur in the thylakoid membrane.
- Photosystems contain pigments.
- The main pigment is chlorophyll.
- Chlorophyll absorbs red and blue light most efficiently.
- Photolysis of water replaces electrons lost by chlorophyll
- Oxygen is a by-product of the light-dependent reactions/photolysis
- Green light is reflected, not absorbed.

c.

- Oxygen production due to increased photolysis of water
- Carbon dioxide uptake for use in carbon fixation in the light-independent reactions
- Oxygen production can be directly measured in aquatic plants by collecting the oxygen bubbles given off
- Carbon dioxide production can be measured indirectly by detecting changes in pH as the CO_2 is taken up from the water/pH increases as CO_2 is taken up during photosynthesis

Genetics

CENTRAL IDEAS

- Chromosomes, genes, alleles and mutations
- Meiosis and inheritance
- Theoretical genetics
- Genetic modification and biotechnology

TERMS

- Allele
- Crossing-over
- Diploid
- Fertilisation

- Gene
- Gene mutation
- Genome
- Haploid

- Homologous chromosomes
- Locus
- Meiosis
- Nondisjunction

INTRODUCTION

All living organisms must pass on genetic information to their offspring. Genetic information is coded in deoxyribonucleic acid (DNA). DNA is a nucleic acid composed of four different nucleotides. The arrangement of the nucleotides in DNA determines the order of amino acids in a protein. Messenger ribonucleic acid (mRNA) is produced from DNA in a process called transcription. Translation is the process of reading an mRNA molecule at the ribosome in order to produce proteins. Different forms of genes (alleles) on DNA code for different arrangements of amino acids, producing thousands of different proteins.

Mitosis creates genetically identical copies of DNA in two new daughter nuclei. These nuclei are then separated during cytokinesis to create two new identical cells. In other words, the parent cell had two copies of each strand of DNA (was diploid). The daughter cells have identical DNA and are therefore also diploid. In contrast, meiosis reduces the chromosome number in gametes (sperm and eggs) to half of the original. The gametes are therefore haploid, which means they have only one copy of each strand of DNA. Since gametes are haploid, fertilisation (the union of the egg and sperm) results in the offspring having a diploid number of chromosomes. Genetics involves the study of how the chromosomes separate and rearrange to form new combinations of genes in a population. Biotechnology has changed the natural process of inheritance by manipulating cells and genes in ways that are not normally seen in nature.

NOS Connections

- Technology developments impact scientific research and exploration (PCR, gel electrophoresis, gene transfer).
- Careful observation and collection of data are vital in assuring the reliability of the scientific process.
- International, national and local collaboration is key to scientific advancements (human genome project).
- Scientific discoveries may have risks associated with them (cloning).

3.1 GENES

Genes, Alleles and Genomes

You need to know the following definitions for the IB Biology Exam:

- **Gene:** A section of DNA that codes for a specific trait. Genes are heritable factors that exist in a specific fragment of DNA and influence the expression of traits.
- **Allele:** Each different form of the same gene. Corresponding alleles are found on homologous chromosomes.
- **Genome:** This is the entire genetic information (coding and noncoding regions) contained in the genes of an organism.

Genes are heritable factors that are passed on to offspring. Each gene determines a specific characteristic seen in an organism. The order of the bases that are found in DNA (A, T, G, C) makes up the genetic code for each gene. A form of a gene is referred to as an allele. Each individual has two alleles for each gene. The two alleles of a gene are found on homologous chromosomes. Figure 3.1 shows the relationship among chromosomes, genes, and alleles.

Each allele for a particular gene differs by one or by just a few bases. Each allele of a gene occupies the same location (locus) on the two DNA strands. Normally, the alleles of a gene are referred to as either the dominant form or the recessive form. The dominant form is expressed with a capital letter, such as *A*. The recessive form is expressed with a lowercase letter, such as *a*.

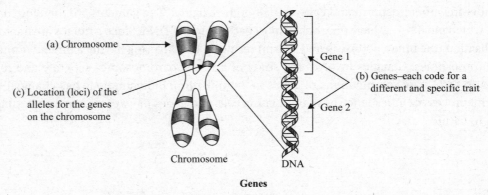

Figure 3.1. (a) Chromosome; (b) genes on DNA;
(c) location of the alleles for the genes

The term genome refers to the total collection of DNA found within an organism. Since only a small percentage of the genomic material codes for proteins (genes), it is incorrect to refer to the genome as the total number of genes an organism possesses. The number of genes an organism possesses does not relate to the complexity of the organism. The lungfish (*Prot-opterus aethiopicus*) has a genome that is 40 times larger than the human genome. Obviously, genome size does not equal complexity in an organism.

Gene Mutation

Mutations may involve the deletion of nucleotides, the addition of nucleotides, or the substitution of nucleotides. Any of these mutations can result in the production of new alleles. The new alleles may code for altered proteins. Although mutations can be harmful, some are beneficial to organisms and may help the organisms survive. Mutations are a major source of variation in populations and can help drive the process of natural selection.

> **Gene mutation:** A permanent change in the sequence of bases that code for a trait in DNA.

Gene mutations can result in several different outcomes:

- **No change**: The protein is unchanged because more than one codon codes for the particular amino acid. Since the amino acid sequence does not change, neither does the protein.
- **Harmful change**: The protein is changed due to a new amino acid being coded. The resulting change is harmful to the organism.
- **Beneficial change**: The protein is changed due to a new amino acid being coded. The resulting change helps the organism survive.

SICKLE CELL ANEMIA

Since more than one codon can code for the same amino acid, some mutations do not result in an altered protein structure. However, if the base change causes a new amino acid to be coded, the protein structure can be altered. The altered structure makes the protein unable to carry out its function. An excellent example of how one base change can affect the function of a protein can be seen in patients who suffer from sickle cell anaemia.

In the DNA of sickle cell patients, one adenine base (A) is replaced by a thymine base (T). This change results in a change in the mRNA codon produced during transcription of the gene that codes for haemoglobin. The altered codon results in the replacement of the amino acid glutamic acid for valine during the process of translation. The presence of valine in the previous location of glutamic acid (the sixth amino acid in the polypeptide chain) causes the structure of haemoglobin to be altered, as shown in Figure 3.2. This single base pair change in the gene that codes for the production of haemoglobin causes red blood cells to lose their ability to carry oxygen. The reason they cannot carry as much oxygen as a normal red blood cell is due to the shape of red blood cells in sickle cell patients. Their red blood cells have a sickle shape, which is rigid and curved like the farm tool known as a sickle. This sickle shape greatly reduces the ability of haemoglobin to carry oxygen to body tissues. In addition, sickle cells are prone to clotting due to their shape.

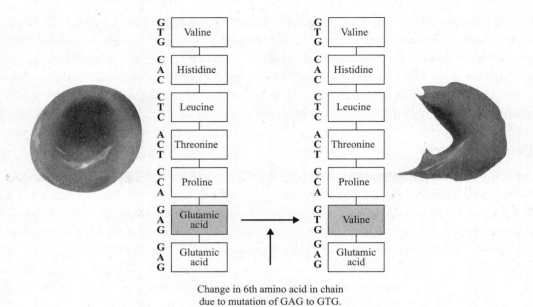

Change in 6th amino acid in chain
due to mutation of GAG to GTG.

Figure 3.2. Sickle cell mutation

FEATURED QUESTION

May 2004, Question #9

In the mutation of the haemoglobin gene that produces the sickle cell allele, CTC is converted to CAC on the DNA strand that is to be transcribed. What will be the anticodon sequence of the tRNA molecule for the translation of the mutated allele?

A. GUG
B. GTG
C. CAC
D. GAG

(Answer is on page 553.)

YOU SHOULD KNOW

Most mutations involve the bases adenine and thymine since they are held together by only two hydrogen bonds. Mutations less frequently involve the bases guanine and cytosine since three hydrogen bonds more forcefully hold these bases together.

Human Genome Project

The human genome project was an international collaborative effort by scientists to produce a complete map of the base sequences found in the human genome (all the genes present in DNA). Although the project was completed in 2003, the full implications of the results have still not been discovered. Outcomes based on the knowledge of the base sequences have allowed scientists to locate specific gene mutations, have allowed them to compare sequences among different organisms, and have impacted pharmaceutical technologies. The identification of genes that can cause disease may allow scientists to correct these genes in the future. Comparing sequences among different species can help in analysing evolutionary relationships. Pharmaceuticals can be produced that are specifically designed for an individual's genetic makeup. This will allow medications to function more effectively for each individual.

REMEMBER

The human genome project was the largest international cooperative project ever to be carried out!

3.2 CHROMOSOMES

TOPIC CONNECTIONS

- 1.6 Cell division

Chromosomes contain the genetic material of an organism. Eukaryotes possess separate and distinct chromosomes that each code for different genes. Prokaryotes possess one circular chromosome and may possess extra circular sections of DNA known as plasmids. Each species possesses a characteristic number of chromosomes. The number of chromosomes an organism possesses does not relate to the complexity of the organism. For example, some ferns possess over 600 chromosomes! Table 3.1 outlines the differences between prokaryotic and eukaryotic chromosomes.

Table 3.1: Prokaryotic and Eukaryotic Chromosomes

	Prokaryotic	Eukaryotic
Chromosome structure	One circular DNA molecule	Many linear DNA molecules
Plasmids	May have plasmids	Do not have plasmids
Packaging	Not packaged with histones	Packaged with histones

Homologous Chromosomes

In the diploid nucleus, a cell possesses chromosomes in pairs. These pairs are termed homologous since they code for the same genes, are the same size, and possess alleles for the same genes in the same location (at the same locus).

> **Homologous chromosomes:** A pair of chromosomes that are the same size, have the same banding pattern, have the centromere located in the same location, and contain genes that code for the same trait.

Most organisms have two alleles for each gene. One allele is received from each parent during sexual reproduction. Each parent contributes half of the genetic makeup to the offspring. The chromosomes from each parent carry the same genes but may have different forms of the genes (alleles) on the matching sets of chromosomes. Figure 3.3 shows a pair of homologous chromosomes.

> **Homo:** same
> **Locus:** position

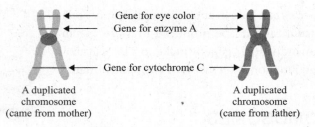

Figure 3.3. Homologous chromosomes

Reproductive cells contain only one of each of a homologous pair. Therefore these cells are referred to as haploid. Haploid cells are created during the process of meiosis. Sperm and eggs are the only haploid cells found in humans. All other cells are referred to as somatic (body) cells and possess the diploid number of chromosomes.

Humans possess 46 chromosomes or 23 homologous pairs of chromosomes. In humans, 22 pairs of chromosomes are referred to as *autosomes* because they do not code for the sex of an individual. One pair of chromosomes is referred to as sex chromosomes (XX or XY) because they determine the sex of an individual.

Karyogram

An organism's karyotype is the total number and appearance of all the chromosomes in the nucleus of the cells. Karyotyping is a process that involves taking a picture of a cell going through mitosis and matching homologous chromosomes as pairs. The picture or diagram of the chromosomes once they are matched up as homologous pairs in decreasing length is termed a karyogram. A cell must be undergoing mitosis in order to obtain a picture of the chromosomes since it is the only part of the cell cycle during which the chromosomes are visible.

Homologous chromosomes are matched by finding those that have the same size and structure. The centromere will be in the same location and the chromatids will be the same height. In other words, pairs of chromosomes are found, one from each parent.

As shown in Figure 3.4, karyotyping can be used to determine whether an individual has an extra chromosome. It can also be used to determine if the individual has an addition or a deletion of a large section of a chromosome.

> **REMEMBER**
>
> A karyogram shows only gross chromosomal changes, not point mutations. Therefore, it will not show the substitution of a single nucleotide—as found in sickle cell anaemia.

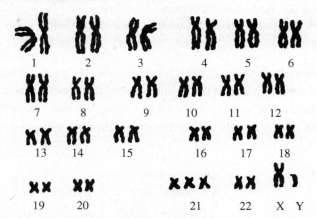

Figure 3.4. Karyogram of a male with trisomy-21 due to nondisjunction

3.3 MEIOSIS

TOPIC CONNECTIONS

- 1.6 Cell division
- 10.1 Meiosis
- 11.4 Sexual reproduction

In order for organisms to carry out sexual reproduction, their gametes must contain half the number of chromosomes found in somatic (body) cells. When the gametes undergo fertilisation, the full chromosome number then results. If the chromosomes were not reduced in the gametes, the number of chromosomes resulting from sexual reproduction would be twice that of the parents.

> **Ploidy** refers to the number of chromosomal sets in the nucleus of a cell.
>
> - **Haploid:** One set of chromosomes
> - **Diploid:** Two sets of chromosomes
> - **Polyploid:** More than two sets of chromosomes

Meiosis occurs only in reproductive cells during the formation of gametes (sex cells). Meiosis results in the production of haploid cells from diploid cells. Just like in mitosis, the chromosomes must go through the S-phase of the cell cycle in order to copy their genetic material. Meiosis involves two divisions of the nucleus. The first division, called meiosis I, separates homologous chromosomes. The second division, called meiosis II, separates the connected chromatids. (See Table 3.2 and Figure 3.5.) The end products of meiosis are sperm and eggs—the reproductive gametes.

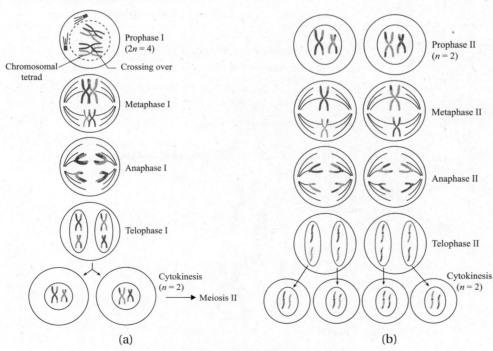

Figure 3.5. (a) Meiosis I; (b) meiosis II

Table 3.2: The Two Divisions of Meiosis

First Division: Separation of Homologous Chromosomes	Second Division: Separation of Sister Chromatids (Replicated Chromosomes)
Prophase I ■ Homologous chromosomes pair up (synapsis occurs). ■ Crossing over is possible. ■ The nuclear membrane breaks down.	**Prophase II** ■ Each haploid set of chromosomes is visible in the two cells.
Metaphase I ■ Homologous chromosomes line up *as pairs* at the metaphase plate.	**Metaphase II** ■ *Individual* chromosomes line up at the metaphase plate.
Anaphase I ■ Homologous chromosomes are pulled apart and move to opposite poles of the cell.	**Anaphase II** ■ Replicated chromosomes (chromatids) separate into individual chromosomes.
Telophase I ■ Individual replicated chromosomes are located at the opposite poles of the cell as haploid sets.	**Telophase II** ■ Nonreplicated chromosomes are located at the poles of the cell.
Key Events ■ Nuclei are reduced from diploid ($2n$) to haploid (n). ■ Crossing over occurs during prophase I. ■ 2 new genetically different haploid cells are produced by cytokinesis.	**KEY EVENTS** ■ Replicated chromosomes are separated into nonreplicated chromosomes by the separation of sister chromatids into individual chromosomes. ■ Four new, genetically different haploid cells are produced by cytokinesis.

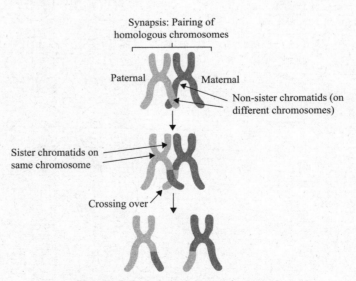

Figure 3.6. Synapsis (crossing over)

During prophase I, homologous chromosomes match up as homologous pairs. The pairing of homologous chromosomes is called synapsis.

The nonsister chromatids of each homologous pair of chromosomes can undergo crossing-over. Crossing-over involves an exchange of genetic material between the homologous chromosomes, as shown in Figure 3.6. Synapsis increases the genetic variation in chromosomes and therefore in the resulting offspring.

Nondisjunction

Nondisjunction occurs when the homologous chromosomes fail to separate. This can occur during anaphase I when the homologous chromosomes fail to separate or during anaphase II when the sister chromatids fail to separate. Figure 3.7 shows the results of nondisjunction in anaphase I and in anaphase II. The failure of homologous chromosomes to separate leads to changes in the chromosome number. One gamete will receive an extra copy of the chromosome, and one gamete will be missing a chromosome. If the gamete with an extra chromosome is fertilised, the resulting zygote will have an extra chromosome. If the gamete missing a chromosome is fertilised, the resulting zygote will be missing a chromosome.

The best example of nondisjunction contributing to genetic diseases is Down syndrome. Down syndrome is also known as trisomy-21 since it results from the fertilisation of a cell with an extra 21st chromosome, creating a cell with 3 copies of chromosome 21. Nondisjunction can involve any of the homologous chromosomes. In most cases, though, the resulting zygote will not survive. An individual with Down syndrome will have 47 chromosomes instead of the normal diploid number of 46.

> **BE PREPARED**
>
> On the IB Biology Exam, be prepared to determine the sex of an individual and any chromosomal abnormalities based on analysing karyotypes.

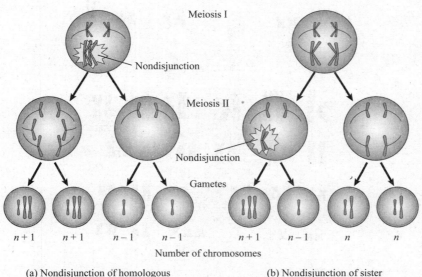

Meiosis I

Nondisjunction

Meiosis II

Nondisjunction

Gametes

$n+1$ $n+1$ $n-1$ $n-1$ $n+1$ $n-1$ n n

Number of chromosomes

(a) Nondisjunction of homologous chromosomes in meiosis I

(b) Nondisjunction of sister chromatids in meiosis II

Figure 3.7. (a) Nondisjunction occurring in anaphase I;
(b) nondisjunction occurring in anaphase II

Amniocentesis and Chorionic Villus Sampling

In order to produce a karyotype of a foetus, the genetic material must be obtained. Amniocentesis and chorionic villus sampling (CVS) are two major methods used to collect cells from the foetus.

Amniocentesis involves inserting a syringe into the amniotic fluid surrounding the foetus and drawing out some of the cells that have been shed from the foetus. This process is guided with an ultrasound in order to avoid injuring the foetus.

CVS also involves the retrieval of foetal cells. However, these cells come from the foetal portion of the placenta. This test involves the insertion of a tool through the cervix (opening of the uterus) into the uterus and taking a small sample of the outer portion of the placenta. This part of the placenta is called the chorion—hence the name of the procedure. The cells of the outer portion of the placenta are produced by the foetus and therefore have the same genetic makeup as the child. Chorionic villus sampling can be done earlier in gestation than amniocentesis but also poses a greater risk of causing miscarriage.

Both amniocentesis and CVS can be used to detect chromosomal abnormalities and identify the sex of the child. After the genetic material is obtained, it is karyotyped, as shown in Figure 3.8.

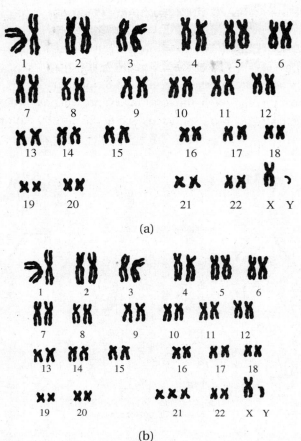

Figure 3.8. (a) Normal karyotype; (b) Down syndrome karyotype

Genetic Variation

Sexual reproduction promotes variation in a species through the processes of meiosis and fertilisation. Meiosis provides variation due to the independent assortment of chromosomes and the process of crossing over. This allows the production of gametes (sperm and eggs) that are different from the parent cell and from each other. The random selection of which egg will be released and which of the millions of sperm released at a particular time also contributes to variation in the population. This is why it is virtually impossible for siblings to look exactly alike unless they are identical twins. (See Figure 3.9.)

Variation in Sexual Reproduction

Independent Assortment

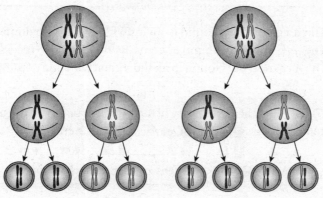

Crossing Over

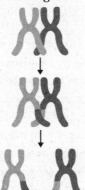

Random Fertilisation

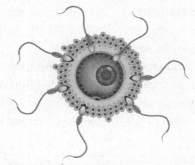

Figure 3.9. Variation with sexual reproduction

3.4 INHERITANCE

TOPIC CONNECTIONS

- 1.6 Cell division

The genotype of an organism consists of the particular alleles present for each gene. Each individual receives one allele from each parent during fertilisation. The genotype is expressed as letters that represent the form of alleles present for a particular gene.

> **Genotype:** The particular alleles found in an organism.

- **Example 1**: The *AA* genotype is found in an individual with two dominant alleles.
- **Example 2**: The *aa* genotype is found in an individual with two recessive alleles.
- **Example 3:** The *Aa* genotype is found in an individual with one dominant and one recessive allele.

The phenotype is the physical appearance of an organism based on the alleles present. You should know the following terms before you carry out genetic crosses:

> **Phenotype:** The characteristics seen in an organism.

- **Dominant allele**: An allele that has the same effect on the phenotype whether it is present in the homozygous or in the heterozygous state. It is symbolized by a capital letter, such as *A*.
- **Recessive allele**: An allele that has an effect on the phenotype only when present in the homozygous state. It is symbolized by a lowercase letter, such as *a*.
- **Codominant alleles**: Pairs of alleles that both affect the phenotype when present in a heterozygote. For example, the genes that code for blood type A and for blood type B in humans are codominant, resulting in the genotype *AB* and in the phenotype *AB*.
- **Locus**: The particular location of a gene on homologous chromosomes.
- **Homozygous**: Having two identical alleles of a gene. An individual can be either homozygous dominant, such as *AA*, or homozygous recessive, such as *aa*.
- **Heterozygous**: Having two different alleles of a gene. For example, an individual with the genotype *Aa* is heterozygous.
- **Carrier**: An individual that has one copy of a recessive allele that causes a genetic disease in individuals that are homozygous for this allele. For example, a person with only one sickle cell gene is a carrier for that trait. The individual's genotype would be *Ss*, where the lowercase *s* is the allele that causes the sickling.
- **Test cross**: Testing an individual expressing the dominant trait in order to determine if a heterozygous (*Aa*) or a homozygous (*AA*) combination of alleles is present. The individual is crossed with an organism known to be homozygous recessive for that trait. The resulting offspring are studied to determine the unknown parent's genotype.

Mendel's Laws

Mendel was a European monk who studied the principles of inheritance involving pea plants during the late 1800s. His studies resulted in the following laws of inheritance that still hold true today:

- **First law**: Mendel's law of dominance states that when organisms exhibit heterozygous (hybrid) combinations of alleles, the dominant allele is always expressed.
- **Second law**: Mendel's law of segregation states that the alleles of a gene will separate during meiosis.
- **Third law**: Mendel's law of independent assortment states that the alleles of a gene will separate independently of each other.

	T	*T*
t	*Tt*	*Tt*
t	*Tt*	*Tt*

Parent (P):   *TT*   ×   *tt*

          Pure tall   Pure dwarf

Offspring (F₁):   *Tt*

          All hybrid tall

Law of dominance
All offspring are tall

The two alleles that exist for each gene are found on the homologous chromosomes in body cells. During meiosis, the separation of the homologous chromosomes results in sex cells containing only one of each of the original two alleles for each gene. Recall that Mendel's law of segregation, as shown in Figure 3.10, states that the allelic combinations of both parents will separate during the formation of the gametes (sex cells). The resulting gametes are haploid. The diploid number of chromosomes is restored when two haploid gametes join during the process of fertilisation. Fertilisation produces zygotes with two alleles for each gene. For each gene, the organisms may have either the same alleles (be homozygous) or have different alleles (be heterozygous).

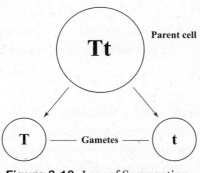

Figure 3.10. Law of Segregation.

The recessive allele is expressed only in organisms that are homozygous recessive for the trait. Homozygous allelic combinations are often referred to as pure breeding. This is because no hidden (nonexpressed) alleles are present.

Monohybrid Crosses

Monohybrid crosses involve crosses of two heterozygous combinations of alleles (hybrids) for a particular gene. They always result in a 3:1 ratio of phenotypes and in a 1:2:1 ratio of genotypes.

		T	t
T		TT	Tt
t		Tt	tt

F$_1$: Tt × Tt

F$_2$: TT, Tt, or tt

Monohybrid cross

The genotype and phenotype for two monohybrid crosses are shown in Figure 3.11. The figure shows that the two alleles present in each parent (Yy) separate into individual alleles (Y and y) in the gametes. The separation of the alleles represents meiosis. Note that only one of the two possible alleles is present in each gamete. The boxes in the Punnett square show the possible combinations of alleles that could occur during fertilisation when the sperm and egg join to make a diploid cell (zygote) during fertilisation.

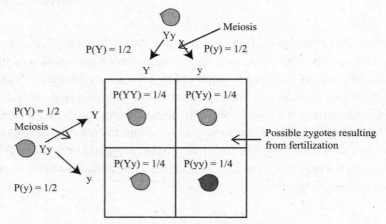

Figure 3.11. Monohybrid cross showing the predicted 3:1 phenotypic ratio (P = probability)

Multiple Alleles

For many genes, only two possible alleles are available in a population to code for a trait. However, some genes have more than two possible alleles available. This leads to a greater variation in allelic combinations. Although more alleles are available in the population, each member of the population can still possess only two of the available alleles at a time. When more than two alleles are available to the population to code for a trait, the condition is referred to as *multiple alleles*.

An excellent example of a trait coded for by more than two alleles is found in blood typing. The three alleles that code for blood typing are *A, B* and *i*. These three alleles code for four different phenotypes for blood: A, B, AB and O. The blood types represent molecules (antigens) that appear on the surface of red blood cells in blood types A, B and AB. Individuals with blood type O do not have any of these antigens on the surface of their red blood cells. (See Figure 3.12.)

	Group A	Group B	Group AB	Group O
Red blood cell type	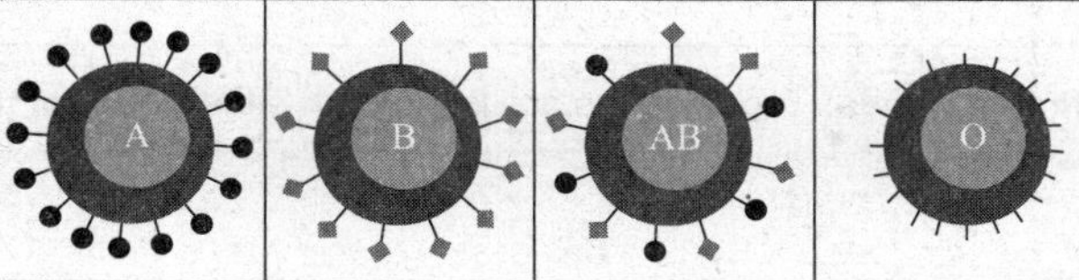 A	B	AB	O
Antibodies in plasma	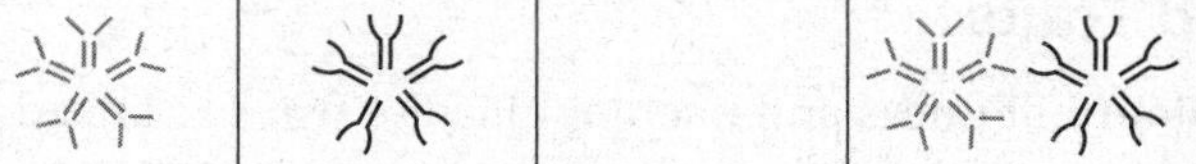 Anti-B	Anti-A	None	Anti-A and Anti-B
Antigens in red blood cell	A antigen	B antigen	A and B antigens	None
Blood genotypes	Homozygous *AA* Heterozygous *Ai*	Homozygous *BB* Heterozygous *Bi*	*AB*	*ii*

Figure 3.12. Blood types

The three alleles that code for blood type can be arranged to produce the four blood types A, B, AB and O. It is customary to use the lowercase *i* to represent blood type O. Individuals who have blood type A or blood type B can be either homozygous or heterozygous for the trait. For example, an individual with blood type A can possess an *A* and an *i* allele (*Ai*) or can have two *A* alleles (*AA*). An individuals who has blood type B can possess either a *B* and an *i* allele (*Bi*) or can have two *B* alleles (*BB*). The alleles for blood types A and B are sometimes represented with an uppercase I (immunoglobin) as I^A and I^B, respectively. When both allele *A* and allele *B* are present (blood type AB), neither allele will mask the other. They will both be expressed on the red blood cell. This is an example of codominance in which both traits (alleles) will appear in the phenotype.

Sex-Linked Genes

Two chromosomes (X and Y) determine the sex of human individuals. The combination of two X chromosomes (XX) results in a female offspring. The combination of one X and one Y chromosome (XY) results in a male offspring. The sperm can carry either an X or a Y. However, the egg can carry only an X chromosome. Therefore, the determination of sex is due to the type of sperm that fertilises the egg. If the sperm carries an X, the offspring will be female. If the sperm carries a Y, the offspring will be male.

Certain genes are found on either the X or the Y chromosome. These are called sex-linked genes. Note, though, that the traits for which they code often have nothing to do with sexual reproduction. For example, the genes for haemophilia (a blood-clotting disease) and colour blindness are sex linked.

The Y chromosome is much shorter than the X chromosome and therefore codes for only a few genes. Males possess only one copy of the genes carried on the X chromosome, while females have the traditional two copies. Only one of the two alleles found on the X chromosomes in females will be expressed for a trait. In contrast, males will always express the one allele they possess on their sole X chromosome. Since only a few genes are on the Y chromosome, females can develop normally without having a Y chromosome.

> **Sex-linked genes:** Those genes are located on either the X or the Y chromosome.

Sex-Linked Traits

Red-green colour blindness and haemophilia are two sex-linked traits that are coded for by alleles found on the X chromosome. Sex-linked traits are more frequently expressed in males than in females. If the male has the allele that codes for the presence of the trait on his sole X chromosome, he will express the trait. In contrast, a female could have an allele that codes for the expression of the trait as well, but her extra X chromosome could contain a dominant allele that masks the trait.

Figure 3.13 shows how a male has to have only one recessive allele to be colour-blind, while females must have two recessive alleles to express the colour-blind trait. Although females can have one recessive allele for colour blindness, having a dominant allele on their second X chromosome that does not code for the trait keeps them from expressing colour blindness. Males do not have this ability since they possess only one X chromosome.

	X^B	X^b
X^B	$X^B X^B$	$X^B X^b$
Y	$X^B Y$	$X^b Y$

½ of the females will be carriers
½ of the females will be normal
½ of the males will be normal
½ of the males will be colour blind

Figure 3.13. Sex-linked inheritance

SEX-LINKED TRAITS AND FEMALES

Females who carry two dominant alleles (homozygous dominant) or a dominant and a recessive allele (heterozygous) for a recessive sex-linked trait will not express the recessive trait. Females who carry two recessive alleles for the trait (homozygous recessive) will express the trait.

Females who are heterozygous for recessive sex-linked traits are referred to as carriers of the trait. They do not express the trait but, instead, possess an allele for the trait that can be passed on to their offspring. Males do not pass sex-linked X chromosome alleles to their sons since males must pass on the Y chromosome to their male offspring.

Here are some important facts about sex-linked traits.

■ Common examples of recessive sex-linked traits are **color blindness**, **haemophilia**, and Duchenne muscular dystrophy**.**

- *All daughters of affected fathers are carriers* (shaded squares).

Punnett square	X–	Y
X	X–X	XY
X	X–X	XY

- Sons cannot inherit a sex-linked trait from the father because the son inherits the Y chromosome from the father.
- A son has a 50 percent chance of inheriting a sex-linked trait from a carrier mother (shaded square).

Punnett square	X	Y
X–	X–X	X–Y
X	XX	XY

- There is no carrier state for X-linked traits in males. If a male has the gene, he will express it.
- It is uncommon for a female to have a recessive sex-linked condition. In order to be affected, she must inherit a mutant gene from *both* parents.

Pedigree Charts

Pedigree charts are used to show how traits have been or can be passed down from one generation to the next. In a pedigree chart, a square represents a male and a circle represents a female. If the shape is shaded in, the individual is affected by (expresses) the trait being studied. If the trait is caused by a recessive allele, those shapes shaded in must possess two recessive alleles (be homozygous) for the trait. If more males are shaded in then females, it is most likely a sex-linked trait. This means the trait is carried on the X chromosome.

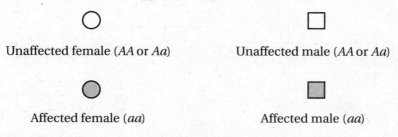

Unaffected female (*AA* or *Aa*)   Unaffected male (*AA* or *Aa*)

Affected female (*aa*)   Affected male (*aa*)

Lines connecting circles to squares indicate a couple that has produced offspring together. Lines branching down from each couple identify the children born to the couple.

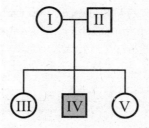

In the pedigree shown above, neither the mother (I) nor the father (II) expresses the trait. The male offspring (IV) is affected, but the female offspring (III and V) are not. So we can conclude that one of the parents must be a carrier for the trait. Since the females do not express the trait, we can assume the trait is sex linked and is carried on the X chromosome.

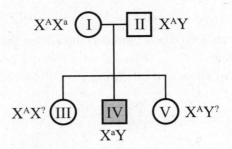

Since the father (II) is unaffected, he must possess a dominant allele. This indicates that the mother (I) must be a carrier (heterozygous) since the male offspring (IV) must receive the recessive allele from the mother (I). Except for knowing the daughters (III and V) must have one dominant allele, we cannot conclude whether or not they are homozygous dominant or heterozygous for the trait.

When interpreting pedigrees, you should always start with what you know. You know that for recessively inherited traits, individuals that are shaded in must be homozygous recessive. Fill in what you know and work backwards. Try to determine the pattern of inheritance from the pedigree chart in Figure 3.14.

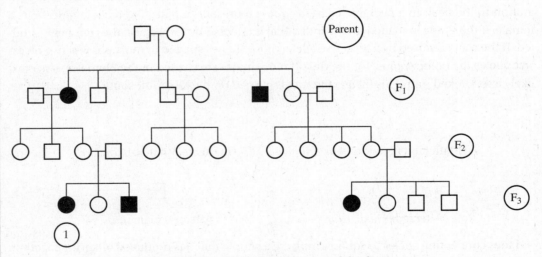

Figure 3.14. Three generations of deafness

The pedigree shown in Figure 3.14 depicts a recessively inherited gene for deafness. All afflicted children have unaffected parents, which means the parents must be carriers (heterozygous) for the gene.

Autosomal Dominant Mutations

Some genetic diseases are due to the presence of a mutated copy of a gene on the dominant allele. Those individuals possessing just one dominant allele or both dominant alleles will exhibit the disease. Huntington's disease is an autosomal dominant disease that causes degeneration of neural tissue. Individuals with Huntington's disease will experience loss of muscle control as well as loss of some cognitive functions. The symptoms of the disease usually begin to appear after the age of thirty. The pedigree in Figure 3.15 shows a possible inheritance pattern for an autosomal dominant disease such as Huntington's.

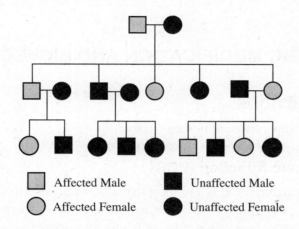

☐	Affected Male	■	Unaffected Male
◯	Affected Female	●	Unaffected Female

Figure 3.15. Inheritance of an autosomal dominant disease

There are several types of genetic diseases. Each type has its own inheritance pattern. Table 3.3 gives an overview.

Table 3.3: Genetic Diseases

Type of Genetic Disease	Example(s)	Inheritance Pattern
Sex-linked	Haemophilia and red-green colour blindness	Inherited on a mutated gene located on the X chromosome. More common in males due to its presence on the X chromosome.
Autosomal dominant	Huntington's disease	Inherited on a mutated gene of a dominant allele, leading to altered protein function.
Point mutation	Cystic fibrosis	Change in a single base, leading to a different amino acid being incorporated into the polypeptide. The altered protein can no longer function.
Nondisjunction	Down syndrome (trisomy-21) Klinefelter's syndrome (XXY)	Failure of the chromosomes to separate properly during meiosis. This leads to a missing or to an extra chromosome in the genotype of the offspring.

Mutagens

A change in the genetic makeup (DNA) of an individual is referred to as a mutation. Mutations can occur from errors during cellular replication and unrepaired DNA damage. Mutations can cause genetic diseases and the development of cancers. If the mutation causes the cell to lose the ability to control the rate of cellular division, cancer develops. Exposure to ultraviolet (UV) light, microscopic pathogens, carcinogenic-chemicals and radiation increase the risk of developing cancers. The radiation released in Hiroshima after the atomic bomb was dropped and in Chernobyl after the nuclear accident were both followed with high rates of cancer in the local population.

3.5 GENETIC MODIFICATION AND BIOTECHNOLOGY

TOPIC CONNECTIONS

- 2.7 DNA replication, transcription and translation

Polymerase Chain Reaction (PCR)

Polymerase chain reaction (PCR) is a process that allows scientists to make many copies of DNA (amplify DNA). The process was developed in 1985 and is considered to be one of the greatest biotechnology tools available today. PCR is important because sometimes only a very small amount of DNA is available (for example, DNA recovered from a crime scene).

The process uses *Taq* polymerase, DNA primers, and nucleotides to produce the new strands of DNA. (Note that *Taq* polymerase is an enzyme that can synthesize DNA. It can survive high temperatures because it is obtained from the thermophilic (heat-loving) bacterium *Thermus aquaticus*.) PCR uses cycles of heating and cooling to break and form new hydrogen bonds between the exposed nucleotides in DNA continually. The steps for PCR are denature, anneal and extend, as shown in Figure 3.16.

1. **Denature:** Heat the solution containing the *Taq* polymerase, DNA and nucleotides in order to separate the double-stranded DNA into two single strands.
2. **Anneal:** Cool the solution so primers can attach to the exposed complementary nucleotides on the single-stranded DNA.
3. **Extend:** Once primers have attached themselves to the DNA, the *Taq* polymerase adds nucleotides to both exposed primed DNA strands.

The process of heating and cooling is repeated many times and can produce over a million copies of the original DNA in just a few hours!

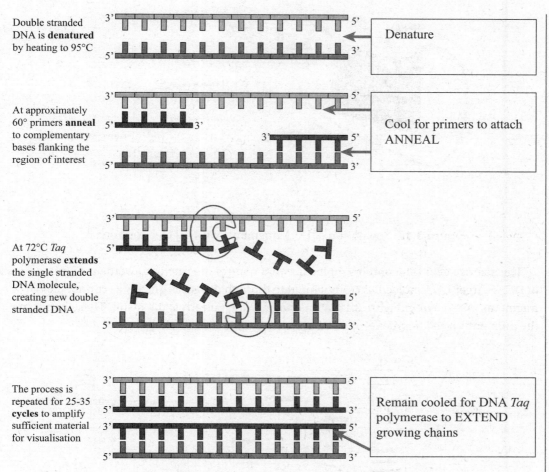

Double stranded DNA is **denatured** by heating to 95°C

Denature

At approximately 60° primers **anneal** to complementary bases flanking the region of interest

Cool for primers to attach ANNEAL

At 72°C *Taq* polymerase **extends** the single stranded DNA molecule, creating new double stranded DNA

The process is repeated for 25-35 **cycles** to amplify sufficient material for visualisation

Remain cooled for DNA *Taq* polymerase to EXTEND growing chains

Figure 3.16. Polymerase chain reaction (PCR)

Gel Electrophoresis

Gel electrophoresis is a process that separates DNA fragments based on their **size** and **charge**. Prior to running DNA through gel electrophoresis, the DNA is cut into small fragments with the use of restriction enzymes. Restriction enzymes always cut DNA at specific base sequence sites. Each individual's base arrangement varies. So cutting the DNA with restriction enzymes creates fragments that are unique in length for each individual's DNA. The fragments are placed into an agarose gel that is submerged in a buffer (a type of liquid) in the electrophoresis chamber. The electrophoresis chamber creates a current that runs through the buffer and through the agarose gel. This current runs from a negative anode (–) to a positive cathode (+), creating a constant, moving charge. Since DNA is negatively charged due to its phosphate groups, the strands migrate through the agarose gel. They move from the negatively charged end of the electrophoresis chamber to the positively charged end. Figure 3.17 shows the separation of DNA fragments via gel electrophoresis.

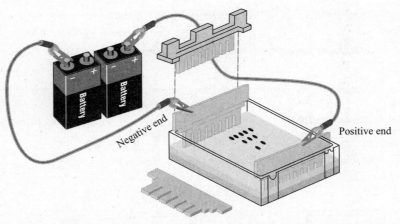

Figure 3.17. Separation of DNA fragments with gel electrophoresis

The agarose gel itself contains molecules that control the rate of movement of the pieces of DNA. Larger DNA molecules cannot migrate through the gel as quickly as can smaller DNA fragments. As shown in Figure 3.18, a unique banding pattern is the result. These bands are the differently sized fragments of DNA, and they can be seen on the gel.

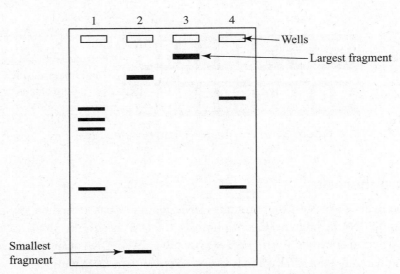

Figure 3.18. Gel electrophoresis results showing
DNA patterns produced by four different sources of DNA

In Figure 3.18, you can see that lane 1 has 4 pieces of DNA, lane 2 contains two pieces of DNA, lane 3 has one large piece of DNA, and lane 4 contains two pieces of DNA. If the banding pattern was the same in any of the wells, we would know that the DNA came from the same organism.

DNA Profiling

The DNA banding patterns that appear on gel electrophoresis are used in DNA profiling. Criminal cases and paternity are often decided based on DNA evidence. The DNA of interest is amplified by PCR in order to obtain enough DNA to run through gel electrophoresis. Each piece of the DNA of interest is cut with the same restriction enzymes.

Identifying individuals based on their DNA banding patterns in gel electrophoresis is known as DNA profiling or DNA fingerprinting. The resulting DNA banding patterns can be compared to determine if any match those from a known crime scene. The DNA profiles will be exact matches if the suspect's DNA is the same as the DNA found at the crime scene. This process has been used to convict felons as well as prove their innocence.

In addition, DNA profiles can be used to determine with very high certainty if a man is the father of a child (paternity). DNA fingerprints from the child and the suspected father are compared for relationship indexes to determine if the child received specific genes from the father. These profiles look for known DNA sequences that must come from the father and are 99.99% accurate.

Analysis of DNA Fragments

The DNA fragments shown on the gel plate in Figure 3.19 are from a crime scene and from three suspects. Look at the plate and determine which suspect was at the crime scene (has matching DNA fragments).

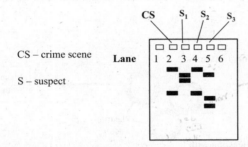

Figure 3.19. DNA fragments from the crime scene and from three suspects

The results shows that suspect 2 produced a banding pattern exactly like the one resulting from the DNA found at the crime scene. This means that the DNA from suspect 2 and the DNA from the crime scene are from the same individual.

Gene Transfer

Every living organism contains DNA that is made of the same four nitrogenous bases: adenine, thymine, guanine and cytosine. For this reason, any organism can read the DNA bases of any other organism and the enzymes used for protein synthesis in one organism will function for all organisms. This allows scientist to transfer DNA from one species to another without the donor DNA being rejected. The protein coded for by the donor DNA will appear the same when the recipient of the foreign DNA produces it.

Gene transfer allows scientist to transfer genes (DNA) from one species into the genome (DNA) of a different species. In order to do this, the gene of interest must be cut from the donor DNA using restriction enzymes. Many different restriction enzymes are available. Each one cuts the DNA at unique base sequence sites. Some restriction enzymes leave sticky ends on the cut DNA, and some leave blunt ends. Restriction enzymes that leave sticky ends are used to obtain DNA fragments that have matching complementary bases. These DNA fragments can then easily form bonds to join together the two pieces of DNA.

A plasmid (small circular piece of DNA) is obtained from a prokaryotic (bacterial) cell and cut with the same restriction enzyme that is used to cut the foreign DNA out of its genome.

REMEMBER

Restriction enzymes are produced by only prokaryotic cells (bacterial cells). However, they will cut the DNA of any organism. This is possible because DNA is universal!

Both the plasmid and foreign DNA must be cut with the same restriction enzyme to ensure complimentary bases on the sticky ends of the DNA pieces. (Refer to Figure 3.20.)

The two cut pieces of DNA (plasmid and foreign) are joined together by the actions of the enzyme DNA ligase. The new DNA is referred to as *recombinant DNA* since it contains both original and foreign DNA.

The recombinant DNA must be placed into an organism in order for the protein coded by the gene to be produced. The organisms of choice for receiving the recombinant DNA are bacteria, yeast and other unicellular organisms. Unicellular organisms make the best hosts for the recombinant DNA since the foreign DNA needs to be taken up by only one cell. The recipient of the recombinant DNA will start producing the protein coded for by the bases. Proteins produced by gene transfer into bacterial cells include human growth hormone (HGH) and insulin. Gene transfer has also been used to modify plants so they can produce insecticides. Figure 3.20 outlines the process of producing plants that are naturally resistant to insects due to the transfer of a gene that codes for an insecticide.

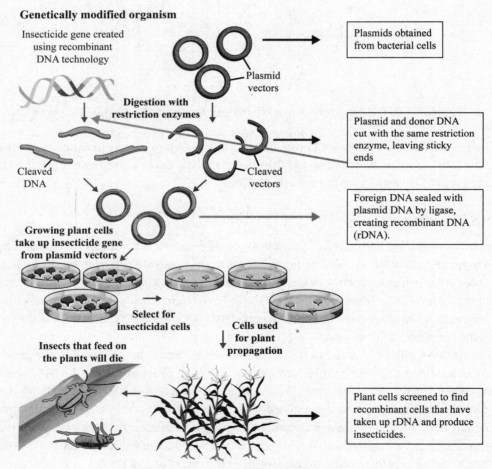

Genetically modified organism

Insecticide gene created using recombinant DNA technology

Plasmid vectors

Plasmids obtained from bacterial cells

Digestion with restriction enzymes

Cleaved DNA

Cleaved vectors

Plasmid and donor DNA cut with the same restriction enzyme, leaving sticky ends

Foreign DNA sealed with plasmid DNA by ligase, creating recombinant DNA (rDNA).

Growing plant cells take up insecticide gene from plasmid vectors

Select for insecticidal cells

Cells used for plant propagation

Insects that feed on the plants will die

Plant cells screened to find recombinant cells that have taken up rDNA and produce insecticides.

Figure 3.20. Gene modification in plants

Genetically Modified Organisms

Genetically modified organisms have been used to help produce crops that are resistant to pests, disease, extreme temperatures and salinity. The addition of genes that code for resistance to salinity and frost (cold temperatures) has allowed the harvesting of plants in environments where they could not otherwise survive. By producing genetically modified plants with new characteristics that enable them to survive in diverse environments, farmers have increased their crop yield and have therefore increased the world's food supply.

When recombinant DNA provides plants with resistance to pests, the need for chemical pesticides decreases. As a result, the use of pesticides in the environment decreases, thereby protecting the ecosystem. The nutritional content of plants can be increased with recombinant DNA technology. For example, the genes that code for the synthesis of beta-carotene can be added to the genome of plants. Nutritionally enhanced plants can help in treating nutritional deficiencies that exist in many parts of the world. An excellent example is "golden rice". Golden rice plants have been genetically modified to produce beta-carotene, which is a precursor to vitamin A. Vitamin A deficiencies in children in poor countries can lead to blindness. This rice has been vital in preventing the loss of sight. When children eat golden rice, they also eat the beta-carotene that the rice now produces due to recombinant DNA technology.

Genetically modified plants possess both potential benefits and possible harmful effects. For example, a plant genetically modified to survive in a new environment may out-compete the natural plants found in the environment. The potential benefits and risks of genetically modified plants are outlined in Table 3.4.

Table 3.4: Benefits and Risks of Genetically Modified Plants

Benefits	Risks
■ Increased crop yield ■ Less land needed for crop growth ■ Increased global food supply ■ Less use of pesticides	■ Other organisms in the ecosystem could be negatively affected. ■ Genetically modified pollen could affect other plants in the ecosystem ■ Individuals could be allergic to the foreign protein

Cloning

Clones are exact genetic copies of each other. They are produced naturally by many plant species as well as by some animal species. Identical twins are naturally occurring clones since they share the same genetic makeup. Asexual reproduction of bacterial cells produces clones. Plants can undergo cloning through asexual vegetative propagation. Artificial cloning is the production of clones in the laboratory. The first animal to be artificially cloned was a sheep named Dolly. Dolly was born in 1996 and lived for about six years. She was cloned from a cell taken from the udder (breast) of a genetic donor sheep (see Figure 3.21). This was the first time an entire organism was produced from a somatic (body) cell.

Clone: A group of genetically identical organisms or a group of cells derived from a single parent cell.

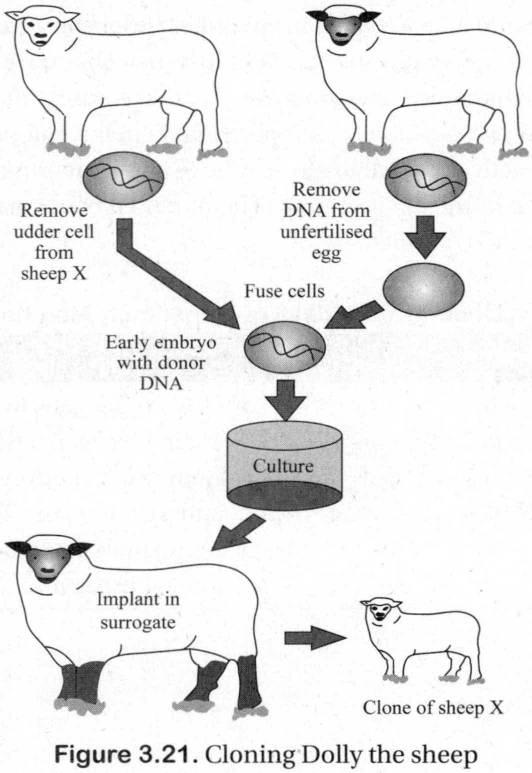

Figure 3.21. Cloning Dolly the sheep

In order to clone an animal, the entire genome (nuclear DNA) must be obtained from the organism chosen for cloning. The diploid set of DNA can be obtained from any body cell. The DNA is removed from the nucleus of the chosen cell and placed into an egg cell that has had its original haploid DNA removed. An electrical shock causes the egg cell to take up the foreign DNA. The egg does not have to be fertilised since it already contains a full set of donor chromosomes. The diploid cell is placed into a nutrient-rich environment and undergoes cellular division until it is a multicellular early embryo. The embryo is placed into the uterus of a surrogate mother for development into a mature offspring. The offspring will be an exact copy of the organism that donated the nuclear DNA.

Ethical Issues of Cloning

Many ethical issues arise related to the process of cloning humans for medical use. The most heavily debated is the use of embryonic stem cells. Embryonic stem cells have the capacity to develop into complete organisms. By studying stem cells undergoing differentiation and development, scientists could make great discoveries involving the treatment and understanding of the disease progression. However, the stem cell must be destroyed in order to carry out the research. For this reason, embryonic stem cell research remains a highly debated and controversial topic. The benefits and risk of embryonic stem cell research are outlined in Table 3.5.

Table 3.5: Benefits and Risks of Embryonic Stem Cell Research

Benefits	Risks
■ Therapies could be developed to treat disease, reduce death, and decrease suffering from pain. ■ Cells can be used solely from embryonic stem cells no longer actively dividing. ■ Stem cells cannot feel pain and do not suffer from the process.	■ Every embryo should be given an equal chance of survival. ■ Many embryos created for research may go unused and be destroyed. ■ There is a risk that embryonic stem cells injected into individuals for treatment could then develop into cancer cells.

1. Eukaryotic chromosomes are made of:

     I. DNA
    II. Protein
   III. Lipids

   A. I only
   B. I and III only
   C. I and II only
   D. I, II and III

2. Which of the following statements describes homologous chromosomes?

   A. They are two chromosomes that code for the same traits and have identical alleles.
   B. They are two chromosomes that code for different traits and have different alleles.
   C. They are two chromosomes that code for the same traits and can sometimes contain different alleles.
   D. They are two chromosomes that code for the different traits and can sometimes contain different alleles.

3. Compared to a haploid cell, a diploid cell has

   A. twice the number of chromosomes.
   B. half as many chromosomes.
   C. the same number of chromosomes.
   D. a varied number of chromosomes.

4. How are chromosomes arranged for karyotyping?

   A. According to their size and charge
   B. According to their size and structure
   C. According to their structure and charge
   D. According to their structure only

5.

The karyotype above is from a:

A. male with Down syndrome.
B. female with Down syndrome.
C. healthy male.
D. healthy female.

6. Black is dominant to white. A black guinea pig mates with a white guinea pig. Although most of the offspring are black, a few white guinea pigs are also produced. What combination of alleles could the black guinea pig possess?

A. Bb or BB
B. Bb
C. bb
D. BB

7. Haemophilia is a sex-linked trait. A child is born from a father who has haemophilia and from an unaffected mother. The child is a female who also suffers from haemophilia. What are the alleles of the three individuals? Note that in the table, H means unaffected (no haemophilia) and h means affected (haemophilia).

	Father	Mother	Daughter
A.	X^hY	X^HX^H	X^HX^h
B.	X^HY	X^hX^h	X^hX^h
C.	X^HY	X^HX^h	X^HX^h
D.	X^hY	X^HX^h	X^hX^h

8. Polymerase chain reaction (PCR) is used to

A. separate DNA based on charge and size.
B. cut individual pieces of DNA.
C. amplify DNA.
D. denature DNA.

9. In order to carry out gene transfer and produce recombinant DNA, which of the following are needed?

   I. Restriction enzymes
  II. DNA ligase
 III. DNA helicase

A. I only
B. I and II only
C. I and III only
D. I, II and III

10. Sickle cell anaemia results from

A. a virus.
B. a single base substitution.
C. recombinant DNA.
D. gene transfer.

11. A heritable factor that controls a specific characteristic is referred to as a(n)

A. genome.
B. allele.
C. gene mutation.
D. gene.

12. The skin cell of an organism contains 30 chromosomes. How many chromosomes will be present in the gametes produced by this organism?

A. 15
B. 30
C. 60
D. 90

13. When nonsister chromatids switch genetic information during meiosis, the process is known as:

A. nondisjunction.
B. fertilisation.
C. crossing-over.
D. a base substitution.

14. The process of collecting cells from the outer portion of the placenta of an embryo in order to produce a karyotype is referred to as

A. an amniocentesis.
B. gene transfer.
C. genetic sampling.
D. chorionic villus sampling.

15. Which of the following allele combination will express the dominant trait?

   I. *Tt*
   II. *TT*
   III. *tt*

   A. I only
   B. I and II only
   C. I and III only
   D. I, II and III

16. Restriction enzymes are used to

   A. cut DNA at specific sites.
   B. cut DNA at random sites.
   C. seal together DNA.
   D. Seal DNA with RNA.

17. The figure below shows DNA fingerprints that were produced for comparison to DNA found at a crime scene.

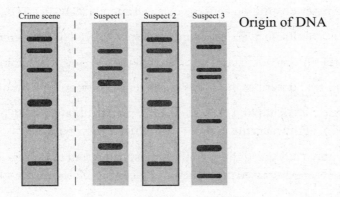

   Origin of DNA

   (a) Deduce which suspect was at the crime scene. (2)
   (b) State which bands contain the largest number of nucleotides. (2)
   (c) Outline how DNA is separated in gel electrophoresis. (3)

18. Explain how Down syndrome develops. (3)

19. Bill (blood type O) and Georgia (blood type A) are the biological parents of Anne (blood type A). Anne marries Ted (blood type B). Anne and Bill have a child who is blood type O. Identify the blood type genotypes of each individual. (4)

20. State two outcomes of sequencing the entire human genome. (2)

21. Define clone. (1)

22. (a) Outline the differences between gene, allele and genome. (4)
   (b) Explain how PCR amplifies DNA. (6)
   (c) Discuss the benefits and risks of genetic engineering. (8)

ANSWERS EXPLAINED

1. **(C)** Eukaryotic chromosomes are made of DNA and protein (histones for packaging).

2. **(C)** Homologous chromosomes code for the same traits, are of the same size and structure, and may or may not possess different alleles.

3. **(A)** Diploid cells have twice the number of chromosomes as do haploid cells.

4. **(B)** During karyotyping, chromosomes are paired based on their size and structure.

5. **(C)** Healthy male. There are no extra chromosomes (2 of each pair are present) and XY codes for a male.

6. **(B)** *Bb*. If the black male was homozygous (*BB*), there would be no white offspring.

7. **(D)** The father is affected, which means he is X^hY. Since the daughter has haemophilia (X^hX^h), the mother must be a carrier X^HX^h.

8. **(C)** PCR is used to amplify (make many copies of) DNA.

9. **(B)** Restriction enzymes are needed to cut the DNA. DNA ligase links together the new recombinant DNA. Helicase is not needed for this process.

10. **(B)** A single base substitution causes sickle cell anemia (GAG to GTG).

11. **(D)** A gene is a heritable factor that controls a specific trait or characteristic.

12. **(A)** Half of the chromosomes will be present in the gametes. The haploid number is 15.

13. **(C)** Crossing-over describes the switching of genes between nonsister chromatids.

14. **(D)** Chorionic villus sampling is the procedure used to obtain a small piece of the outer layer (chorion) of the placenta in order to produce a karyotype.

15. **(B)** I and II only. Only one dominant allele must be present to express a dominant trait. Recessive traits are expressed only when two recessive alleles are present.

16. **(A)** Restriction enzymes cut DNA at specific sites.

17. (a)
    - Suspect 2 was at the crime scene.
    - Suspect 2 and the crime scene DNA have matching bands.
    - Bands are from the same source since they travel the same distance .
    - Bands are of the same length.

    (b)
    - Bands closest to the origin contain the most nucleotides.
    - Bands closest to the origin travelled the least distance due to their increased size.
    - Smaller bands travel farther and are found at a greater distance from the origin.

    (c)
    - Gel electrophoresis separates DNA based on charge and size.
    - DNA is placed into the gel chamber with a buffer.
    - Electrical charge is run through the chamber.
    - DNA migrates from a negative to a positive pole.
    - DNA is negatively charged due to the phosphate groups.

18.

- Down syndrome results from nondisjunction.
- Chromosomes fail to separate.
- Chromosomes should separate during anaphase I/anaphase II.
- The resulting gamete has an extra chromosome.
- At fertilisation, there are three copies of the chromosome instead of the normal two.
- Down syndrome is also known as trisomy-21.

19.

- Bill is *ii*.
- Georgia is A–. We don't have enough information to know if she is homozygous or heterozygous.
- Anne must be *Ai* since her child is *ii* and must receive an *i* allele from each parent.
- Ted must be *Bi* since his child is *ii* and must receive an *i* allele from each parent.

20.

- Production of designer drugs/pharmaceuticals/tailor drugs to an individual's DNA
- Treatment of disease
- Diagnosis of disease
- Evolutionary relationships

21. A clone is a group of genetically identical organisms or a group of cells derived from a single parent.

22. (a)

- Gene: heritable factor that controls a specific trait
- Gene: found on chromosomes
- Gene: found at specific loci
- Allele: one specific form of a gene
- Allele: differs from other alleles by one or a few bases
- Allele: occupies the same locus as other alleles for the gene
- Allele: example of dominant and recessive alleles
- Genome: the whole of the genetic information of an organism
- Genome: mention of human genome project

(b)

- Developed in 1985
- Used to amplify/make many copies of DNA
- Heat is used to split the DNA/break hydrogen bonds/denature the strands
- Cooling cycle allows the primers to attach/primers attach
- DNA polymerase adds nucleotides
- DNA polymerase is *Taq*/from *Thermus aquaticus;* produces millions of copies in a few hours/copies very fast
- Amplified DNA can be used for gel electrophoresis

(c) Benefits:

- Higher crop yield/more plants survive
- Can feed the growing population
- Less land needed for same yield
- Use less pesticides/less toxins in the environment

Risks:

- Consequences of consuming genetically modified organisms (GMOs)
- Other organisms affected by the presence of the GMO
- Pollen from GMO plant could contaminate other organisms
- Cross-pollination could create new types of plants
- GMO could outcompete the natural organisms in the ecosystem
- One point is awarded for any named example of a GMO

Ecology

CENTRAL IDEAS

- Ecosystems, food chains and food webs
- Trophic levels
- Greenhouse effect
- Precautionary principle
- Populations

TERMS

- Carrying capacity
- Species
- Habitat
- Population

- Community
- Ecosystem
- Ecology
- Consumers

- Detritivores
- Saprotrophs
- Trophic level
- Evolution

INTRODUCTION

The biotic and abiotic features of ecosystems impact an organism's survival, its behaviours and its adaptions. Ecosystems rely on energy transfer and nutrient recycling in order to continue to thrive. In order for an ecosystem to be sustained, energy must flow through it and nutrients must be recycled. The carbon cycle can be impacted by human intervention. Changes in the carbon cycle can have profound effects on ecosystems.

NOS Connections

- Patterns, trends and discrepancies are seen in nature (species, populations and ecosystems).
- Scientific claims must be assessed in order to obtain validity (climate change).
- Accuracy in data collection is important in evaluating scientific phenomena (climate change, energy conversions).
- Theories are used to explain natural processes (energy flow, climate change).
- Scepticism exists within the scientific world (the human impact on climate change).
- Causation and correlation must be established in order to support scientific theories (climate change).

ECOLOGY

4.1 SPECIES, COMMUNITIES AND ECOSYSTEMS

TOPIC CONNECTION

- 4.1 Molecules to metabolism

Important Terminology

- **Species**: A group of organisms that can interbreed and produce fertile offspring.
- **Habitat**: The environment in which a species typically lives or the location of a living organism.
- **Population**: A group of organisms of the same species that live in the same area at the same time.
- **Community**: A group of populations living and interacting with each other in an area.
- **Ecosystem**: A community and its abiotic environment
- **Ecology**: The study of relationships among living organisms and among living organisms and their environment

> **REMEMBER**
>
> Autotrophs produce organic molecules from inorganic molecules.
>
> Heterotrophs must ingest or absorb organic molecules.

Autotrophs and Heterotrophs

Organisms that can produce their own organic molecules from simple inorganic substances are referred to as autotrophs. The best examples of autotrophs are found in the plant kingdom. Plants (and almost all algae) build their own organic molecules through the process of photosynthesis using radiant energy from the sun. Some organisms that live in the deep hydrothermal vents on the bottom of the ocean floor can also produce their own organic molecules. These organisms are chemosynthetic and can convert simple molecules such as methane into organic molecules.

In contrast, heterotrophs must obtain organic molecules from other organisms. Food (energy) enters an ecosystem through autotrophs and flows through the ecosystem from one consumer to another.

Decomposers

Detritivores and saprotrophs are both decomposers. Decomposers are as vital to an ecosystem as are producers. Without decomposers, dead organic matter would not be able to be broken down (decomposed) and returned to the ecosystem. Without the recycling of the organic material found in dead organisms, living organisms would run out of the necessary nutrients that they must ingest in order to survive.

Organisms that obtain energy to build their organic molecules from autotrophs are referred to as consumers. Consumers ingest other organic matter that either is living or has recently been killed. Detritivores, such as earthworms and vultures, ingest dead organic matter. Saprotrophs, such as fungi (for example, mushrooms) and some types of bacteria, live in or on dead organic matter. Saprotrophs secrete digestive enzymes into the dead organic matter in order to digest the material before absorbing the products. Since saprotrophs lack a large entrance for material to enter into their tissues, they must predigest the organic material into small molecules that can be absorbed across their cell membranes. Decomposers are vital to maintaining a supply of inorganic material, such as carbon, in the environment through nutrient cycling. The process of decomposition of organic material releases inorganic material back into the ecosystem. Table 4.1 compares and contrasts detritivores and saprotrophs.

Table 4.1: Decomposers

Decomposer	Method of Feeding	Examples
Detritivore	Ingests and then digests	Earthworm and vulture
Saprotroph	Digests externally and then ingests	Fungi and bacteria

Food Chains

A food chain shows the direct flow of energy from one feeding level to the next. Each feeding level is referred to as a trophic level. Arrows drawn in food webs represent the direction of energy transfer. Energy passes from the producer to the consumer up the food chain. Most of the energy consumed is lost, mainly as heat. As a result, only about 10 percent of the ingested food is converted to new organic molecules. For this reason, the number of feeding levels is limited. Rarely will a food chain exceed four or five feeding levels.

Primary consumers feed directly on producers, secondary consumers feed on primary consumers and tertiary consumers feed on secondary consumers.

Food Webs

Food chains do not exist in isolation of each other. Many organisms feed on more than one organism and may feed at several different feeding levels. The complex feeding interactions that exist in ecosystems can be shown by use of a food web, as in Figure 4.1.

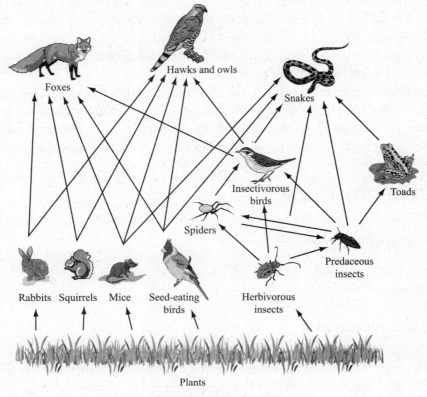

Figure 4.1. Food web

Trophic Levels

A trophic level is a feeding level in a food chain or food web. It represents an organism's position in the food chain or food web. Trophic levels include producers (autotrophs—most commonly plants), primary consumers (herbivores—organisms that feed on plants), secondary consumers (carnivores—organisms that feed on meat), and tertiary consumers (carnivores—organisms that feed on meat). Rarely, higher trophic levels can be seen. However, these are limited due to the small amount of energy that is passed on from each trophic level. (See Figure 4.2.)

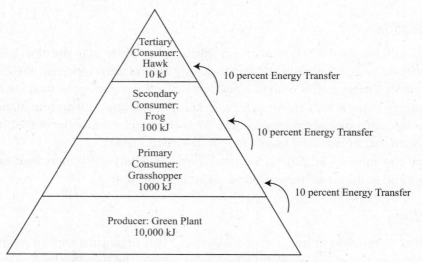

Figure 4.2. Food pyramid showing trophic levels

Identify the trophic levels occupied by the mice, plants, spiders, and hawks and owls in Figure 4.1. You should be able to identify the plants as autotrophs (producers), the mice as primary consumers (herbivores), the spiders as secondary consumers (carnivores), and the hawks and owls as both secondary and tertiary consumers. Notice that the hawks and owls are members of more than one trophic level.

When asked to construct a food web, you must include at least ten organisms and show at least three identified trophic levels. Drawings can be used to show each organism, but its name must be stated as well. State the name of each organism using either genus and species name or common name. Include arrows to show the direction of energy flow through the food web.

4.2 ENERGY FLOW

TOPIC CONNECTIONS

- 2.8 Cellular respiration
- 2.9 Photosynthesis

Energy in Ecosystems

Although radiant energy (light) is the main energy source for most communities, we must take into account the communities that exist in the deep ocean trenches. These communities rely on chemosynthesis for their initial energy source. For most ecosystems, energy from

ECOLOGY

light is converted into chemical energy stored in carbon compounds. The energy stored in carbon compounds travels through ecosystems as organisms feed on organic material. During cellular respiration, energy is released from organic compounds. Note that excess energy is released as heat. This heat is released to the environment and represents lost energy since organisms cannot convert the heat into other forms of energy.

Energy Transfer

Energy transformations between trophic levels are never 100 percent efficient. Recall that only about 10 percent of the energy stored in organic material is transferred on to the next trophic level. Energy lost between trophic levels includes material not consumed or material not assimilated. Heat energy is lost through the process of cellular respiration.

Some of the organic material produced at each trophic level is never consumed. This means the energy fails to be passed on to the next trophic level. The organic material that is consumed is never fully digested and cannot be absorbed by the consumers, resulting in the loss of more of the energy available for transfer. Most of the energy stored in organic molecules is lost as heat during chemical digestion. This is one way that endothermic organisms can create body heat, but it represents a large loss of energy from the ingested organic matter.

> Biomass, as well as energy, decreases along food chains due to the loss of:
>
> - Carbon dioxide
> - Water
> - Waste products (such as urea)

ENERGY PYRAMIDS

Energy pyramids represent energy transfer through trophic levels. Since only about 10 percent of the available energy is transferred to the next trophic level, each level of the pyramid is about 1/10 the size of the level it follows. Energy produced by autotrophs (producers) is calculated in energy per unit area per unit in time, for example $kJm^{-2}yr^{-1}$ (kilojoules per square meter per year). The amount of energy transferred and the relative size of each of the parts of the pyramid representing the energy available at each level can be seen in Figure 4.3.

> **REMEMBER**
>
> Biomass is the total weight of living organisms in a specified area at a specific time.

Producers are constantly supplied with new energy (mainly from the sun), but nutrients are limited in ecosystems. Decomposers are vital for recycling nutrients from dead organic material back into the soil for use by living organisms. If the nutrients in dead organisms were not returned to the ecosystem, the ecosystem would be quickly depleted of many of the molecules necessary for life processes (nutrients). In order to exist, every ecosystem must possess decomposers as well as producers. Fungi and bacteria are the main decomposers in ecosystems. They break down dead organic matter and are responsible for returning nutrients to the soil and water for use by other living organisms.

- Energy supply in terrestrial ecosystems in the form of sunlight may vary but is continuous.
- Nutrient supply in terrestrial ecosystems is limited and finite.

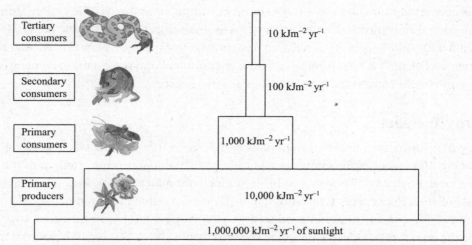

Figure 4.3. Energy pyramid showing energy transfer

4.3 CARBON CYCLING

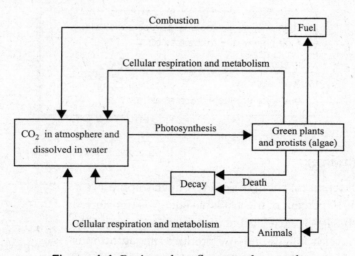

Figure 4.4. Basic carbon flow—carbon cycle

Living organisms interact with the biosphere as they incorporate carbon into organic molecules, as seen in Figure 4.4. Carbon dioxide enters ecosystems through the process of photosynthesis. The carbon in carbon dioxide is used in the production of organic compounds. Terrestrial organisms obtain carbon directly from carbon dioxide. Aquatic organisms can obtain carbon from both carbon dioxide dissolved in water and from hydrogen carbonate ions. The process of cellular respiration in both aquatic and terrestrial organisms releases carbon dioxide back into the environment.

Carbon is recycled back to the atmosphere through the processes of cellular respiration and combustion. Biomass and fossilized organic matter (coal, oil and natural gas) release carbon dioxide during combustion.

Carbon Fluxes

The amount of carbon dioxide in the atmosphere has been increasing since the earliest recorded levels in the 1700s, as shown in Figure 4.5. The early levels were deduced based on the analysis of air bubbles found in ice at different levels of ice sheets in Antarctica. In recent years, the levels have been increasing much faster than seen in recorded history. Many scientists argue that human interaction with the environment is causing the increase in carbon dioxide levels.

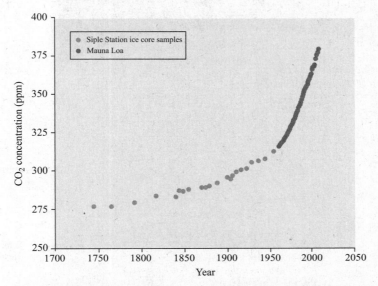

Figure 4.5. Atmospheric concentration of carbon dioxide

4.4 CLIMATE CHANGE

The Greenhouse Effect

The two most significant contributors to greenhouses gases are water vapour and carbon dioxide. Rising levels of methane and nitrogen oxides in the atmosphere have a similar but smaller impact. Although the greenhouse effect is a natural phenomenon, the increased production of greenhouse gases has enhanced the greenhouse effect. (See Figure 4.6.)

The earth receives short-wave radiation from the sun and reflects some of this energy back into the atmosphere as long-wave radiation. The greenhouse gases naturally block much of the long-wave radiation from leaving the lower atmosphere. These wavelengths are responsible for warming Earth's surface and the lower atmosphere. The problem with too many greenhouse gases being produced is that more of the long-wave radiation becomes trapped in the atmosphere, enhancing the warming effect.

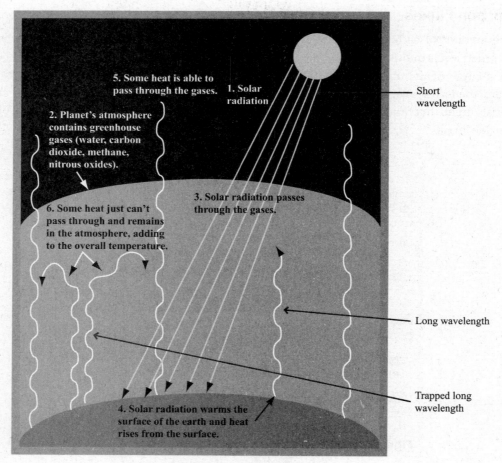

Figure 4.6. Greenhouse effect

The following labels appear in the figure:

5. Some heat is able to pass through the gases.

1. Solar radiation

Short wavelength

2. Planet's atmosphere contains greenhouse gases (water, carbon dioxide, methane, nitrous oxides).

3. Solar radiation passes through the gases.

6. Some heat just can't pass through and remains in the atmosphere, adding to the overall temperature.

Long wavelength

Trapped long wavelength

4. Solar radiation warms the surface of the earth and heat rises from the surface.

GREENHOUSE GASES

Carbon dioxide (CO_2) is released from organic molecules during the combustion of fossil fuels and the process of cellular respiration. Partially decomposed organic matter can become fossilized due to conditions that inhibit total decomposition of the organic material. For example, peat forms from partially decomposed organic matter in wetlands due to the acidity of the soil, anaerobic conditions or the presence of both inhibiting further decomposition. Fossilized organic matter from past geologic eras that was converted into fossil fuels (coal, oil and natural gas) contains stored organic carbon. This carbon is released during combustion of the fossil fuels and enters the atmosphere as carbon dioxide.

> Carbon fluctuations in the atmosphere are measured in gigatonnes (billions of tonnes, or tons).

Methane (CH_4) traps more heat than does carbon dioxide but is much shorter lived than is carbon dioxide. Once in the atmosphere, methane is oxidized into water vapour and carbon dioxide (the two most significant greenhouse gases). Methane is released into the atmosphere and ground from processes such as volcanic eruptions and from organic matter during the metabolic processes carried out by anaerobic methanogenic bacteria (domain Archaea). Some mammals possess methanogenic microbes in their digestive tract, producing methane that is released along with digestive waste.

Processes that incorporate calcium carbonate into reefs and mollusk shells absorb carbon dioxide from the atmosphere. The carbon stored in these organisms can become fossilized as limestone. The formation of limestone decreases the amount of free carbon dioxide in the atmosphere. Conversely, the decomposition of limestone can release carbon dioxide into the atmosphere increasing the levels of greenhouse gases. (See Table 4.2.)

Table 4.2: Release and Storage of Greenhouse Gases

Processes Releasing (Increasing) Greenhouse Gases	Processes Absorbing (Decreasing) Greenhouse Gases
■ Cellular respiration (releases CO_2) ■ Combustion of fossil fuels ■ Methanogenic bacteria in wetlands (rice paddies) and mammalian digestive tracts ■ Industrial processes and fertiliser use release nitrous oxides ■ Deforestation	■ Photosynthesis (stores CO_2) ■ Coral reef formation stores calcium carbonate ■ Calcium carbonate incorporated into mollusc shells ■ Carbon stored in fossilized limestone

FEATURED QUESTION

November 2008, Question #15

Which of the following gases contribute to the greenhouse effect?

  I. Methane
 II. Water vapour
III. Nitrogen

A. I only
B. I and II only
C. II and III only
D. I, II and III

(Answer is on page 553.)

The Precautionary Principle

The precautionary principle holds that if the effects of a human-induced change would be very large, those responsible for the change must prove that it will do no harm before proceeding. In other words, if a company wants to increase the production of a product that would result in the emission of a greenhouse gas, it must first prove that the emissions will not harm the environment before being allowed to proceed with production. Since the consequences of the greenhouse effect could be catastrophic, the precautionary principle applies.

The precautionary principle affects the economy due to limiting many industrial pursuits by preventing production of products that could lead to devastating results due to the increase in greenhouse gases. The loss of economic gain is counteracted by the protection gained by protecting the environment from greater harm that would impact future generations. Ethical issues arise when balancing economic pursuit with the devastation and possible extinction of species. Greenhouse gases could cause global temperatures to rise, resulting in changing weather patterns and loss of land. The coastal regions would be greatly affected due to loss of land and intensification of storm strength. Fertile land may become infertile due to the climate change. Diseases not previously observed in areas may become prevalent.

The greenhouse effect is a global issue that rests on the shoulders of some countries more than others. There exists an unequal global contribution to the production of greenhouse gases. Developed countries far outdo undeveloped countries in the production of greenhouse gases. There is inequity between the countries contributing the most to the enhanced greenhouse effect and those countries that will be most harmed. International cooperation to control the enhanced greenhouse is the best solution to this problem.

Consequences of Global Warming

Consequences of the enhanced greenhouse effect include increased rates of decomposition of detritus trapped in permafrost, expansion of the range of habitats available to temperate species, loss of ice habitats, changes in distribution of prey species affecting higher trophic levels, and increased success of pest species, including pathogens. As more detritus is available for decomposition, cellular respiration by decomposers will result in even more carbon dioxide being released. The melting of the ice caps will result in loss of habitat for many species and will open up niches for foreign species to enter the ecosystems. This could result in the extinction of existing species. Pathogens will be able to spread into previously uninhabitable niches and bring disease. In addition, as the polar ice caps melt, the sea level will rise, thereby covering land that is currently used for farming or living space for organisms. (See Table 4.3.)

Table 4.3: Consequences of Global Warming

Effect of Global Warming	Consequences
Rising sea level	■ Loss of habitat for terrestrial organisms ■ Changes in habitat for aquatic organisms
Melting of permafrost	■ Increased decomposition of detritus trapped in permafrost, which will increase carbon dioxide production (further enhancing the greenhouse effect) ■ Loss of habitat for organisms adapted to the permafrost
Change in weather patterns	■ Increased percentage of hazardous weather patterns ■ Serious flooding and droughts may be widespread ■ Fertile land may change and lose its ability to support plant life ■ Disease-causing organisms could survive in previously uninhabited areas

1. Which of the following must be present in an ecosystem?

   I. Producers
   II. Consumers
   III. Decomposers

   A. I only
   B. I and II only
   C. I and III only
   D. I, II and III

2. What processes(s) return carbon dioxide to the atmosphere?

   A. Photosynthesis only
   B. Combustion, decomposition and cellular respiration
   C. Combustion, decomposition, cellular respiration and photosynthesis
   D. Combustion only

3. Which of the following actions contribute to the enhanced greenhouse effect?

   A. Planting trees in the rain forest
   B. Transporting produce in trucks across the country
   C. Buying locally grown produce
   D. Decreasing the use of fossil fuels

4. What term describes a group of members of the same species located in the same area?

   A. Community
   B. Species
   C. Population
   D. Ecosystem

5. Which of the following are considered to be greenhouse gases?

   I. Methane
   II. Carbon dioxide
   III. CFCs

   A. I only
   B. II only
   C. I and II only
   D. I, II and III

6. Which actions remove carbon dioxide from the atmosphere?

   A. Photosynthesis, reef building and limestone formation
   B. Photosynthesis, combustion of fossil fuels and limestone formation
   C. Cellular respiration, combustion of fossil fuels and limestone formation
   D. Cellular respiration, reef building and combustion of fossil fuels

ECOLOGY

7. An ecosystem consists of:

   A. populations only
   B. communities and abiotic factors
   C. populations and communities
   D. abiotic factors only

8. Which of the following are heterotrophs?

   I. Fungi
   II. Detritivores
   III. Algae

   A. I only
   B. I and II only
   C. I and III only
   D. I, II and III

9. Members of the same species:

   A. interbreed, producing fertile offspring.
   B. interbreed without producing fertile offspring.
   C. do not interbreed.
   D. are structurally prevented from interbreeding.

10. What correctly describes saprotrophs?

    A. Decomposers with internal digestion.
    B. Decomposers with external digestion.
    C. Autotrophic decomposers with internal digestion.
    D. Autotrophic decomposers with external digestion.

The graph below represents global carbon emissions from fossil fuel burning between the years of 1751 and 2003. Refer to the graph as you answer Questions 11 (a)–11 (c).

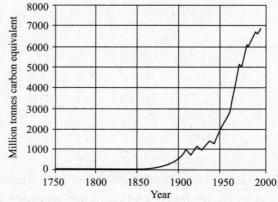

GLOBAL CARBON EMISSIONS FROM FOSSIL FUEL BURNING, 1751–2003

Source: Earth Policy Institute, based on Worldwatch, ORNL, BP figures.

ECOLOGY

11. (a) State the range of the time period in which global carbon emissions began to increase. (1)

    (b) Calculate the percent change in global carbon emissions between 1900 and 2000. (2)

    (c) Suggest two reasons why global carbon emissions increased between 1950 and 2000. (2)

12. Identify the trophic level for organisms I and III. (2)

A simple food web

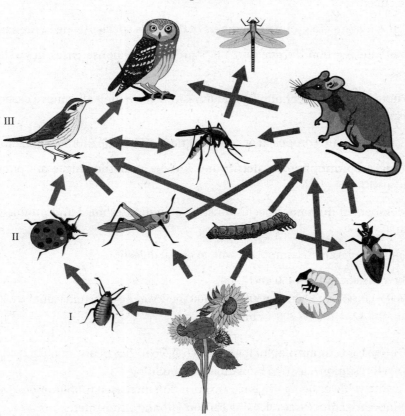

13. Distinguish between detritivores and saprotrophs. (2)

14. Explain the enhanced greenhouse effect. (4)

15. List three greenhouse gases. (3)

16. (a) Draw a food web that includes 8 different organisms and at least 3 trophic levels. (8)

    (b) Explain how energy travels through trophic levels. (6)

    (c) Describe how carbon cycles through the environment. (6)

ECOLOGY

ANSWERS EXPLAINED

1. **(C)** Producers must be present to provide the initial source of energy (mainly through photosynthesis). Decomposers must be present in order to ensure the recycling of limited nutrients in the environment.

2. **(B)** Combustion (burning fossil fuels), decomposition (through cellular respiration) and cellular respiration all return carbon dioxide to the atmosphere.

3. **(B)** Transporting produce across the country requires the burning of fossil fuel that contributes carbon dioxide to the atmosphere, contributing to the enhanced greenhouse effect.

4. **(C)** A population is a group of members of the same species living in the same area.

5. **(D)** Methane, carbon dioxide and CFCs are all greenhouse gases (as well as nitrous oxides).

6. **(A)** Photosynthesis, reef building and limestone formation all remove carbon from the atmosphere.

7. **(B)** Ecosystems are composed of communities and abiotic (nonliving) factors.

8. **(B)** Fungi (saprotrophs) and detritivores are decomposers. Algae are photosynthetic autotrophs.

9. **(A)** Members of the same species must be able to interbreed and produce fertile offspring.

10. **(B)** Saprotrophs are decomposers with external digestion.

11. (a) The range between 1850 and 1900
    (b) 1900 = 500 GT and 2000 = 6500 GT/6500 (new value) − 500 (old value) = 6000/500 (old value) = 12 × 100 = 1200% change
    (c)
    - Increased combustion of fossil fuels (oil, coal, gas, peat)
    - Deforestation/clearing of plant life for building
    - More vehicles being driven/transportation increased (public and private)
    - Industrial processes releasing carbon into the atmosphere

12. Organism I is a primary consumer, and organism III is both a secondary and a tertiary consumer.

13.
    - Detritivores and saprotrophs are both decomposers
    - Detritivores ingest and then digest dead organic material
    - Saprotrophs digest and then ingest dead organic material
    - Both feed on dead organic material
    - Vital for ecosystems in order to ensure recycling of nutrients
    - Named example of detritivore (example, earthworm, vulture)
    - Named example of saprotroph (example, mushroom)

14.

- Enhanced greenhouse effect is the increasing amount of greenhouse gases in the atmosphere
- Human contribution by burning of fossil fuels (oil, coal, natural gas)
- Traps UV light
- Short-wave radiation enters and long-wave radiation is reflected from Earth
- Some long-wave radiation is trapped by gases, increasing Earth's temperature
- Can lead to damage to DNA
- Mutation of DNA
- Named example of damage (example, skin cancer, retina damage)

15.

- CFCs (chlorofluorocarbon)
- Carbon dioxide
- Water vapour
- Methane (CH_4)
- Nitrous oxides

16. (a)

- Food web contains at least 8 organisms
- Food web contains at least 4 trophic levels
- Producers at bottom level
- Primary consumers at second level
- Secondary consumers above primary
- Tertiary consumers above secondary
- Top consumers above tertiary
- Appropriate organism chosen for each level
- Arrows flow in correct direction (show energy transfer going up the trophic levels)
- Shows multiple arrows to some organisms (weblike)
- Labeled percentage of energy transferred
- Included a method to show energy lost as heat

(b)

- Energy travels from bottom level up to top
- Only 10% of energy is transferred
- Most energy is lost as heat
- Some energy lost because not all is consumed
- Some energy lost due to not being digested (as faeces)
- Since energy is lost at each level, the number of levels is limited (usually to a maximum of 4 or 5)
- Examples of organisms at each level
- Use of values for energy transfer ($kJm^{-2}y^{-1}$)

(c)

- Carbon enters ecosystem through the process of photosynthesis/autotrophs
- Deforestation increases the level of carbon dioxide in the atmosphere due to loss of photosynthetic organisms
- Carbon is released as carbon dioxide through cellular respiration
- Carbon is released due to decomposition of organic material/by decomposers

- Carbon is released due to the combustion of organic matter and fossilized organic matter/burning of fossil fuels/coal/oil/gas/peat
- Carbon is released from volcanic eruptions
- Carbon is removed from the atmosphere by the process of formation of mollusc shells/calcium carbonate incorporated into living structures
- Methane can be oxidized into carbon dioxide and water, increasing carbon levels in the atmosphere/water
- Dissolved carbon dioxide/hydrogen carbonate ions exist in aquatic environments
- One point is awarded for a correctly drawn representation of carbon cycle

Evolution and Biodiversity

CENTRAL IDEAS

- Evidence for evolution
- The process of natural selection
- Classification and biodiversity
- Cladistics

TERMS

- Analogous structures
- Clade
- Cladogram

- Convergent evolution
- Divergent evolution
- Evolution

- Homologous structures
- Taxonomy

INTRODUCTION

Although the exact origin of life on Earth still remains a mystery, several theories exist as to how living organisms evolved. The scientific theory of evolution is one such theory. Evolution refers to genetic changes in organisms that occur over time as a species evolves. Environmental factors, mutations, natural selection, meiosis and sexual reproduction all contribute to the evolution of a species.

Classifying organisms allows scientists to identify newly discovered organisms easily, to identify common characteristics shared among members of a group and to construct evolutionary links among different groups of organisms. Newly discovered organisms can be easily identified when using a classification system simply by finding the group with which they share the most characteristics. Evolutionary timelines can be constructed based on the number of shared characteristics among different groups. The use of an international classification system allows scientists to share information as they sort, identify and add new organisms to the global classification system. The classification of newly discovered organisms can be based on evolutionary relationships, structural similarities, embryological evidence and genetic similarities in DNA.

NOS Connections

- Patterns, trends and discrepancies exist in nature (evolution, classification).
- Theories can explain natural phenomena (evolution).
- Collaboration and cooperation among scientists leads to scientific discoveries and explanations for observed phenomena (classification system, evolutionary theory, scientific nomenclature).

5.1   EVIDENCE FOR EVOLUTION

> **Evolution:** The cumulative change in the heritable characteristics of a population.

New species arise through the evolution of existing species. Members of a population exhibit variation among the traits they possess. Those organisms with the best traits for survival in the environment are considered to be the most "fit" organisms. Organisms possessing these favourable characteristics (the "fittest" organisms) will reproduce, passing these favourable traits on to their offspring.

Fossil Record and Evolution

Comparisons among common features seen in different organisms provide evidence for the theory of evolution. The fossil record supports the theory based on the changes in structures of organisms that can be seen when comparing older fossils with those from more recent times. Fossils allow scientist to reconstruct the timeline of evolution based on where the fossils were discovered. Deeper layers of earth contain older fossils unless uplifting has occurred (layers have been flipped due to some process). Fossils can be laid out based on the layer of earth that they were discovered in. In addition, carbon dating allows scientists to get a relatively accurate age for rocks that many fossils are preserved in. These processes allow scientists to reconstruct an evolutionary timeline in which changes in the organisms over time can be seen.

Selective Breeding and Evolution

Selective breeding of domesticated animals has shown how the artificial selection of specific organisms for breeding can lead to changes in traits in a population. This happens naturally in the environment as organisms that possess favourable traits for surviving reproduce and pass those traits on to their offspring. Artificially selecting the traits desired in animals and plant species allows the favoured alleles to stay in the population. Selective breeding can lead to evolution as new varieties and breeds of organisms are produced. (Refer to Figure 5.1.)

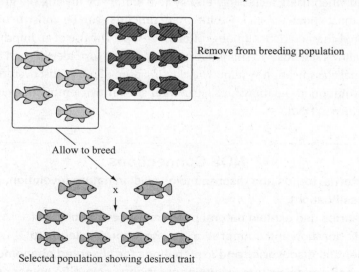

Figure 5.1. Selective breeding

HOMOLOGOUS STRUCTURES BY ADAPTIVE RADIATION

- **Homologous characteristics:** Similar characteristics that can be seen in organisms that evolved from a common ancestor. These are often referred to as divergent characteristics since they have evolved to suit the habitat of each new organism but maintain similar characteristics.

Homologous structures, such as those seen in the bones of mammals, bats, birds and amphibians, support the theory of evolution. (Refer to Figure 5.2.) All of these organisms possess a pentydactyl limb consisting of a hand with five digits, a lower limb with two bones and an upper limb with only one bone. This suggests that they all evolved from a common ancestor and the primitive basic structure was modified (evolved) in each of the organisms as they adapted to their environments. Although the limbs are of the same basic structure, they have adapted to perform different and specific functions to allow for survival in diverse environments (adaptive radiation).

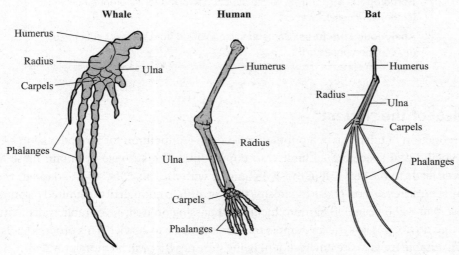

Figure 5.2. Pentydactyl limb

ANALOGOUS STRUCTURES

- **Analogous characteristics:** Similar structures that can be seen in organisms that have adapted to similar environments. They are often referred to as convergent characteristics since the organisms may not show common evolutionary lineage (ancestors). The characteristics of the organisms become more similar as they evolve and adapt to similar environments.

Analogous structures such as those seen in the bat, butterfly and bird wing have evolved in order to allow these organisms to adapt to their environments. Although these organisms are not thought to share direct evolutionary relationships, the evolution of wings for flight can be seen in each of them. The similar function of the wing suggests that it evolved to allow flight in order for the organisms to best adapt to their environments. Similar functions seen in analogous structures that have evolved from unrelated organisms support the theory that these structures evolved based on their ability to help the unrelated organisms adapt to similar environments. (Refer to Figure 5.3.)

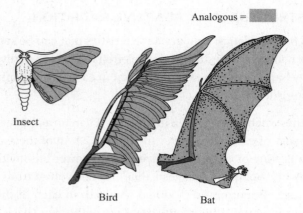

Analogous =

Insect

Bird

Bat

Figure 5.3. Analogous structures

> **Homologous structures:** Same structure evolved for different functions (common ancestor)
>
> **Analogous structures:** Same function evolved from different structures (no common ancestor)

Survival of the "Fittest"

Overproduction of offspring in a population leads to competition for limited resources. Food, water, shelter and space are all limited in populations. The struggle to obtain these limited resources leads to the survival of those organisms with the most "fit" (best adapted) traits for obtaining these resources. Those organisms that can obtain enough of the limited resources will be those that will be capable of reproducing and passing on their favourable traits ("fitness"). Those that do not possess the favourable traits will struggle to survive. This process leads to the most favourable traits in the environment being the ones that will be more prevalent.

Divergent Evolution

As members of populations gradually change to adapt to different environmental conditions, new species can evolve. Darwin's finches of the Galapagos Islands are an excellent example of divergent evolution. Over time, groups from the original population of finches became geographically separated from each other as new islands formed. The isolated birds evolved to be best adapted to the environment they inhabited. The different shapes of beaks seen on each species of finch evolved to feed on specific types of food available in the environment. Variations seen in organisms over geographical regions lend support to the theory of divergent evolution. This type of divergent evolution is often referred to as adaptive radiation, as shown in Figure 5.4.

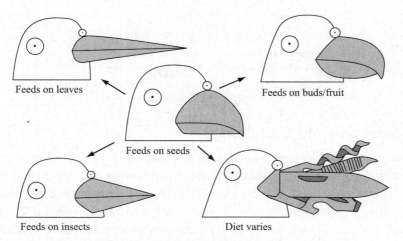

Figure 5.4. Adaptive radiation and Darwin's finches

Industrial Melanism (Melanistic Insects)

Industrial melanism in the peppered moth is an example of the process of evolution based on survival of the fittest. The peppered moth population consisted mainly of light-coloured moths and a few darker ones. The peppered moths rested on trees that possessed white bark. Naturally, the dark-coloured moths were more easily seen by birds and were eaten more often than the light-coloured moths. Pollution from the industrial revolution left soot on the white bark of the trees and made the white moths much easier for the birds to see and consume. Over a short period of time, the dark moths became the dominant phenotype in the population. This was due to them being able to survive, reproduce and pass on the favourable characteristic for the changing environment. Laws were passed to clean up the pollution that was producing the soot on the tree bark. The trees became less polluted with soot, and the dominant phenotype switched back to the light-coloured moths again. The peppered moth is an excellent example of modern-day microevolution and provides evidence that supports the theory of evolution (evolution that occurs at a rate that allows humans to witness the phenomenon).

5.2 NATURAL SELECTION

Darwin (1809–1882) and Wallace (1823–1913) were scientists who outlined the theory of evolution by natural selection. The theory of evolution by natural selection consists of four main parts.

- Populations produce more offspring than the environment can support.
- There will be a struggle for survival.
- Variation will exist among the offspring produced.
- Natural selection will favour those in the population that possess the best traits (are the most "fit") for survival. Those that do not possess the best traits will not survive.

The surviving members of the population will reproduce and pass on the "fit" traits to their offspring.

REMEMBER

1. Natural Selection
2. Overproduction
3. Variation
4. Survival of the "fittest"

Passing on "fit" traits to offspring through reproduction

Sources of variation in a population include:

- Mutation
- Meiosis (crossing-over and independent assortment)
- Sexual reproduction (union of random sperm and egg)

Natural Selection and Evolution

The process of natural selection leads to evolution though the adaptation of organisms to changing environments. Adaptations are features an organism possesses that allow the organism to carry out its life processes most effectively. Those organisms that possess favourable traits for adapting to the environment will be naturally selected for survival. The greater survival rate and reproductive success of individuals with favourable traits can lead to changes in the characteristics of the population. Variation must exist in the population in order for populations to evolve with changing environments. Variation can be due to sexual reproduction or random mutations.

Evolution of Antibiotic Resistance

Bacteria can gain resistance to antibiotics through the process of evolution. Within a population of bacteria, a few members may possess resistance. This antibiotic resistance is coded for in plasmids. When these bacteria are exposed to antibiotics, they will survive and pass along the resistance (coded for in plasmids) to their offspring. Rarely, bacteria can undergo a form of sexual reproduction known as conjugation. In conjugation, the bacteria share their plasmids through connections in their cytoplasm made through the joining of their pili. Both asexual reproduction (binary fission) and conjugation pass the resistant plasmid to the future generations. The future generations of the bacteria populations will be resistant to the antibiotic. They have "evolved" to have this resistance.

Antibiotic resistance develops when individuals take too many antibiotics, take antibiotics when they don't need them or fail to finish all of the antibiotics prescribed. Each of these situations provides bacteria with opportunities to develop resistance. Some bacteria are naturally resistant to antibiotics and pass their antibiotic resistance to their offspring. Other previously susceptible bacteria acquire resistance to antibiotics when their genes mutate to code for proteins that provide antibiotic resistance. Antibiotic resistant bacteria reproduce and pass the antibiotic resistance to their offspring. The resulting antibiotic resistant bacterial populations are not affected by antibiotic use. For this reason, it is very important to finish all antibiotics in order to ensure no bacteria survive and mutate to develop antibiotic resistance. Antibiotic resistance is a major concern worldwide, since without antibiotics, the number of people dying from bacterial infections would greatly increase. Refer to Figure 5.5.

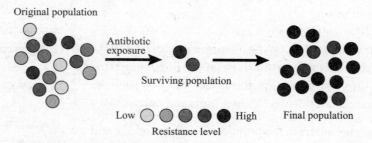

Figure 5.5. Antibiotic resistance

CLASSIFICATION OF BIODIVERSITY

Binomial Nomenclature

The Swedish botanist Carolus Linnaeus (1707–1778) developed a universal system for naming organisms that is known as binomial nomenclature (two-name system). The two-naming system consists of a genus and species name for each organism that has been discovered. When new species are discovered, they are named using the same binomial nomenclature method. It is customary to capitalize the first letter of the genus name and use lowercase for the species name. In addition, the genus and species name must be underlined or italicized. The use of a universal scientific naming system ensures that scientists can make reference to any organism anywhere in the world and everyone will understand which organism is being referenced.

> **REMEMBER**
>
> *Homo sapiens*: human
> *Canis familiaris*: dog
> *Zea mays*: corn

Hierarchy of Taxa

All organisms are classified as being members of one of three domains. The three domains are Eubacteria, Archaea and Eukarya. The domains Eubacteria and Archaea are both composed of prokaryotic organisms. The domain Eukarya is composed of all eukaryotic organisms. (Refer to Figure 5.6.)

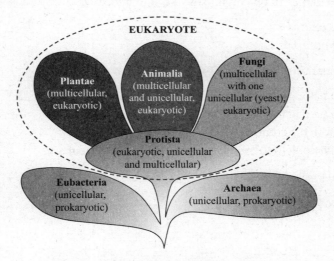

Figure 5.6. Three domains of classification.

Seven ranks of classification are used to classify eukaryotic organisms. The most inclusive rank is that of kingdom. Each rank below kingdom becomes more exclusive than the previous, with species being the most exclusive rank of all. In order to be a member of the same species, an organism must be able to reproduce with other members of the species and produce viable (fertile) offspring. An example of the complete classification of corn, the orca whale and humans are outlined in Table 5.1.

Table 5.1: Classification of Corn, Orca and Humans

Organism	Corn	Orca	Human
Domain	Eukarya	Eukarya	Eukarya
Kingdom	Plantae	Animalia	Animalia
Phylum	Magnoliophyta	Chordata	Chordata
Class	Lilopsida	Mammalia	Mammalia
Order	Poales	Cetacea	Primates
Family	Poaceae	Delphinidae	Hominidae
Genus	*Zea*	*Orcinus*	*Homo*
Species	*mays*	*orca*	*sapiens*

FEATURED QUESTION

May 2004, Question #17

The scientific names of two organisms are shown below.

Lathyrus palustris
Angelica palustris

What is the relationship between these organisms?

A. They both belong to the same genus but they are different species.
B. They both belong to the same species but different genera.
C. They are both different species and different genera.
D. They both belong to the same species and the same genus.

(Answer is on page 553.)

The orca and human are both members of the same kingdom, phylum and class. However, they differ at the level of order. Since they are in different orders, they must be in different ranks for the rest of their classification. You cannot be in the same species if you are not in the same genus as well. Be careful when evaluating relationships for taxa because you may see the same species name with a different genus name. These organisms would not be in the same species; they just happen to share the same species name.

> Use the phrase "King Phillip Came Over For Good Soup" to remember the order of classification: K, P, C, O, F, G, S (Kingdom, Phylum, Class, Order, Family, Genus, Species)

Tables 5.2 and 5.3 list the various plant and animal phyla, respectively.

Table 5.2: Plant Phyla

Phyla	Features	Example
Bryophyta	■ Lack vascular tissue (xylem and phloem) ■ Must live near water and close to the ground (limited in height) ■ No roots; rootlike structures: rhizoids ■ Reproduce by spores	Mosses
Filicinophyta	■ Possess vascular tissue ■ Reproduce by spores on sporangia located on the underside of the leaves ■ Possess roots ■ Curled divided leaflets	Ferns
Coniferophyta	■ Possess vascular tissue ■ Trees or shrubs ■ Possess roots ■ Narrow leaves (example, pine needles) ■ Possess cones with pollen (male cone) or seeds on the surface (female cone)	Pine tree
Anigospermophyta	■ Possess vascular tissue ■ Possess roots, stems, leaves ■ Produce flowers ■ Possess ovaries where ovules are located ■ Fertilised ovules form seeds encased in fruit	Rose Sunflower

Table 5.3: Animal Phyla

Phyla	Features	Example
Porifera	Symmetry: none (asymmetrical) ■ Body consists of pores ■ Attached to a surface ■ No anus or mouth	Sponges
Cnidaria	Symmetry: radial ■ Possess stinging tentacles ■ No anus but have mouth	Jellyfish
Platyhelminthes	Symmetry: bilateral ■ Flat, unsegmented bodies (*platy-* means "flat") ■ No anus but has mouth	Tapeworm Planarian
Annelid	Symmetry: bilateral ■ Segmented, round bodies ■ No anus but has mouth	Earthworm Leech
Mollusca	Symmetry: bilateral ■ Possess a foot and a mantle ■ Unsegmented ■ Some possess shells ■ Possess anus and a mouth	Snail Clam Octopus
Arthropoda	Symmetry: bilateral ■ Segmented, jointed appendages ■ Possess hard exoskeleton ■ Has anus and mouth	Insects Spiders Crustaceans (example, crabs)
Chordata	Symmetry: bilateral ■ Segmentation ■ Notochord (rod supports a central nerve cord) ■ Possess anus and a mouth	Birds Mammals Amphibians Reptiles Fish

Symmetry describes the general structural arrangement of body parts in an organism. Animals possess body plans that can be asymmetrical, radially symmetrical, or bilaterally symmetrical. Asymmetrical organisms have no defined body plan. Radially symmetrical organisms possess parts that extend from a central core. Bilaterally symmetrical organisms possess a body that, when split down the centre, forms two halves that are mirror images of each other. (Refer to Figure 5.7.)

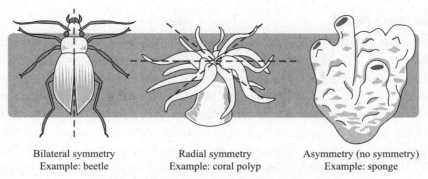

Bilateral symmetry
Example: beetle

Radial symmetry
Example: coral polyp

Asymmetry (no symmetry)
Example: sponge

Figure 5.7. Types of symmetry in organisms

Members of the animal phyla are further classified into separate classes on the basis of shared characteristics. The main classes of the animal phyla include birds, mammals, amphibians, reptiles and fishes. The main distinguishing characteristics of the animal classes are identified in Table 5.4

Table 5.4: Characteristics of the Animal Classes

Animal Class	Distinguishing Characteristics
Birds	■ Warm-blooded ■ Possess hollow bones ■ Feathers ■ Egg layers (hard shell)
Mammals	■ Warm-blooded ■ Live birth ■ Possess hair or fur ■ Most live on land (not all) ■ Mammary glands (nurse young) ■ Breathe with lungs
Amphibians	■ Cold-blooded ■ Eggs laid in water ■ Possess moist, tough skin ■ Live on land (adult) or water (larvae/young) ■ Breath with gills (water) and lungs (land)
Reptiles	■ Cold-blooded ■ Egg layers (leathery shell) ■ Live on land only ■ Possess dry, scaly skin
Fish	■ Cold-blooded ■ Possess scales and gills (for breathing) ■ Lay slimy eggs ■ Live in water (possess fins and tails for swimming)

Dichotomous Keys

Dichotomous keys are used to identify organisms based on structures and features that organisms possess. Each level in a dichotomous key includes two statements about an organism that refer to the same feature. One statement will be that the organism possesses the feature, and the other will state that the organism lacks the feature. Once the decision is made as to whether the organism possesses the feature, the key will either identify the organism or take you to another level in the key in order to help correctly identify the organism. Dichotomous keys are often used in the field and laboratory to identify or verify the scientific name of an organism. Use the sample key shown in Figure 5.8 to identify the genus of bird based on the beak structure shown.

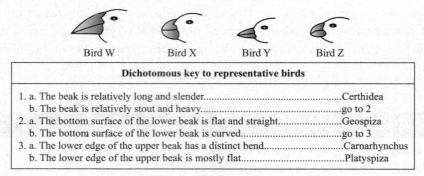

Figure 5.8. Dichotomous key

5.4 CLADISTICS

Evolutionary timelines can be constructed based on the number of shared characteristics among different groups. The use of an international classification system allows scientists to share information as they sort, identify and add new organisms to the global classification system. The classification of newly discovered organisms can be based on evolutionary relationships, structural similarities, embryological evidence and genetic similarities in DNA.

Biochemical Evidence

All organisms contain DNA (and RNA) and produce proteins made from the same 20 amino acids. The basic structure of DNA, composed of nucleotides, is the same in all organisms, suggesting a common ancestry for all life-forms. The development of a method to store and transfer information is thought to have evolved in a primitive cell and was transferred to all organisms as they evolved from this primitive life-form. The genetic code consisting of triplet bases that code for specific amino acids will always code for the same amino acid regardless of the organism. Enzymes that act on the DNA of one species of an organism can be used to catalyse reactions on the DNA of any other species of organisms due to the universality of DNA.

Phylogeny

The system of classifying organisms based on evolutionary origins and changes over time is known as phylogeny. Changes in DNA that code for specific proteins can be used to help construct phylogenic trees. Phylogenic trees show stems and branches that indicate evolutionary relationships among organisms, as shown in Figure 5.9.

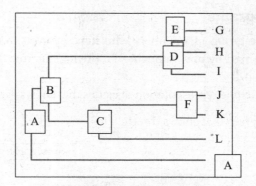

Figure 5.9. Phylogenic tree

In Figure 5.9, the most primitive organism is represented by *A*. Organisms *B* through *L* evolved from the same common life-form represented by *A*. The first organisms to evolve from *A* was *B*, followed by *C* that evolved into *F*, that evolved into *J* and *K*. Organisms *G* through *L* each represent a new organism that evolved from the primitive organism *A* along various paths. Organism *B* evolved into organisms *D* and *E* that further evolved into organisms *G*, *H* and *I*.

Biochemical Variations

By looking at specific proteins coded for by different species of organisms, scientists can make suggestions as to how the evolution of the organisms occurred. The changes in proteins have been shown to occur at constant rates, allowing the changes to serve as evolutionary clocks. The changes can be analysed and placed in order as to when they evolved. Organisms that exhibit similarities in protein structure are most likely more closely related than organisms that show greater variation in protein structure. Since almost all living organisms contain the protein cytochrome c, it is often used to compare amino acid sequences in order to determine phylogenic relationships. (Refer to Table 5.5.)

Table 5.5: Similarities in Amino Acid Sequences of Cytochrome c

Organism	Number of Amino Acids that Differ from Humans
Yeast	42
Wheat germ	37
Fruit fly	24
Bullfrog	20
Pigeon	12
Cow	10
Rabbit	9
Rhesus monkey	1
Chimpanzee	0

Clades and Cladograms

- **Clade:** A complete group of members who have evolved from a common ancestor. Clades are represented by one "branch" of a phylogenic tree. Each branching point is referred to as a "node".
- **Cladogram:** A branching representation of clades that shows evolutionary relationships.

Figure 5.10 is a cladogram that shows three clades.

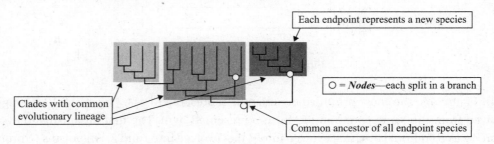

Figure 5.10. Cladogram showing three clades

HOMOLOGOUS AND ANALOGOUS STRUCTURES

Homologous and analogous structures are both considered when constructing cladograms. Cladograms show relationships among organisms and are constructed based on similarities seen between different groups of organisms. Molecules (DNA, proteins) as well as morphological structures (forms) are compared when constructing a cladogram. Any group of organisms can be compared and used to construct a cladogram. Consider the following information provided in Table 5.6 about organisms *A, B, C* and *D*.

Table 5.6: Sample Organisms and Their Characteristics

Characteristic	Organism			
	A	*B*	*C*	*D*
Multicellular	X	X	X	X
Backbone	X		X	X
4-chamber heart	X		X	
Wings	X			

CONSTRUCTING CLADOGRAMS

Figure 5.11 is a cladogram made from the information presented in Table 5.6.

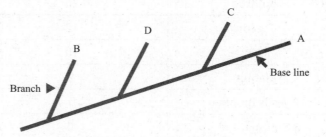

Figure 5.11. Simple cladogram.

EVOLUTION AND BIODIVERSITY

The term "clade" in Greek refers to a branch.

The simple cladogram shown in Figure 5.11 shows that all the organisms evolved from a common ancestor represented by the base line. The first branch to reach off of the base line is considered to be the first one to have evolved, followed by the next branch up and so on to the top of the cladogram. The top of the base line represents the most recently evolved organism, in this case organism A. The cladogram also shows that organisms A and C share the most characteristics and that A and B share the least.

Cladograms can be created vertically, as shown in Figure 5.11, or horizontally, as shown in Figure 5.12.

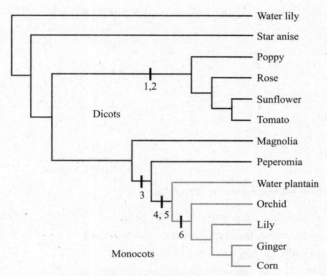

Figure 5.12. Horizontal cladogram.

Each node (split) in a branch represents the evolution of two new lineages of organisms from a common ancestor. Figure 5.12 shows many nodes resulting from the evolution of the water lily (still existing as shown) into many new lineages of plant forms. The most recently evolved plants are the sunflower and tomato (evolving directly from a common ancestor) and the ginger and corn (evolving directly from a common ancestor).

Cladograms are important tools to use when comparing and classifying organisms. Cladograms can show morphological or biochemical relationships and have often been found to mirror what scientists believe to be evolutionary relationships. The use of cladograms for interpreting evolutionary relationships does have limitations. The suggested evolutionary branches used in cladistics show branching with nodes that are based solely on selected characteristics. In addition, cladograms include only selected organisms, making the entire picture of evolution difficult to interpret. Oftentimes, cladograms that include the same organisms based on different characteristics contradict each other when showing evolutionary relationships. Regardless of existing scepticism about the usefulness of cladograms, scientists find them helpful in the process of interpreting the history of evolution.

PRACTICE QUESTIONS

1. Which description correctly describes a clade?
   A. Organisms that share a common ancestor
   B. Organisms with similar feeding patterns
   C. Organisms that do not share a common ancestor
   D. Organisms of the same species

2. Which of the following are members of the domain Eukarya?

   I. Bacteria
   II. Fungi
   III. Amoeba

   A. I only
   B. I and II only
   C. II and III only
   D. I, II and III

3. Amphibians possess the following features:

   A. Live on land only, lay eggs and are warm-blooded
   B. Live on land or water, lay eggs and are cold-blooded
   C. Live on land only, lay eggs and are cold-blooded
   D. Live in water only, lay eggs and are cold-blooded

4. The finches of the Galapagos Islands possess different beak shapes due to:

   I. Adaptive radiation
   II. Divergent evolution
   III. Mutations

   A. I only
   B. I and II only
   C. II and III only
   D. I, II and III

5. Which of the following contribute to the process of evolution?

   I. Variation
   II. Reproduction
   III. Struggle for existence

   A. I only
   B. I and II only
   C. I and III only
   D. I, II and III

6. Which level of taxa is the most inclusive?

   A. Species

   B. Genus
   C. Order
   D. Family

   K P C O F G S

7. An organism that possesses stinging cells and radial symmetry is a member of phylum:

   A. Porifera
   B. Cnidaria — *jelly fish*
   C. Annelida
   D. Arthropoda

8. A plant was discovered living near water. It lacked vascular tissue. What phylum could this plant belong to?

   A. Filicinophyta
   B. Angiospermophyta
   C. Bryophyta
   D. Coniferophyta

9. Which of the following scientific names are written correctly?

   A. *Homo Sapiens*
   B. *homo Sapiens*
   C. *homo sapiens*
   D. *Homo sapiens*

10. Which of the following observations supports the theory of evolution?

    A. Bacteria dividing by binary fission
    B. Bacteria gaining resistance to antibiotics and passing the resistance on to future generations
    C. Bacteria living in the human intestine
    D. Bacterial colonies being destroyed by exposure to antibiotics

11. What is true of the following two organisms?

    Zea mayas
    Rhea mayas

    A. They are members of the same genus and the same species.
    B. They are members of a different genus but are of the same species.
    C. They are members of the same genus but not of the same species.
    D. They are members of a different genus and a different species.

12. The graph below displays data concerning the length of the big toe as observed in the year 1800 and in the year 2000.

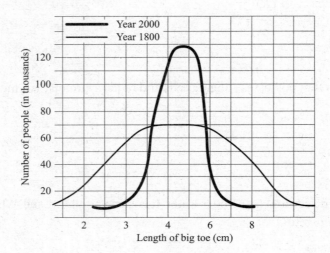

(a) State the number of people who had big toe lengths of 5 centimetres in the year 2000. (1)

(b) Calculate the percent change between the number of people with 5-centimetre big toe lengths in the year 1800 and in the year 2000. (2)

(c) Suggest reasons for the differences between the lengths of the big toe in 1800 and in 2000. (2)

13. Explain how sexual reproduction promotes variation among members of a species. (4)

14. Outline the taxa of classification for eukaryotic organisms. (4)

15. (a) Explain the Darwin–Wallace theory of evolution. (4)
   (b) Outline evidence that exists to support the theory of evolution. (6)
   (c) List the complete classification of humans. (6)

ANSWERS EXPLAINED

1. **(A)** A clade includes organisms that share common ancestry.

2. **(C)** Fungi and amoeba are members of the domain Eukarya. All members of domain Eukarya are eukaryotic. Bacteria are prokaryotic and are members of domain Eubacteria or Archaea.

3. **(B)** Amphibians live on land or in water, lay eggs and are cold-blooded.

4. **(D)** I, II and III. Mutations allow organisms to evolve. When organisms evolve to become less alike, they are undergoing divergent evolution. Adaptive radiation is a type of divergent evolution.

5. **(D)** Variation allows organisms to adapt to changing environments. Reproduction allows genes to be passed on to offspring. The struggle for existence selects for the "fittest" traits.

6. **(C)** Order is the most inclusive (kingdom, phylum, class, order, family, genus, species).

7. **(B)** Cnidarians possess stinging tentacles and have radial symmetry.

8. **(C)** Bryophytes lack vascular tissue (xylem and phloem) and must live near water.

9. **(D)** *Homo sapiens*. Genus is always capitalized, and species is always lowercase. The genus and species name must be italicized or underlined.

10. **(B)** Bacteria must gain resistance and pass it on to the offspring in order to exhibit the process of evolution.

11. **(D)** The represented organisms are members of a different genus and a different species. You cannot be a member of the same species if you are not a member of the same genus (although you may have the same species name).

12. (a)   130 thousand people (+/− 2)

    (b)
    - 1800 = 70 centimetres and 2000 = 130 centimetres
    - 130 − 70 = 60/130 = 0.46 × 100 = 46% change

    Percent change = (New value − Original value)/Original value × 100

    (c)
    - Environment selected for longer big toes
    - Longer big toes provide better stability/ability to walk/grasp items with feet
    - Shorter toes were not selected for in the environment
    - Individuals with longer toes reproduced more and passed on the longer toe trait/ short toes were not reproducing as often

13.
    - The process of meiosis produces variation in gametes
    - Sexual reproduction promotes variation in gametes through crossing-over/chiasmata formation/synapsis of chromosomes
    - Independent assortment leads to variation in gametes
    - The production of egg and sperm by different parents promotes variation
    - The random union of sperm and egg from the many produced promotes variation

14.

- Eukaryotic organisms are members of the domain Eukarya.
- Eukaryotic organisms include the kingdoms Protista, Fungi, Plantae and Animalia.
- Eukaryotic organisms are classified by the taxa: kingdom, phylum, class, order, family, genus, and species.
- All members of the same species must be able to interbreed and produce fertile offspring.
- Kingdom is the most inclusive level, and species is the least inclusive level.
- All eukaryotes have a true nucleus/membrane-bound organelles.
- Classification into taxa allows scientists to compare organisms/predict characteristics of organisms/identify species/compare evolutionary relationships.
- Organisms can be reclassified into new taxa based on new information/classification taxa can change with new information.

15. (a)

- Variation exists within populations.
- Organisms produce more offspring than can survive.
- Those organisms with traits best adapted ("fit") to the environment will survive.
- Surviving organisms can reproduce.
- Reproducing organisms pass on their traits ("fitness") to offspring.
- Offspring will possess the best traits for survival in the environment.
- Variation in the population allows populations to adapt to changes in the environment.
- One point is awarded for example of evolution.
- One point is awarded for evidence of evolution.

(b)

- Fossil record shows changes over time in structures of organisms
- DNA/nucleic acid comparisons show evolution of changes
- Universality of DNA suggests evolution from a common ancestor
- Homologous structures in different organisms
- Continuous variation among organisms over geographic ranges supports evolution to adapt to various habitats/gradual divergence
- The microevolution seen in melanistic insects/peppered moth
- The development of antibiotic resistance shows evolution of traits
- Any acceptable example of evolution of an organism

(c)

- Domain Eukarya
- Kingdom Animalia
- Phylum Chordata
- Class Mammalia
- Order Primates
- Family Hominidae
- Genus *Homo*
- Species *sapiens*

Human Physiology

CENTRAL IDEAS

- Digestion and absorption of nutrients
- Transport of nutrients and waste
- Defence mechanisms and immunity
- Gas exchange
- Maintaining homeostasis
- Regulation of blood sugar
- Type I and type II diabetes
- Human reproduction and infertility treatment

TERMS

- Absorption
- Assimilation
- Villi
- Erythrocytes
- Leucocytes

- Pathogen
- Antigen
- Antibody
- Ventilation
- Gas exchange

- Cellular respiration
- Resting potential
- Action potential

INTRODUCTION

Human health depends on the ability of the body systems to function and maintain the body at homeostatic levels. The digestive system breaks down food, releasing its nutrients into the digestive cavity to be absorbed by body cells. The respiratory system brings fresh air into the body to supply oxygen to and allow the removal of carbon dioxide from the body. The circulatory (transport) system delivers products to cells and picks up waste for removal from body tissues. The immune system is vital in preventing pathogens from destroying healthy tissue. The body cells communicate chemically with hormones and electrically with action potentials. The integration and communication within and between body systems assists in the many processes required by the human body to maintain homeostasis.

NOS Connections

- Models are used to represent structures and processes (human systems, physiological processes).
- Scientific research leads to new discoveries and theories (physiology of body systems and disease progression).
- Cooperation and collaboration lead to scientific discoveries and understandings (brain and memory research, disease progression, epidemics and pandemics, antibiotic discoveries).
- Scientists play a role in educating the public about scientific phenomena and discoveries (human disease and physiological processes).
- Accidental discoveries contribute to scientific knowledge and progress (discovery of penicillin).

6.1 DIGESTION AND ABSORPTION

TOPIC CONNECTIONS

- 2.1 Molecules to metabolism
- 2.5 Enzymes

Digestion

If they did not digest food, organisms would be unable to absorb nutrients necessary for growth, repair and metabolic processes. In order for food to be absorbed, it must be broken down into molecules small enough to pass through cell membranes and enter into the circulatory system to be distributed throughout the body. In addition, if polymers were not digested prior to absorption, they could potentially block the circulatory system due to their large size. The digested monomers can be rearranged back into new polymers inside of cells and then be used by the cells or sent out for extracellular use. Without digestion of the macromolecules, monomers resulting from digestion would not be available for cellular use.

In order to break down polymers into monomers, enzymes must be able to function. Each enzyme has an optimal pH and temperature at which it functions most effectively. The temperature at which most enzymes in the human body function efficiently is 98.6 degrees Fahrenheit or 37 degrees Celsius, which is why this is the average body temperature. Recall from objective 3.6 that pH, temperature and substrate concentration can all affect the rate at which enzymes can catalyse reactions. If enzymes were not available to speed up the rate of reactions, the rate of reactions would be too slow to be accomplished before the polymers exit the body. This would result in the loss of nutrients needed for the body to function.

THE DIGESTIVE SYSTEM

Food enters the digestive system through the oral cavity and travels down the oesophagus through the process of peristalsis. Peristalsis involves the involuntary rhythmic contraction of the oesophagus and intestines in order to move food through the digestive system. The cardiac sphincter (circular muscular structure that can dilate and contract) controls the movement of food into the stomach. When food presses on the cardiac sphincter, the sphincter is stimulated to open. Once the food passes into the lumen of the stomach, the sphincter closes. The stomach holds the food for a short period of time and constantly churns and mixes the food into a solution called chyme. The chyme passes through another sphincter (the pyloric) and enters into the small intestine. Finally, the food enters the large intestine, reaches the rectum (holding site until the faeces are ready to be expelled) and exits through the anus. (See Figure 6.1.)

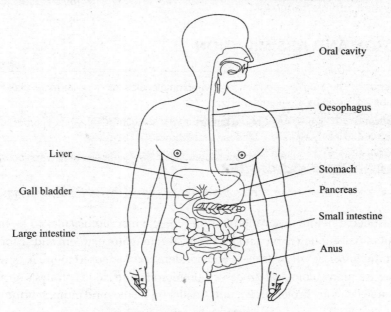

Figure 6.1. The digestive system

The salivary glands, liver, gall bladder and pancreas are all accessory organs of the digestive system, as shown in Figure 6.2. Accessory organs contribute products to the digestive process, but food never enters these tissues. The salivary glands produce salivary amylase, which enters the oral cavity to begin the chemical digestion of amylase (starch) and glycogen (carbohydrates). The pancreas produces several digestive enzymes that enter the small intestine through the pancreatic duct. The liver produces bile, which is stored in the gall bladder. Bile leaves the gall bladder through the common bile duct and enters into the small intestine.

Liver—
produces bile

Gall bladder—
stores bile until
it is released
into the small
intestine

Pancreas—
produces
enzymes that
are sent to the
small intestine

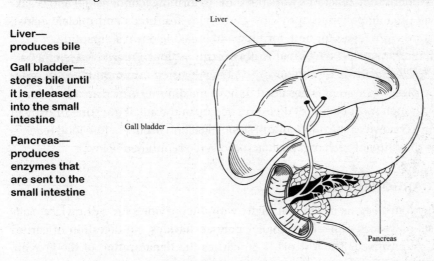

Figure 6.2. Accessory organs

The organs of the digestive system all work together in order to carry out the functions of digestion and absorption. After absorption of monomers, the body assimilates the monomers into new molecules that are used for human body functions. The phrase "you are what you eat" is literally true—the digested polymers are just rearranged into new polymers that are ideal for human body use. The amino acids in the protein from the steak you ate last week may now be part of an enzyme functioning in your digestive system. The amino acids have just been digested and rearranged.

ABSORPTION AND ASSIMILATION

> **Digestion:** The breaking down of large macromolecules into smaller molecules that can be absorbed by the human body.
>
> **Absorption:** The movement of monomers across cellular membranes in order to be incorporated into human tissue or used for metabolic processes.
>
> **Assimilation:** The incorporation of absorbed food (monomers) into new molecules (polymers) to be used by the body.

Digestion must occur before absorption in order for the macromolecules to be broken down into monomers small enough to be transported by the circulatory system and absorbed by cells.

Recall that the processes of hydrolysis and condensation are used to break down and build new molecules for many biological processes. (Refer to Figure 2.6.) Hydrolysis occurs continually as macromolecules are broken down into smaller molecules and ultimately into monomers.

> **Hydrolysis:** Sucrose + water → Glucose + Galactose
>
> (Disaccharide) → (Monosaccharides)
>
> Hydrolysis breaks dimers or polymers into monomers with the addition of water.
>
> **Condensation:** Glucose + Galactose → Sucrose + water
>
> (Monosaccharides) → (Disaccharide)
>
> Condensation forms dimers or polymers with the removal of water.

Hydrolysis and condensation reactions work the same in forming (condensing) and breaking down (lysing) all organic polymers (proteins, carbohydrates, lipids and nucleic acids). Refer to Figure 2.6 for the processes for both forming and breaking down a dipeptide.

The digestion of macromolecules occurs at various locations in the digestive system. Carbohydrates, lipids, nucleic acids and proteins must all be digested into monomers in order to be absorbed. The digestive system is designed to have maximum surface area in order to achieve this goal. Although most chemical digestion occurs in the small intestine (mostly in the first 25 centimetres known as the duodenum), some digestion occurs before food reaches the duodenum. The digestion of each organic macromolecule is outlined below.

DIGESTION OF CARBOHYDRATES

Digestion of carbohydrates begins in the oral cavity with the enzyme salivary amylase. Salivary amylase digests polysaccharides into smaller polysaccharides. No digestion of carbohydrates occurs in the stomach. The low pH (2–3) causes the denaturation of the enzyme salivary amylase. Denaturation changes the enzyme's structure so salivary amylase no longer functions. Digestion of carbohydrates continues in the small intestine with pancreatic amylase, which breaks polysaccharides into disaccharides. The final digestion of carbohydrates occurs at the brush border (the villi of the small intestine) of the duodenum. Disaccharidases (sucrase, lactase, maltase) are produced by the villi cells and digest the disaccharides into monosaccharides. Monosaccharides are small enough to cross the villi and enter into the circulatory system. (Refer to Figure 6.3.)

Mouth—
Salivary amylase
(polysaccharides to
smaller polysaccharides)

Small intestine—

- **Lumen: Pancreatic amylase**
  (smaller polysaccharides
  to disaccharides

- **Brush border (villi)**
  (disaccharides to
  monosaccharides)

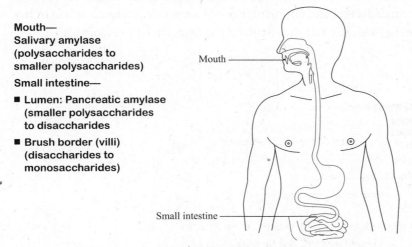

Figure 6.3. Digestion of carbohydrates

DIGESTION OF PROTEINS

Digestion of proteins begins in the stomach. Pepsin is the only enzyme that can function at such a low pH (2–3). The precursor to pepsin, pepsinogen, is produced by the cells of the stomach wall and becomes pepsin in the presence of hydrochloric acid. Stomach cells produce hydrochloric acid as well as a mucous lining. The mucus protects the stomach tissue from the low pH and inhibits the stomach from digesting itself. Pepsin digests large polypeptides into smaller polypeptides. Digestion continues in the duodenum with the enzyme trypsin from the pancreas. Trypsin digests polypeptides into dipeptides. Dipeptidases produced by the brush border digest dipeptides into amino acids. Amino acids can cross the villi and be absorbed by the nearby capillaries, allowing them to enter the circulatory system. (Refer to Figure 6.4.)

Stomach—
Pepsin (polypeptides to
smaller polypeptides)

Small intestine—

- **Lumen: Trypsin**
  (polypeptides to dipeptides)

- **Brush border: (dipeptides**
  to amino acids)

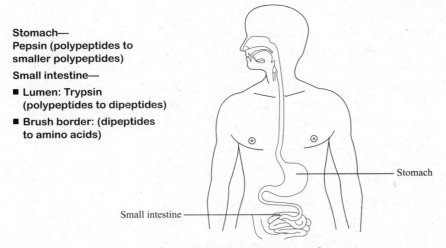

Figure 6.4. Digestion of proteins

DIGESTION OF LIPIDS

Digestion of lipids occurs solely in the duodenum. Bile from the gall bladder enters the duodenum and emulsifies lipids (breaks large fat droplets into smaller fat droplets). This increases the surface area of the lipids for easier digestion. Emulsification is mechanical digestion, not chemical digestion that is accomplished by enzymes.

The pancreas contributes lipase to the duodenum to begin the chemical digestion of lipids into glycerol and fatty acids. Glycerol and fatty acids can cross the villi and be absorbed by the nearby capillaries, allowing them to enter the circulatory system. (Refer to Figure 6.5.)

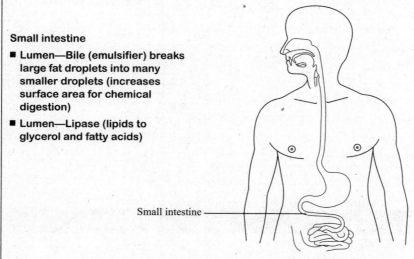

Small intestine

- **Lumen—Bile (emulsifier) breaks large fat droplets into many smaller droplets (increases surface area for chemical digestion)**
- **Lumen—Lipase (lipids to glycerol and fatty acids)**

Small intestine

Figure 6.5. Digestion of lipids

DIGESTION OF NUCLEIC ACIDS

Nucleases from the pancreas digest nucleic acids into nucleotides that can cross the villi and be absorbed by the nearby capillaries. (Refer to Figure 6.6.)

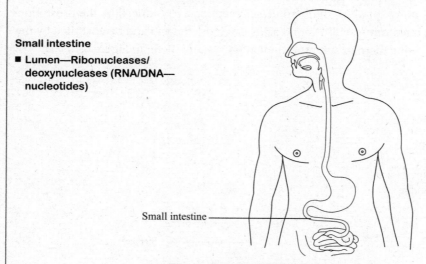

Small intestine

- **Lumen—Ribonucleases/ deoxynucleases (RNA/DNA— nucleotides)**

Small intestine

Figure 6.6. Digestion of nucleic acids.

The complete digestion of the organic macromolecules is outlined in Table 6.1. Table 6.2 outlines the source, substrate, products and optimal pH for Amylase, Protease and Lipase.

Table 6.1: Digestion of Organic Macromolecules

	Carbohydrates	Proteins
Site of Digestion	**Mouth:** ■ Enzyme: salivary amylase Resulting molecules: smaller polysaccharides **Small intestine** (Duodenum) ■ Enzyme: pancreatic amylase Resulting molecules: disaccharides ■ Enzyme: disaccharidases: (sucrase, maltase, lactase) Resulting molecules: monosaccharides	**Stomach:** ■ Enzyme: pepsin Resulting molecules: Smaller polypeptides **Small intestine** (Duodenum) ■ Enzyme: trypsin (endopeptidase) Resulting molecules: dipeptides ■ Enzyme: dipeptidases Resulting molecules: amino acids
Resulting Monomers	Monosaccharides: glucose, galactose, fructose	Amino acids
	Lipids	Nucleic Acids
Site of Digestion	**Small intestine** ■ Enzyme: lipase ■ Emulsified with bile (from gall bladder) Resulting molecules: glycerol and fatty acids	**Small intestine** ■ Enzyme: nucleases Resulting molecules: nucleotides
Resulting Monomers	Glycerol and fatty acids	Nucleotides

Table 6.2: Source, Substrate, Products and Optimal pH for Amylase, Protease and Lipase

Enzyme (Site of Production)	Substrate (Site of Digestive Action)	Products	Optimal pH
Amylase ■ Salivary amylase (salivary glands) ■ Pancreatic amylase (pancreas) ■ Disaccharidases (brush border/villi)	Polysaccharides (oral cavity)	Smaller polysaccharides	7
	Polysaccharides (small intestine)	Disaccharides	7
	Disaccharides (small intestine/brush border)	monosaccharides	7
Protease ■ Pepsin (stomach cells) ■ Trypsin (endopeptidase/pancreas) ■ Dipeptidases (brush border/villi)	Polypeptides (stomach)	Smaller polypeptide	2–3
	Polypeptides (small intestine)	Dipeptides	7
	Dipeptides (small intestine/brush border)	Amino acids	7
Lipase ■ Pancreatic lipase (pancreas)	Lipids (small intestine)	Glycerol and fatty acids	7

FUNCTIONS OF THE MOUTH, STOMACH, SMALL INTESTINE AND LARGE INTESTINE

- **Mouth**
  - □ Chemical digestion of carbohydrates with salivary amylase
  - □ Mechanical digestion with teeth and tongue
- **Stomach**
  - □ Chemical digestion of proteins with pepsin
  - □ Mechanical digestion due to low pH (2–3) and churning

The stomach functions for both mechanical and chemical digestion of polymers. Mechanically (no chemical bonds are broken between polymers) the stomach churns the food, which helps break large chucks of food into smaller pieces. This allows more surface area to be exposed in order for enzymes to digest the food chemically (break chemical bonds between polymers). The stomach maintains a pH level ranging between 2 and 3. This helps with the immune system because many pathogens cannot survive in this acidic environment. This pH also helps digestion by mechanically breaking down proteins. Remember that proteins will denature (lose their 3-D structure) at pH levels that are not optimal to their functioning. The only enzyme that can function in the stomach's acidic environment is pepsin. Hydrochloric acid is produced by the stomach. The presence of hydrochloric acid causes the precursor to pepsin (pepsinogen) to become the active enzyme pepsin.

■ **Small intestine**
  □ Chemical digestion of carbohydrates, proteins, lipids and nucleic acids
  □ Absorption of the monomers of carbohydrates, proteins, lipids and nucleic acids

The main roles of the small intestine are the digestion and absorption of nutrients. Most digestion of macromolecules occurs here. Different methods of transport across the intestinal membrane are used to absorb the many nutrients present. The small intestine is very long and twisted in order to maintain a high surface area to help support the need for absorption. Longitudinal and circular muscle layers contract to move food along the digestive tract. The mucosa of the small intestine keeps food moist and facilitates the smooth movement of food through the small intestine. (Refer to Figure 6.7.)

■ **Large intestine**
  □ Absorption of water and minerals

The term that describes the rhythmic contractions of the longitudinal and circular muscles lining the intestines in order to move digested food is called *peristalsis*.

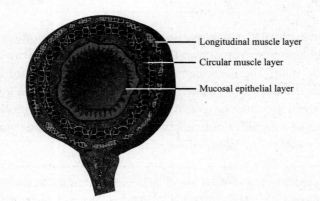

Longitudinal muscle layer

Circular muscle layer

Mucosal epithelial layer

Figure 6.7. Cross section of small intestine

Folds of the small intestine (villi) and folding of the cell membrane of the villi (microvilli) further increase the surface area. Due to all this folding, the tissue resembles the brushes on a comb and is often referred to as the "brush border" of the small intestine.

The large intestine functions mainly for absorption of water. The large intestine is much shorter than the small intestine and is often referred to as the colon.

VILLI STRUCTURE

The function of the villi in the small intestine is to increase the surface area for maximum absorption of nutrients. Only monomers are small enough to cross the villi and enter into nearby capillaries. The villi are only one-cell thick, as are the capillaries, which allows for monomers to move efficiently. Once the monomers enter the capillaries, they can be transported throughout the body. The lacteal (a fingerlike projection that reaches into each villus) functions for the absorption of lipids and contains numerous white blood cells. The white blood cells allow for the destruction of foreign antigens that may enter the villus along with digested food material, making lacteals part of the immune system as well as part of the digestive system. Each villus contains microvilli to increase the surface area further. (Refer to Table 6.3.)

Table 6.3: Functions of Villi Structures

Structure	Function	Role
Villi and microvilli	Increase surface area	Increased surface area allows for the efficient absorption of monomers.
Blood capillaries	Absorption of digested nutrients	Numerous capillaries can absorb a large percentage of nutrients.
Lacteals	Absorption of lipids (fats) and immune support	■ Absorption of lipids helps keep lipids out of the bloodstream. ■ White blood cells in lacteals help with immunity

As shown in Figure 6.8, polymers must be digested into monomers in order to be absorbed by the villi. The monomers cross over the villi and enter into the capillaries to be transported throughout the body. The monomers resulting from the digestion of macromolecules are transported by the circulatory system throughout the body. The absorbed monomers can be used directly in their monomer form (for example, glucose for cellular respiration) or the monomers can be assembled into polymers (for example, glycogen or DNA) that can be used for cellular processes or to build cellular structures.

> The small intestine contains over 3 million villi to allow for maximum surface area. The presence of villi and microvilli make the surface area of the small intestine about the size of a tennis court!

REMEMBER

Polymers: Large molecules formed from many monomers

Dimers: Small molecules made of only 2 monomers

Monomers: Smallest subunit of macromolecules

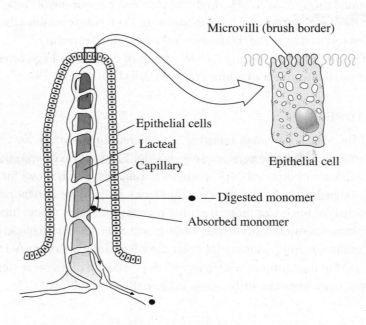

Figure 6.8. Villus absorbing monomer

6.2 THE BLOOD SYSTEM

TOPIC CONNECTIONS

- 2.2 Water
- 2.3 Carbohydrates and lipids
- 6.4 Gas exchange
- 6.6 Hormones, homeostasis and reproduction

The transport system consists of the heart and associated blood vessels. The heart pumps the blood to the lungs (the pulmonary circuit) to carry out gas exchange and out to the body to supply the body cells with nutrients and to pick up waste (the systemic circuit). The transport system is vital in order for each of the billions of body cells to receive required nutrients and get rid of waste products.

The Heart

The human heart consists of four chambers and four valves that completely separate oxygenated blood from deoxygenated blood. The top chambers include the right and left atria; the bottom chambers include the right and left ventricles. Blood in the right atria enters the right ventricle through an atrioventricular valve. Blood in the left atria enters the left ventricle through an atrioventricular valve. Blood is pumped from the right ventricle into the pulmonary artery to be sent to the lungs to carry out gas exchange (pulmonary circulation). Blood is pumped from the left ventricle to the aorta and out to capillaries in order to deliver nutrients and pick up waste material from the body cells (systemic circulation). The structure of the heart is shown in Figure 6.9.

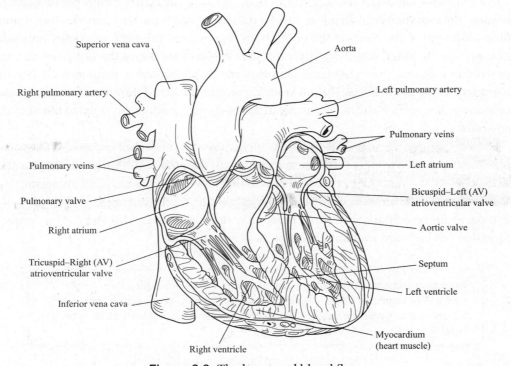

Figure 6.9. The heart and blood flow.

> **Two Circulatory Paths**
>
> 1. **Systemic:** Blood flows from the heart to the body (out the aorta and back through the vena cava).
> 2. **Pulmonary:** Blood flows from the heart to the lungs (out through the pulmonary artery to the lungs and back to the heart through the pulmonary vein).

CORONARY ARTERIES

The heart is a muscle made of many cells that must be supplied with nutrients and cleared of waste. The coronary arteries are located on the outside of the heart and supply the heart cells with nutrients and clear the heart cells of metabolic waste products. If blood flow to the coronary arteries were to become blocked, the individual would suffer a heart attack. A heart attack results when blood is blocked from bringing nutrients and from taking waste away from the heart muscles.

ACTIONS OF THE HEART

The heart consists of four chambers and four valves. Two atria receive blood returning to the heart from the body and from the lungs. The right atrium receives blood returning to the heart from the body, and the left atrium receives blood returning to the heart from the lungs. The superior vena cava receives blood from the upper torso, and the inferior vena cava receives blood from the lower torso. These veins bring blood back to the heart from the body. The pulmonary vein brings blood back to the heart from the lungs. The ventricles receive blood from the atria through the atrioventricular (AV) valves. The ventricles pump blood out to the lungs (right ventricle) and to the body (left ventricle).

The ventricles are much more powerful than the atria since they pump blood a greater distance. The most powerful chamber (largest muscular wall) is the left ventricle since it must pump blood out of the heart to the entire body. In contrast, the right ventricle pumps the blood only to the lungs, which is a shorter distance. The blood leaves the ventricles through the semilunar valves. The right ventricle pumps the blood through a semilunar valve to the pulmonary artery to the lungs. The left ventricle pumps the blood through a semilunar valve to the aorta. The aorta is the largest artery in the body and provides oxygenated blood to the body cells.

When the atrioventricular (AV) valves are open, the semilunar valves are closed. The valves open due to pressure from the contracting muscles of the atria and ventricle walls. The atria contract before the ventricles in order to ensure that blood exits the atria and enters the ventricles prior to the ventricles contracting. The valves control the direction of blood flow. If a valve does not close all the way, some blood may flow backwards. This leaky valve causes a sound that can be heard with a stethoscope as a heart murmur.

> vena cava (superior and inferior) → right atrium → atrioventricular valve → right ventricle → semilunar valve → pulmonary artery → lungs → pulmonary vein → left atrium → atrioventricular valve → left ventricle → semilunar valve → aorta → body cells → back to heart to vena cava . . .

CONTROL OF THE HEART

The heart sets its own pace for beating and can do so without influence from the nervous system. In fact, a person who is brain dead will have a heart that continues to beat but must be on a respirator in order to breathe. The term to describe the fact that the heart can beat on its own is *myogenic*.

The pacemaker that controls the rate at which the heart beats is a collection of neurons (node) located in the sinus (cavity) of the right atrium. This bundle of neurons is known as the sinoatrial (SA) node. The SA node controls the contraction of the atria and indirectly the contraction of the ventricles. The SA node sends an impulse to the atrioventricular node (AV) after it triggers the contraction of the atria. The AV node is located in the right atrium as well. When the AV node receives electrical impulses from the SA node, it triggers the contraction of the ventricles by sending out more electrical impulses. Electrical impulses travel through the walls of the heart, initiating contractions. There is a very small (0.1 second) delay after the SA node triggers the contraction of the atria before it sends impulse to the AV node. This ensures that the atria contract and empty before the ventricles do so. (Refer to Figure 6.10.)

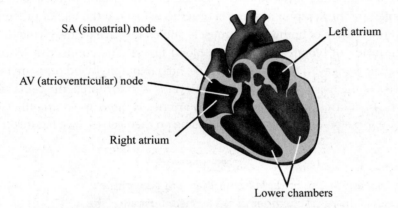

Figure 6.10. The sinoatrial (SA) node and atrioventricular (AV) node

The rate at which the heart beats is set by the SA node and is a constant rate. Sometimes the heart needs to beat faster in order to deliver more oxygen to body cells and get rid of excess carbon dioxide. The nervous system can influence heart rate through impulses delivered by nerves that branch from the central nervous system (CNS) to the SA node (pacemaker). The medulla oblongata is the part of the brain that communicates with the pacemaker via nerve impulses that impact (increase and decrease) both breathing and heart rate. The endocrine system can influence heart rate as well by releasing epinephrine (adrenaline). This hormone causes several physiological effects, including increasing heart rate.

> **REMEMBER**
>
> The nodes are named based on their location in the heart.
>
> **Sinoatrial node:** Located in the sinus of the right *atrium*.
>
> **Atrioventricular node:** Located in the right atrium near the *atrioventricular* valve.

- Two nerves from the medulla oblongata communicate with the heart (SA node) and can change the rate (increase or decrease) at which the heart contracts.
- Epinephrine from the adrenal glands can increase heart rate during stress or physical exercise.

Arteries, Veins and Capillaries

The largest vessels in the circulatory system are the arteries. Arteries take blood away from the heart and must withstand the high pressure exerted by the ventricles in order to pump the blood out to great distances. The aorta is the largest artery and must withstand the highest pressure since it receives the blood directly from the heart and delivers it to other arteries throughout the body. Most arteries carry oxygenated blood. The exception to this rule is the pulmonary artery that carries blood away from the heart to the lungs to pick up oxygen and get rid of carbon dioxide (gas exchange).

Veins bring blood back to the heart, and most carry deoxygenated blood with the exception of the pulmonary vein. The pulmonary vein carries blood high in oxygen since it is bringing blood back to the heart from the lungs. The superior vena cava brings blood back to the heart from the upper regions of the body, and the inferior vena cava brings blood back to the heart from the lower regions of the body. Both vena cava return deoxygenated blood to the right atria.

Arteries and veins consist of three tissue layers, including the endothelium (inner layer consisting of only one layer of cells), smooth muscle (middle layer allowing arteries to help pump blood), and an outer layer of connective tissue. The smooth muscle and connective tissue in arteries are thicker than in veins in order to withstand the higher pressure exerted. Veins have thinner layers of both smooth muscle and connective tissue, giving them a wider lumen (space inside of blood vessels where blood flows). The connective tissue contains elastic fibres that, along with the smooth muscle, help maintain blood pressure between the pumping cycles of the heart. When blood flow to veins is interrupted, the veins will collapse on themselves. In contrast, when blood flow to arteries is interrupted, the thicker layers of arteries prevent these blood vessels from collapsing on themselves. (See Figure 6.11.)

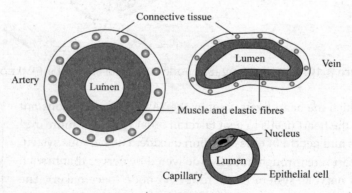

Figure 6.11 Blood vessels

Capillaries are the most numerous of all blood vessels and consist of only the endothelium. Since the endothelium is only one cell thick, capillaries have the ability to carry out the exchange of material between blood and body cells.

Arteries take blood away from the heart. Arteries branch to form smaller vessels known as arterioles. Arterioles branch to form capillaries that exchange material with body cells. Capillaries converge to form venules. Venules converge to form veins that bring blood back to the heart. (Refer to Table 6.4.)

> heart → arteries → arterioles → capillaries → venules → veins → heart

Table 6.4: Structure and Function of Blood Vessels

Vessel	Tissues	Size	Function
Artery	Endothelium, smooth muscle, connective tissue (with elastic fibres)	Largest (thick layer of smooth muscle and connective tissue)	Takes blood away from the heart and helps maintain blood pressure (pumps blood)
Vein	Endothelium, smooth muscle, connective tissue (with elastic fibres)	Second largest (thin layer of smooth muscle and connective tissue)	Brings blood back to the heart and helps maintain blood pressure (pumps blood)
Capillary	Endothelium only	Thinnest and most numerous	Allows exchange of material with body cells

FEATURED QUESTION

November 2008, Question #18

What are the characteristics of blood flowing through most arteries and veins?

	Arteries	Veins
A.	Slow velocity	Fast velocity
B.	High pressure	Low pressure
C.	Deoxygenated	Oxygenated
D.	Greater than 37 degrees	Less than 37 degrees

(Answer is on page 553.)

Composition of Blood

The liquid portion of blood containing dissolved substances and suspended blood cells is known as plasma. Erythrocytes (red blood cells), leucocytes (white blood cells) and platelets are found in blood. Erythrocytes function mainly to carry oxygen. Leucocytes are the white blood cells that function for immunity. Platelets aid in blood clotting. Plasma, which is composed of mostly water (the universal solvent), dissolves and transports nutrients, carbon dioxide, hormones, antibodies, urea and heat.

6.3 DEFENCE AGAINST INFECTIOUS DISEASE

TOPIC CONNECTIONS

- 5.2 Natural selection

The body is constantly being challenged by disease-causing organisms and viruses (pathogens) that want to gain entry to the body. The body is designed to prevent entry of these pathogenic organisms. The immune system has evolved to protect the body from these dangerous pathogens.

> **Pathogen:** Any disease-causing antigen (nonself substance)

Viruses and Antibiotics

Viruses are nonliving structures that must gain entry into a host cell in order to reproduce. Bacteria are living organisms that possess metabolic activities that allow them to survive and reproduce on their own. Antibiotics are chemicals that interrupt metabolic processes performed by bacteria. Since viruses do not carry out metabolic processes, antibiotics are not effective against them.

Primary Defences

The skin and mucous membranes are part of the primary (first-line) defence against pathogenic invasion. First-line defensive mechanisms are nonspecific and aim to keep out any and all types of pathogenic organisms. The skin consists of many keratinized (tough) layers of cells that form a barrier to entry. The acidity of the skin also prevents many opportunistic pathogens from being able to invade the body.

Mucus traps pathogens, preventing them from gaining entry into cells. Mucus also holds pathogens for ease of expelling them out of the body. Mucus that contains pathogens can be expelled from the body by coughing up the mucus, by sneezing or by blowing the nose.

Phagocytic White Blood Cells (Lymphocytes)

Once inside the body, pathogens can be engulfed by phagocytic lymphocytes that make up the major part of the body's secondary defence. Phagocytic lymphocytes (macrophages) are large cells that migrate around the body, engulfing any foreign molecules in the blood or body tissues. They take in the foreign material by endocytosis, forming a vesicle containing the pathogen. Fusion of the vesicle with a lysosome allows the hydrolytic digestion of the pathogen. After digestion of the pathogen, the macrophage displays parts of the pathogen on its surface to alert other immune cells to the presence of the pathogen in the body.

Antigens and Antibodies

Phagocytic leucocytes ingest any substance that is foreign to the body or needs to be removed because the substance is no longer needed by the body. Foreign substances are referred to as *antigens.* The immune system produces *antibodies* that recognize specific antigens and prevent them from entering into body cells. (Refer to Figure 6.12.)

- **Antigens:** Foreign substances that enter the body and promote the formation of antibodies.
- **Antibodies:** Globular proteins produced by the immune system to recognize and rid the body of antigens.

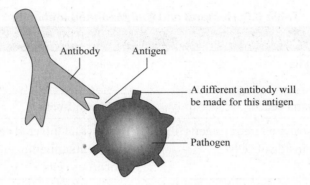

Figure 6.12. Antibody and antigen.

Specific Immunity

The immune system consists of two types of lymphocytes: B-cells and T-cells. These cells recognize and destroy specific pathogens. Both B-cell and T-cell progenitors are produced in the bone marrow and have distinct functions in the immune system. B-cells function to destroy any free (not inside of a cell) pathogen. T-cells function to destroy any body cells infected with pathogens, multicellular pathogens, and cancerous cells.

T-cells need to be carefully regulated since they destroy cells and the body is made of cells. For this reason, T-cell progenitors leave the bone marrow and migrate to the thymus (a gland located right above the heart) to complete their development and to be screened to make sure they destroy only foreign cells or infected body cells. Unlike the T-cells, B-cells are fully functional when they leave the bone marrow.

- **B-cells:** Originate and fully mature in the bone marrow (centre of long bones). B-cells are named for their location of maturation—the bone marrow.
- **T-cells:** Originate in the bone marrow but must migrate to the thymus to mature. T-cells are named based on their location of maturation—the thymus.

> T-cells that do not recognize the body's own cells cause autoimmune diseases. In other words, the immune system destroys healthy cells. Examples of autoimmune diseases include lupus, rheumatoid arthritis and multiple sclerosis.

Humoral and Cell-Mediated Immunity

The immune system is divided into two types of responses based on the function of the B- and T-cells. B-cells are involved in the humoral immune response, and T-cells are involved in the cell-mediated immune response. Both responses are controlled by T-cells known as helper T-cells. Helper T-cells activate both humoral and cell-mediated responses when presented with an invading pathogen. Helper T-cells cannot recognize a pathogen unless it is presented by an antigen-presenting cell (APC). Both macrophages and B-cells act as APC and present antigens to helper T-cells. Activation of the humoral and cell-mediated immune responses involves the production of memory cells as well. Once the pathogen has been contained, memory B-cells and memory T-cells will remain and can quickly identify the pathogen if it invades the body again. (Refer to Table 6.5.)

Table 6.5: Humoral and Cell-Mediated Immunity

Feature	Humoral Immunity	Cell-Mediated Immunity
Lymphocyte involved	B-cells	T-cells
Effector cell	Plasma cells	Cytotoxic T-cells
Target	Destroys all free antigens (those not inside of cells)	Destroys all infected cells, multicellular pathogens and cancerous cells
Mode of action	Plasma cells produce antibodies that surround and neutralize the pathogen	Cytotoxic T-cells release chemicals that trigger the apoptosis (destruction) of the infected cells

ANTIBODY PRODUCTION

The humoral immune response is responsible for activating B-cells that produce antibodies that will destroy any free (not inside of a cell) pathogens. The B cells will divide and become plasma cells (as well as long-lived memory B-cells) that will produce antibodies specific to the invading pathogen. The antibodies produced by the plasma cells are the same as those that are imbedded in the cell membrane of the B-cells. Antibodies are very important in stopping the invasion of body cells. Antibodies neutralise pathogens. (Refer to Figure 6.13.)

T-Helper Cells

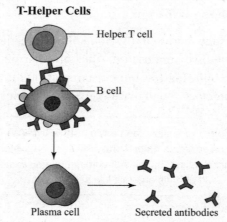

Helper T cell

B cell

Plasma cell

Secreted antibodies

Figure 6.13. Antibody production

Blood Clotting

Blood clotting is necessary in order to prevent loss of blood from the circulatory system as well as to help prevent the entry of pathogens into the body. Damaged tissue (cells) releases clotting factors that attract platelets (thrombocytes) to the injured area. If the damage is minor, the platelets can seal the wound and stop blood loss. However if the damage is more extensive, fibrin will be produced to secure the clot. The protein prothrombin is present in blood in a soluble form. When enough clotting factors are released, the prothrombin will be converted into thrombin. Thrombin catalyses the conversion of fibrinogen, a protein also present in a soluble form in the blood, to fibrin. Fibrin is a protein that weaves into the platelets and secures the clot. Since damaged cells and active platelets release clotting factors, the more cells that are damaged along with increased active platelets results in higher levels of clotting factors. High levels promote the cascade of reactions that produce fibrin. (See Figure 6.14.)

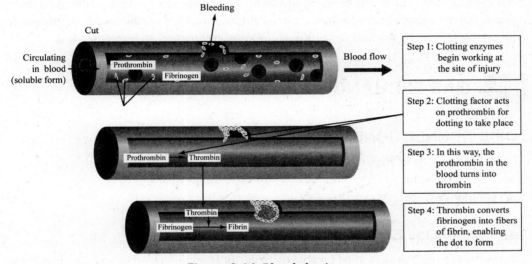

Figure 6.14. Blood clotting

As shown in Figure 6.19, blood clotting occurs in the following steps:

1. Injury occurs, resulting in the release of clotting factors by the damaged cells and platelets.
2. Clotting factors activate prothrombin to be converted into thrombin.
3. Thrombin catalyses the conversion of fibrinogen into fibrin.
4. Fibrin weaves into platelets, thereby securing the clot.

HIV and the Immune System

Human immunodeficiency virus (HIV) is effective in destroying the human immune system because it invades and stops the helper T-cells from functioning. Since helper T-cells activate both the humoral and cell-mediated responses, individuals infected with HIV have weakened immune systems. The number of lymphocytes present is reduced as well as the number of antibodies present when a pathogen invades the body. Individuals with HIV often die from opportunistic infections that they cannot fight effectively or efficiently.

Individuals infected with HIV in which the virus is actively replicating are diagnosed with having acquired immunodeficiency disease (AIDS). The virus invades lymphocytes, decreasing the ability of the body to fight off pathogens.

Direct contact between specific body fluids of an infected individual and an uninfected individual can transfer the virus to the uninfected individual. The virus can be acquired through unprotected sexual contact with an infected individual, by sharing a needle from an infected individual and across the placenta from an infected mother to a foetus. The virus must be able to enter the bloodstream of an uninfected individual in order to be transferred from one individual to another. For this reason, you cannot get HIV from touching or kissing an infected individual.

Since there is no cure for this deadly disease, many infected individuals are treated poorly in social situations. Infected individuals may be shunned by others and isolated from society. Employers and insurance companies may unfairly discriminate against HIV-positive individuals and block them from employment or limit their access to medical treatment. The number of infected individuals in the world is not evenly distributed. Most of the reported AIDS cases are in undeveloped countries. Southern Africa has the most severe problems with the disease. Debate exists about whether or not wealthy countries have the responsibility to provide aid, such as medicine and education about preventing AIDS, to poorer regions.

6.4 GAS EXCHANGE

TOPIC CONNECTIONS

- 1.4 Membrane transport
- 1.6 Cell division
- 6.2 Blood system

Ventilation, Gas Exchange and Cellular Respiration

In order to survive, humans must be able to carry out efficient gas exchange. The respiratory system allows the ventilation of the lungs by expanding the chest cavity to bring in air and by contracting the chest cavity to push out air. The term to describe the movement of air into and out of the lungs is ventilation. Ventilation maintains a concentration gradient between oxygen and carbon dioxide in the alveoli and the capillary blood. The concentration gradient allows for simple diffusion of these gases as they move into and out of the body.

Gas exchange occurs between the capillaries and alveoli in the lungs where carbon dioxide is exchanged for oxygen. Oxygen passes from the alveoli into the capillaries surrounding the alveoli, and carbon dioxide passes from the capillaries into the alveoli. The body needs oxygen in order to carry out cellular respiration. Ventilation brings air into the body to allow for gas exchange between capillaries and body cells.

- **Ventilation:** Inhalation and exhalation of air into and out of the lungs. Ventilation maintains concentration gradients between carbon dioxide and oxygen in the air, in the alveoli of the lungs, and in the air in the capillaries.
- **Gas exchange**: The exchange of carbon dioxide for oxygen that occurs in the alveoli of the lungs.

Cellular respiration uses the oxygen provided by the ventilation system to create energy (ATP). It also produces carbon dioxide that must be removed from the body. The movement of oxygen (O_2) and carbon dioxide (CO_2) between the alveoli and capillaries allows these gases to be exchanged with the environment.

- **Cellular respiration**: The controlled release of energy from organic molecules coupled with the production of ATP.

A ventilation system is vital in order to maintain a high concentration gradient across the alveoli to facilitate efficient gas exchange. If fresh air were not constantly being brought into the lungs and if stale air were not being forced out, the exchange of gases would slow to a rate that would not be efficient to support cellular respiration. The high concentration gradient is maintained by the ventilation system between the amount of oxygen and carbon dioxide in the alveoli. Maintaining a high concentration of oxygen in the alveoli facilitates the movement of oxygen from the alveoli into the bloodstream. Maintaining a low concentration of carbon dioxide in the alveoli facilitates the movement of carbon dioxide from the capillaries into the alveoli.

Two types of pneumocytes are found in the alveoli. Type I pneumocytes are elongated and very thin cells. Their shape helps them facilitate gas exchange. Type II pneumocytes are thick and short cells. They secrete a solution containing surfactant to ensure the membranes of the alveoli cells remain moist. The presence of moisture helps ensure that the sides of each alveolus do not stick together. In addition, the moisture helps to facilitate a high rate of diffusion of respiratory gases. (Refer to Figure 6.15 and Table 6.6.)

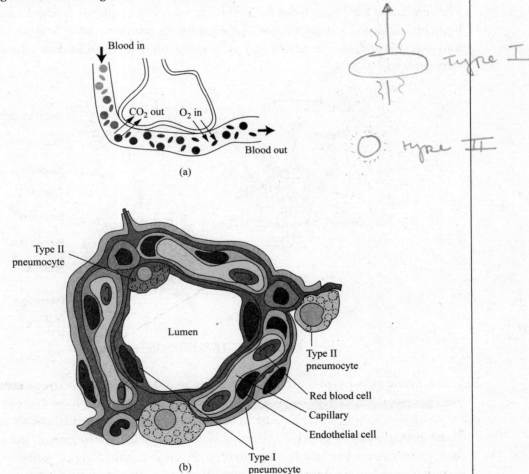

Figure 6.15. (a) Alveoli gas exchange; (b) type I and type II pneumocytes

Table 6.6: Features of Alveoli

Feature	Significance
Many alveoli are present	Increases surface area for more efficient gas exchange
Capillaries surround the alveoli	Close association with the capillaries allows efficient gas exchange between the capillaries and the alveoli since the gases need to diffuse only a short distance
Alveoli are only one-cell thick	Allows ease of diffusion across a small distance
Alveoli are moist	Moisture increases the rate of diffusion by dissolving the gases and also prevents the alveoli from sticking together
Alveoli are flexible	The single cell layer can stretch and become even thinner to facilitate more rapid gas exchange

The Ventilation System

The ventilation system includes the mouth, nasal cavity, epiglottis, trachea, lungs, bronchi, bronchioles, alveoli and diaphragm. These respiratory structures allow fresh air to be brought into the body and stale air to be removed from the body. Figure 6.16 shows the structures of the ventilation system.

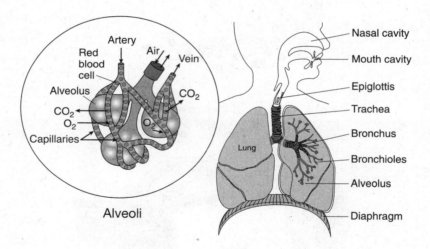

Figure 6.16. The ventilation system

Gas exchange is based on changing pressure in order to force air to move. Air will always move from an area of high pressure to an area of low pressure. As volume decreases, pressure increases. The lungs change volume in order to change pressure and force air into and out of the alveoli. In order to exhale (force air out of the lungs), the volume of the chest cavity must be reduced to increase pressure that causes air to exit the lungs. In order to exhale, the diaphragm and external intercostals must relax and both the internal intercostals and the abdominal muscles must contract. In order to inhale, the diaphragm and external intercostals must contract and both the internal intercostals and abdominal muscles must relax. (Refer to Table 6.7.)

Table 6.7: Ventilation

Muscle	Inhalation	Exhalation
Diaphragm	Contracted (pulls downwards)	Relaxed (bends upwards)
Internal intercostals*	Relaxed	Contracted (pulls ribs down and inward)
External intercostals*	Contracted (pulls ribs up and outward)	Relaxed
Abdominal muscles	Relaxed and pushed out	Contracted and pull inwards

*The internal and external intercostal muscles are antagonistic muscles since they carry out opposite movements of the ribs during the respiratory cycle. (The internal muscles contract, pulling the ribs downwards. The external muscles contract, pulling the ribs upwards.)

Lung Cancer and Emphysema

Lung cancer develops from cells that divide uncontrollably and destroy healthy lung tissues. Lung cancer is the most prevalent and deadliest form of cancer among both men and women. The cancerous cells in the lungs can take over tissues involved in the exchange of respiratory gases. Inefficient gas exchange can be life threatening. Individuals with lung cancer may exhibit symptoms such as a persistent cough, coughing up blood, fatigue and weight loss. The number one cause of lung cancer is exposure to cigarette smoke, although other factors can contribute to the development of the cancer. (Refer to Table 6.8.)

Emphysema is a chronic respiratory disorder caused by damage to the alveoli in the lungs. The damage can result from exposure to cigarette smoke, pollutants in the air and exposure to asbestos. Symptoms include wheezing, shortness of breath and decreased gas exchange. Emphysema and lung cancer are thought to have possible genetic causes as well. (See Table 6.8.)

Table 6.8: Lung Cancer and Emphysema

	Causes	Consequences
Lung cancer	■ Cigarette smoke ■ Asbestos exposure ■ Radon gas ■ Exposure to air pollutants ■ Genetics	■ Persistent cough ■ Coughing up blood ■ Fatigue ■ Weight loss ■ Shortness of breath
Emphysema	■ Cigarette smoke ■ Exposure to air pollutants ■ Asbestos exposure ■ Genetics	■ Wheezing ■ Coughing ■ Fatigue ■ Genetics

6.5 NEURONS AND SYNAPSES

TOPIC CONNECTIONS

- 1.4 Membrane transport

The Nervous System

The nervous system is divided into two parts that include the central nervous system (CNS) and the peripheral nervous system (PNS). The CNS consists of the brain and spinal cord. The PNS consists of all the nerves that branch from the CNS and go to the rest of the body. Nerves are composed of cells called neurons as shown in Figure 6.17. Neurons transmit messages by sending rapid electrical impulses.

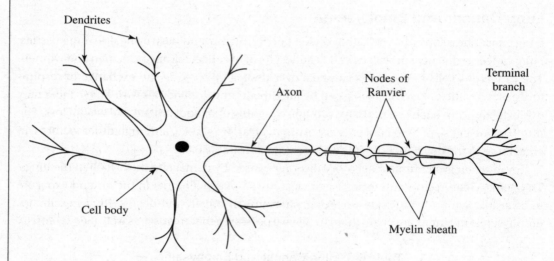

Figure 6.17. The motor neuron

Receptors respond to stimuli and send messages to the CNS using electrical impulses known as action potentials. The electrical impulse must be changed into a chemical message in order to cross the synapse (space between the neurons) and carry the message to a relay neuron inside the CNS. The relay neuron receives the chemical message and turns it into an electrical impulse that travels through the relay neuron from the CNS to the effector cell (the cell that will elicit a response to the initial stimuli) by another chemical messenger. If the effector is a muscle cell, the neuron that delivers the message to the muscle is a motor neuron. This pathway of information transformation from a receptor to an interneuron and out to an effector is known as a reflex arc. Notice that the reflex arc does not involve the brain. Hence reflexes, such as the pain reflex shown in Figure 6.18, are involuntary behaviours. The message will be delivered to the brain but not before the response occurs. It is much faster to elicit a response to stimuli by allowing the spinal cord, which is part of the CNS, to handle the initiation of the response. This is an adaptive behaviour for survival. It allows the body to respond to harmful stimuli by allowing the spinal cord to carry out the response at the same time that the message is being delivered to the brain. For example, you will pull your hand away from a hot stove before you are actually aware of the response.

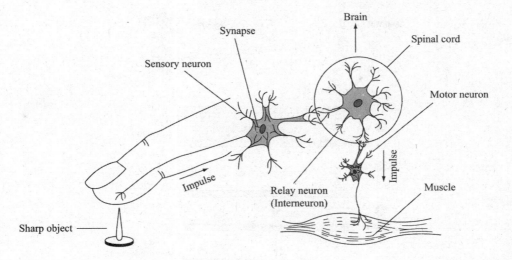

Figure 6.18. Reflex arc

Resting Potential and Action Potential

A difference in electrical charge exists between the inside (cytoplasm) of a cell and the outside (extracellular fluid) of a cell. This difference in charge is known as the membrane potential. Although several ions are responsible for this difference, the main ions involved in maintaining the membrane potential are sodium (Na^+) and potassium (K^+). When at rest (when not sending an action potential), the charge inside of the cell is more negative than the charge outside of the cell by –70 millivolts. This difference (–70 mV) is known as the resting potential.

The resting potential is maintained by the sodium-potassium pump, which uses ATP to pump sodium out of the cell while at the same time pumping potassium into the cell. The pump actively transports more sodium out than potassium to maintain the resting potential of –70 mV. Since some of the potassium ion channels are "leaky", the sodium potassium pump is always working to maintain the resting potential.

Some cells, such as neuron and muscle cells, possess sodium-gated ion channels that can open in response to a stimulus and allow sodium to diffuse into the cell. Potassium-gated ion channels can open as well and allow potassium to diffuse out of the cell. When a stimulus is detected by a receptor, the sodium-gated ion channels open. Sodium diffuses into the cell, causing the membrane potential to become less negative. (The addition of a positive charge to the inside of the cell makes the cell less negative.) This process alters the membrane potential and depolarizes the cell. If the cell depolarizes to a charge difference of around –50 mV, an action potential is propagated. This charge is known as the threshold potential. A cell must have enough sodium channels open in order to reach threshold before an action potential is sent down the axon. An action potential is a moving wave of depolarization (less negative inside due to sodium ions entering the cell) and repolarization (back to a more negative state inside due to potassium ions exiting the cell). This is accomplished by the initial opening of sodium channels followed by the opening of potassium channels down the axon of a neuron. (See Figure 6.19.)

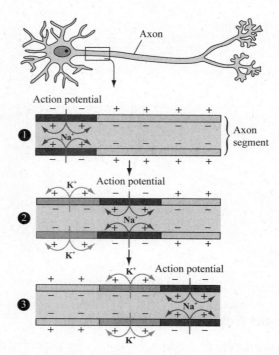

Figure 6.19. Propagation of an action potential

- **Resting potential:** The charge of a cell at "rest". The neuron is not sending an action potential. Resting potential is maintained by the sodium-potassium pump.

- **Action potential:** The changing charge in a cell due to the opening of sodium-gated ion channels (depolarization) followed by the opening of potassium-gated ion channels (repolarization).

NERVE IMPULSES

Once a cell reaches threshold (–50 mV), an action potential is propagated down the axon towards the terminal buttons. The axon will experience depolarization followed by repolarization as the sodium gates open followed by the opening of the potassium gates. Each action potential propagates an action potential in the immediate adjacent membrane. This creates a wave of travelling membrane charge changes (depolarization followed by repolarization) known as the action potential down the axon.

Myelinated axons send action potentials (electrical impulses) at greater speeds than do nonmyelinated axons. The action potential simply needs to be generated at the nodes (spaces between the myelination). The "jumping" of the impulse from one node to another (across the Schwann cells, which are cells that make up the myelination) is known as saltatory conduction.

As the wave of depolarization and repolarization travels down the axon, the sodium-potassium pump follows and reinstates the resting potential by pumping sodium ions back out of the cytoplasm and potassium ions back into the cytoplasm. If not for the sodium-potassium pump, the resting membrane potential would not be reinstated and the cell could no longer send an action potential. (Refer to Figures 6.19 and 6.20.)

> ### REMEMBER
>
> In Spanish, *saltar* means "to jump". During saltatory conduction, the impulse is "jumping" from node to node.

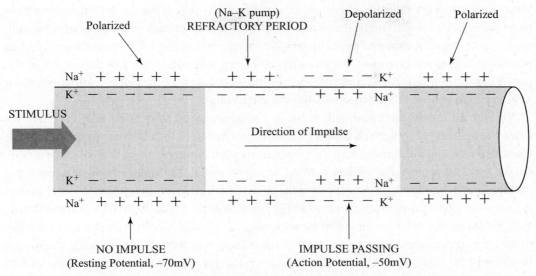

Figure 6.19. Propagation of an action potential along an axon.

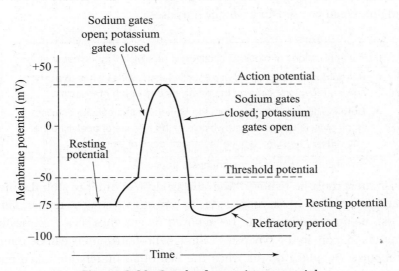

Figure 6.20. Graph of an action potential.

SYNAPTIC TRANSMISSION

In order for an action potential to continue from one neuron to another or from a neuron to a muscle cell, the action potential must be able to cross the space between the cells. The space between two neurons or a neuron and a muscle cell is called a synapse. In order for the message to cross the synapse, it must be changed from an electrical signal (action potential) into a chemical signal. The molecules that are responsible for delivering the electrical message from a neuron across the synapse in a chemical form are appropriately called neurotransmitters. Neurotransmitters are stored in the terminal buttons of neurons inside of membrane-bound vesicles.

The arrival of an action potential at the terminal buttons triggers the opening of calcium-gated ion channels in the presynaptic membrane, allowing calcium to diffuse into the presynaptic membrane. The presence of calcium triggers the release of the neurotransmitters into the synapse by exocytosis. The neurotransmitters diffuse across the synaptic gap (space) and bind to receptors on gated ion channels, causing them to open. If the neurotransmitters bind to sodium-gated ion channels, the channels will open and sodium will diffuse into the cell, resulting in depolarization of the postsynaptic membrane. If the neurotransmitters bind to potassium-gated ion channels, the channels will open and potassium will diffuse out of the postsynaptic membrane, resulting in hyperpolarization (inside of the cell becomes even more negative than the outside) of the postsynaptic membrane.

Neurotransmitters are identified as being excitatory when they open sodium channels since the opening of sodium channels can initiate an action potential. Neurotransmitters are identified as being inhibitory when they open potassium channels since this will hyperpolarize the cell and prevent the initiation of an action potential. Examples of neurotransmitters include acetylcholine, dopamine, serotonin, epinephrine and norepinephrine. Acetylcholine is the most abundant neurotransmitter in the body. Synapses that use acetylcholine for synaptic transmission are considered to be cholinergic synapses. Cholinergic synapses are the main type of synapses involved in controlling muscle contractions. Many toxic chemicals function by blocking the action of acetylcholine at the synapse. Modern science has used its knowledge of cholinergic synapses to develop methods of insect control. Science has also used its knowledge of these synapses so a toxin produced by the bacterium *Clostridium botulinum*, called Botox, can be used for cosmetic purposes.

Useful applications of blocking cholinergic (acetylcholine) synapses:

- **Pesticides:** Neonicotinoid (neuroactive insecticide) binds to acetylcholine receptors, causing the paralysis and death of insects.

- **Botox:** This toxin derived from the bacterium that causes botulism stops muscle contractions by blocking the release of acetylcholine. The relaxed muscles reduce the appearance of wrinkles.

Neurotransmitters must be removed from the synapse in order to stop the message from being sent and to allow the postsynaptic cell to return to resting potential. This occurs in two ways. First, neurotransmitters can be removed by enzymes. The enzymes break down neurotransmitters that are in the synapse. Second, neurotransmitters can be removed by the process of reuptake. Reuptake occurs when the neurotransmitters are taken back into the terminal buttons through endocytosis. This process allows the neurotransmitters to be available to be released again when another action potential arrives at the terminal button. (Refer to Figure 6.21.)

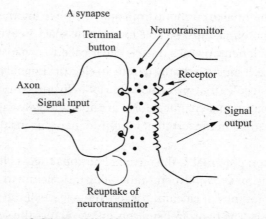

Figure 6.21. Terminal branch of neuron at synapse.

6.6 HORMONES, HOMEOSTASIS AND REPRODUCTION

TOPIC CONNECTIONS

- 3.2 Chromosomes
- 3.3 Meiosis
- 10.1 Meiosis

Communication between cells that are in close association with each other occurs electrically with action potentials and occurs chemically with neurotransmitters. Sometimes information needs to be communicated between cells that are a great distance from each other. The endocrine system accomplishes this type of long-distance communication by producing and releasing hormones into the circulatory system from endocrine glands. The hormones travel in the blood to their target tissue. The target tissue consists of cells that have receptors for the hormone and that will elicit a specific response.

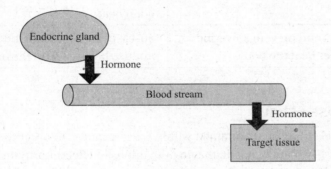

Homeostasis

Homeostasis is the ability of the body to maintain stable internal conditions regardless of changing external conditions. The body (specifically the hypothalamus of the brain) monitors blood pH, carbon dioxide concentration, blood glucose concentration, body temperature and water balance. If the homeostatic level of any of these variables is altered, the body will respond by counteracting any changes in set homeostatic levels. The counteraction that occurs to bring back a set value when it is altered is known as *negative feedback*. In other words, when a change is detected, the body elicits responses to bring the altered level back to where it was before the change.

THERMOREGULATION

Maintaining the temperature of the body close to its homeostatic level of 98.6 degrees Fahrenheit (37 degrees Celsius) is vital in order to maintain homeostasis. The hypothalamus monitors the body temperature and elicits specific negative feedback mechanisms to counteract any change from the optimal body temperature. The hormone thyroxin from the thyroid gland increases metabolic rate and helps to control body temperature as well. An increase in metabolic activities leads to the generation of heat that elevates the core body temperature.

If the body temperature drops too far below 98.6 degrees Fahrenheit (37 degrees Celsius), the body will begin to shiver to produce heat. Blood transports heat. When the body is too cold, the arterioles that take blood to the extremities will constrict to send more blood to the body's core to keep vital organs warm. If the body temperature rises too far above 98.6 degrees Fahrenheit (37 degrees Celsius), the body will begin to sweat to allow for evaporative cooling. In addition, the arterioles to the skin will dilate, sending more blood to the surface of the body for cooling. (See Table 6.9.)

Table 6.9: Thermoregulation

Cooling Mechanisms	Warming Mechanisms
Sweat glands release sweat for evaporative cooling.	Sweat glands do not release sweat in order to limit evaporative cooling.
Arterioles leading to the skin's surface dilate and bring more blood to the surface for cooling. (Remember that blood transports heat!)	Arterioles leading to the skin's surface constrict and keep blood in the body's core to warm vital organs.
The body spreads out (wide arms and legs) to increase the surface area for cooling .	The body pulls inwards (arms and legs very close to body) to decrease surface area and retain heat.
Muscle cells relax and prevent shivering that would further heat the body.	Muscle cells contract to induce shivering to produce heat to warm the body.

BLOOD GLUCOSE LEVELS

Blood glucose levels must be maintained within narrow ranges in order to keep the body at homeostatic levels. Two antagonistic (having the opposite effect) hormones are involved in maintaining blood glucose levels. Both of these hormones are produced in the pancreas by the pancreatic islet cells. The alpha (α) islet cells of the pancreas produce glucagon, and the beta (β) islet cells produce insulin. The hypothalamus monitors the blood glucose levels and elicits hormonal responses to any change (above or below the set value) in blood glucose levels. This is another example of negative feedback. Any change in a set homeostatic level is counteracted by a reaction to bring the level back to the homeostatic level. (See Table 6.10.)

Table 6.10: Blood Glucose Regulation

Rise in Blood Glucose Levels	Drop in Blood Glucose Levels
Insulin is released from beta cells (β).	Glucagon is released from alpha cells (α).
Liver cells store glucose as glycogen, removing glucose from the bloodstream.	Liver cells break down stored glycogen into glucose and release the glucose into the bloodstream.
Cells take up glucose from the bloodstream (capillaries).	Cells take up very little glucose from the bloodstream (capillaries).
Blood sugar levels fall to homeostatic levels.	Blood glucose levels rise to homeostatic levels.

When blood sugar levels fall below the homeostatic level, glucagon is released. Glucagon stimulates the liver to break down stored glycogen into glucose and release it into the bloodstream. When blood sugar levels rise above the homeostatic level, insulin is released. Insulin stimulates body cells to take up glucose and the liver to store glucose as glycogen. The release of insulin when blood sugar levels are high and glucagon when blood sugar levels are low maintains blood glucose levels at the homeostatic value.

Type I and Type II Diabetes

Diabetes is a disease associated with the loss of the ability to control blood glucose levels. Type I diabetes results when the islet cells lose their ability to produce insulin. Insulin acts as the "key" that opens up protein channels that allow glucose to leave the bloodstream and enter into body cells. Without insulin, body cells cannot take up glucose and blood glucose levels become dangerously high. In addition, cells are starved of glucose and cannot undertake cellular respiration to make ATP. Although currently no cure is available for type I diabetes, the disease is controlled with insulin injections and low-sugar diets. If they did not receive daily insulin injections, individuals with type I diabetes would not survive. Although scientists do not fully understand the cause of type I diabetes, it is speculated to have several causes. These include an autoimmune response that destroys the islet cells, viral infection and possibly a genetic link.

Type II diabetes results when the cellular receptors for insulin fail to respond to the hormone. Insulin recognition can be related to a lock and key model. While insulin serves as the "key" that opens the protein channels for glucose, the cellular receptors serve as the "keyhole" that recognizes insulin. In type II diabetes, the "keyhole" is misshapen and fails to allow insulin to open the protein channels that allow glucose to enter cells. Type II diabetes, just like type I, results in high blood sugar levels that disrupt homeostasis. Type II diabetes is caused by obesity and may also have a genetic link. Unlike those with type I diabetes, individuals with obesity-related type II diabetes can reverse the disease with weight loss by controlling their diet and exercising. Insulin injections can help manage type II diabetes by stimulating the receptors that are still responding to insulin to take in more sugar. (Refer to Table 6.11.)

Table 6.11: Type I and Type II Diabetes

	Type I Diabetes	Type II Diabetes
Onset	Sudden	Gradual
Prevalence	Less common	More common
Age of onset	Mostly in young children	Usually in adults but due to childhood obesity becoming more common in children
Insulin levels	No insulin production or very low levels	Normal insulin or altered (higher or lower) levels
Insulin needs	Must have insulin injections for survival	Vary with individual
Treatment	Insulin injections/diet changes	Weight loss with diet and exercise/ may use insulin injections

Applications of Hormone Use

Hormones can be used to treat or detect disorders involving the endocrine system. Table 6.12 outlines a few of the current uses involving hormone therapy and the use of hormone levels as indicators of disorders of the endocrine system.

Table 6.12: Hormone Applications

Hormone	Source	Mode of Action	Application
Insulin	Beta cells of the pancreas	Lowers blood glucose levels	Used to treat type I diabetes
Thyroxin	Thyroid gland	Increases metabolic processes and thermo-regulation (increases body temperature)	Used as replacement therapy for patients with reduced thyroid function (hypothyroidism)
Leptin	Adipose (fat) tissue	Acts on the hypothalamus, inhibiting appetite	Used to help diagnose causes of obesity by testing levels of leptin in obese individuals
Melatonin	Pineal gland	Circadian rhythms (sleep cycles)	Used to treat jet lag

REPRODUCTION

MALE AND FEMALE REPRODUCTIVE SYSTEMS

The male and female reproductive systems are designed to produce, nourish and prepare gametes for fertilisation. The structures of the male and female reproductive systems are shown in Figures 6.22 and 6.23, respectively.

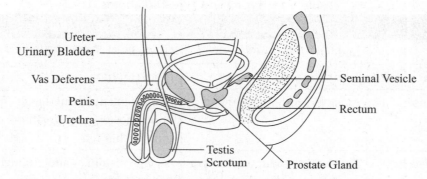

Figure 6.22 Male reproductive system

- The **scrotum** holds the testes outside of the body for optimal sperm development.
- **Seminiferous tubules** are located inside of the testes and are the site of spermatogenesis (sperm development).
- The **epididymis** is where the sperm are stored as well as where they gain motility (the ability to swim).
- The **prostate gland** is the largest gland that contributes to semen.
- The **seminal vesicles** secrete a fluid rich in fructose to provide energy for the sperm to function.
- The semen travels through the **urethra** and out of the male's body. Urine is also released through the urethra.

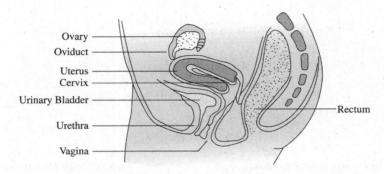

Figure 6.23 Female reproductive system

- The **ovaries** are the site of oogenesis (egg development).
- The **oviducts** (fallopian tubes) transport the eggs to the uterus and are the site where fertilisation takes place.
- The **uterus** is a muscular organ that is lined with the endometrium (lining of the uterus that will thicken and provide a place for the embryo to implant and grow).
- The **cervix** is the opening to the uterus. The menstrual blood (the endometrium) flows through the cervix when exiting the uterus. The baby exits from the uterus through the cervix and into the birth canal (**vagina**) when ready to be born.

HORMONES AND THE MENSTRUAL CYCLE

The menstrual cycle is under the control of several hormones, as shown in Table 6.13. Each one plays a distinctive role in the regulation of the monthly cycle. Follicle-stimulating hormone (FSH) and luteinizing hormone (LH) are released from the pituitary gland in response to gonadotropin-releasing hormone (GnRH) from the hypothalamus. FSH triggers the immature follicles in the ovary to begin to develop into mature eggs. Although both FSH and LH are released at the same time, the developing egg initially has receptors for only FSH and develops receptors for LH as it matures. One (or rarely more) immature follicles will be stimulated by FSH to develop into a mature egg and be released in response to LH around day 14 of the menstrual cycle. Oestrogen and progesterone are released from the ovaries and stimulate the thickening of the endometrium (oestrogen) and the maintenance of the endometrium (progesterone).

Hormonal Control of the Menstrual Cycle

Hypothalamus

Releases

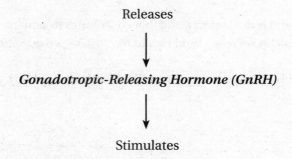

Gonadotropic-Releasing Hormone (GnRH)

Stimulates

Anterior Pituitary

Releases

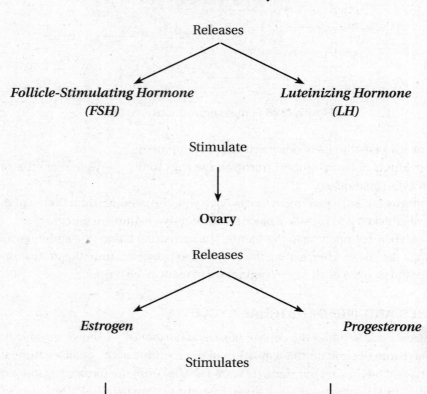

Follicle-Stimulating Hormone
(FSH)

Luteinizing Hormone
(LH)

Stimulate

Ovary

Releases

Estrogen

Progesterone

Stimulates

Thickening of the lining of the uterus

Table 6.13: Hormones of the Menstrual Cycle

Hormone	Source	Role
Follicle-stimulating hormone (FSH)	Pituitary gland (simulated by GnRH from the hypothalamus)	Development of immature follicles into mature oocytes (eggs)
Luteinizing hormone (LH)	Pituitary gland (stimulated by GnRH from the hypothalamus)	Release of the mature egg from the ovary, resulting in the formation of the corpus luteum
Oestrogen	Ovaries	Development (thickening) of the endometrium
Progesterone	Ovaries	Maintenance of the endometrium

Menstrual Cycle

Day 1 of the menstrual cycle is marked by the first day of bleeding, which usually lasts about 4 to 6 days. The days of bleeding are known as menstruation. FSH will stimulate the development of a mature egg between day 6 and day 14 (follicular phase). LH will stimulate ovulation (the release of a mature egg from the ovary) around day 14. After ovulation between days 14 and 28 (luteal phase), oestrogen continues to stimulate the development of the endometrium and progesterone keeps the endometrium from being shed. When the egg is released, follicles that originally surrounded the egg are left behind. These follicles are referred to as the corpus luteum. The corpus luteum is responsible for secreting progesterone that sends messages to the uterus to maintain the endometrium. The corpus luteum slowly disintegrates and eventually can no longer release progesterone. The lack of progesterone stimulates the shedding of the endometrium, marking the onset of menstruation. (Refer to Figure 6.24.)

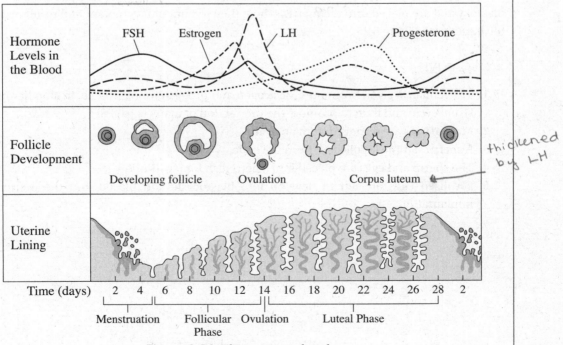

Figure 6.24. The menstrual cycle.

TESTOSTERONE

During embryonic development, a gene located on the Y chromosome causes the development of the testes from embryonic gonads. The hormone testosterone is released from cells in the testes known as Leydig cells. Luteinizing hormone (LH) is released at puberty and stimulates the Leydig cells to release testosterone. Testosterone serves several roles in males:

- Prenatal development of male genitalia
- Development of secondary sexual characteristics
- Maintenance of sex drive

OESTROGEN AND PROGESTERONE

Oestrogen and progesterone are responsible for the development of female reproductive organs during embryonic development. Oestrogen and progesterone are responsible for the development of secondary sexual characteristics in females at the time of puberty.

IN VITRO FERTILISATION (IVF)

In vitro fertilisation is fertilisation that occurs outside of the body in a laboratory environment. The female is given hormones that suspend the normal release of reproductive hormones. After suspension of the normal release of hormones, the female is given artificial doses of reproductive hormones to induce the development of many eggs (superovulation) at once. The ovaries are periodically checked with an ultrasound in order to determine when the eggs are ready for retrieval. The eggs are retrieved in the doctor's office with use of a guided needle that pierces the vaginal wall to reach the mature eggs. The mature eggs are placed into a nutrient solution along with collected sperm.

Fertilised eggs must be left in the laboratory to develop into multicellular embryos known as blastocysts before they can be placed into the uterus. Depending on how many eggs are retrieved and fertilised, 1 to 3 (usually not more) are used for possible implantation. The blastocyst(s) are placed into the uterus, where they will hopefully implant and result in a live birth. (Refer to Figure 6.25.)

Steps to IVF:

1. The female is given hormones (oestrogen and progesterone), initially to stop her menstrual cycle and then to promote the production of many eggs (FSH).
2. Mature eggs are retrieved from an ovary.
3. A sperm sample is collected from the male.
4. The sperm and eggs are placed into a petri dish for fertilisation.
5. Fertilised eggs are left to develop into blastocysts and placed into the uterus for implantation.

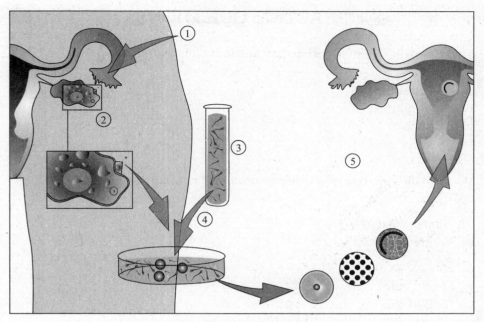

Figure 6.25. In vitro fertilisation

Ethical Issues and IVF

Many ethical issues surround in vitro fertilisation. The most heavily debated issue revolves around what will become of the embryos that were not chosen for implantation. Since multiple embryos may form during the process and not all viable embryos may be selected for implantation, embryos are often left for future use. Many couples choose not to use these embryos, and the fate of these embryos is a source of great controversy.

Scientists would like to use these embryos to study the embryonic stem cells they possess. Embryonic stem cells are capable of becoming an entire organism through the process of cell differentiation. How cells differentiate into many different tissue types is a scientific question that holds possible answers for questions about many human developmental processes. Using the embryonic stem cells for scientific research could lead to discoveries that could help in finding treatments and possibly cures for life-threatening developmental diseases and disorders.

Individuals who oppose the use of embryonic stem cells for research argue that a living human organism is being destroyed. Some people believe that the process is unnatural or against God's plan. In addition, some argue that the risks of the procedure to the female are dangerous to her health and that the process may create embryos that are not as healthy as naturally created embryos—although this has not been found to be true.

Embryos can also be screened for genetic disorders or screened to determine the sex of the embryo before implantation. The information gained from screening of embryos could potentially be used by couples to make decisions about the embryos they select for implantation. The ability to screen and select specific embryos for implantation over other available embryos is yet another highly debated ethical issue related to IVF.

1. The part of the digestive system that functions mainly for the absorption of water is the:
    A. Small intestine
    B. Large intestine
    C. Stomach
    D. Liver

2. Which of the following help in the absorption of nutrients in the small intestine?

    I.   Villi
    II. Protein channels
    III. Veins

    A. I only
    B. I and II only
    C. I and III only
    D. I, II, and III

3. Which of the following are transported in blood?

    I.   Nutrients
    II. Hormones
    III. Heat

    A. I only
    B. I and II only
    C. II and III only
    D. I, II and III

4. What is the result of HIV in the human body?

    A. Reduced number of erythrocytes
    B. Reduced number of thrombocytes (platelets)
    C. Reduced number of lymphocytes
    D. Increased number of lymphocytes

5. Which of the following correctly depicts the condition of the valves of the heart when the **ventricles** are contracting?

	Atrioventricular (AV) Valve	Semilunar valve
A.	Open	Closed
B.	Open	Open
C.	Closed	Open
D.	Closed	Closed

6. What are produced by the immune system that helps destroy pathogens?

   A. Antigens
   B. Antibodies
   C. Erythrocytes
   D. Platelets

7. What ion channels will open when a cell is undergoing depolarization?

   A. Sodium
   B. Potassium
   C. Calcium
   D. Magnesium

8. Which of the following are associated with type II diabetes?

   I. Obesity
   II. Loss of cell sensitivity to insulin
   III. Absence or low levels of insulin

   A. I only
   B. I and II only
   C. II and III only
   D. I, II and III

9. Where in the male reproductive system do sperm gain motility?

   A. Prostate gland
   B. Seminal vesicles
   C. Epididymis
   D. Urethra

10. What is the role of luteinizing hormone (LH) in females?

    A. Maintaining the endometrium
    B. Thickening of the endometrium
    C. Development of mature eggs
    D. Ovulation

11. Which cells produce insulin?

    A. Beta cells of the liver
    B. Beta cells of the pancreas
    C. Alpha cells of the liver
    D. Alpha cells of the pancreas

12. The male reproductive system and excretory system share which structure?

    A. Bladder
    B. Seminal vesicles
    C. Ureter
    D. Urethra

13. What is the condition of the external intercostal muscles and the diaphragm when inhalation is occurring?

	External Intercostal Muscles	Diaphragm
A.	Contracted	Relaxed
B.	Contracted	Contracted
C.	Relaxed	Relaxed
D.	Relaxed	Contracted

14. What is responsible for facilitating the release of neurotransmitters from the synaptic button into the synapse?

    A. The opening of sodium channels
    B. The opening of potassium channels
    C. The opening of calcium channels
    D. The closing of sodium channels

15. Which of the following shows the direction a nerve impulse travels in a neuron?

    A. From dendrite to terminal button
    B. From axon to dendrite
    C. From terminal button to dendrite
    D. From dendrite to synapse

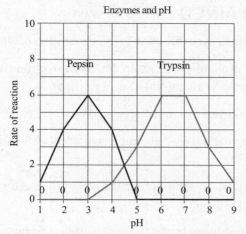

Enzymes and pH

The graph above shows the effect of pH on enzyme action.

16. (a) State the optimal pH for pepsin. (1)
    (b) Deduce the pH level in which both pepsin and trypsin would be able to have at least partial functional activity. (1)
    (c) Describe why pepsin no longer functions at pH levels above 5. (2)

17. State four substances that are transported in blood. (4)

18. State the components of the central nervous system (CNS). (2)

19. Describe three features of the alveoli that make them perfect for gas exchange. (3)

20. Outline the process of in vitro fertilisation. (6)

21. (a) Draw a motor neuron. (4)
    (b) Explain the principle of synaptic transmission. (8)
    (c) Outline homeostatic mechanisms that control body temperature. (6)

ANSWERS EXPLAINED

1. **(B)** The large intestine's main function is the absorption of water.

2. **(B)** Villi and protein channels function for the absorption of nutrients. The vessels involved are capillaries. Veins cannot absorb nutrients.

3. **(D)** Blood transports nutrients, hormones and heat.

4. **(C)** HIV infects lymphocytes (B-cells and T-cells), reducing their numbers in the human body.

5. **(C)** When the ventricles are contracting, the atrioventricular valves (AV) are closed and the semilunar valves are open.

6. **(B)** Antibodies are produced by the immune system and help destroy pathogens.

7. **(A)** In order for a cell to depolarize, the sodium channels must open.

8. **(B)** Type II diabetes is associated with obesity and loss of sensitivity to insulin. Absence or low levels or insulin are associated with type I diabetes.

9. **(C)** Sperm gain motility (the ability to swim) in the epididymis.

10. **(D)** Luteinizing hormone (LH) causes ovulation in females.

11. **(B)** Beta cells in the pancreas produce insulin.

12. **(D)** The male reproductive system and excretory system share the urethra (both semen and urine exit through the urethra).

13. **(B)** During inhalation (breathing in), the external intercostal muscles pull up and out (contract) and the diaphragm pulls down (contracts) in order to increase the volume of the chest cavity, decreasing pressure so air moves into the lungs.

14. **(C)** The opening of calcium channels facilitates the movement of calcium into the synaptic button. This causes the vesicles with the neurotransmitters to undergo exocytosis, releasing the neurotransmitter into the synapse.

15. **(A)** Action potentials travel from dendrites to the terminal buttons.

16. (a) pH level of 3 (highest peak in rate of reaction)
    (b) pH levels of 3–5 (overlap of bell curve)
    (c)
    - Enzyme denatures/loses its 3-D structure
    - Change in enzyme structure blocks its ability to catalyse reactions
    - Active site no longer fits with substrate

17.
    - Blood cells/or named type of blood cell (this answer can score a maximum of one point)
    - Heat
    - Nutrients
    - Carbon dioxide
    - Oxygen

- Minerals
- Hormones
- Salts
- Urea (waste)

18. Brain and spinal cord

19.

- Alveoli are close to capillaries so gases have to travel only a short distance. Alveoli are only one cell think, so diffusion can occur rapidly over a small area.
- Alveoli are moist, which facilitates diffusion.
- Alveoli are numerous, so a large surface area is available for gas exchange.
- Alveoli are flexible to become even thinner and closer to capillaries to facilitate gas exchange. (Each feature must be explained.)
- Type I pneumocytes in the alveoli are thin and flexible to facilitate gas exchange.
- Type II pneumocytes are thick and excrete surfactant to maintain moisture in the alveoli membranes/prevent the sides of alveoli from sticking together.

20.

- In vitro fertilisation is fertilisation that occurs outside the body/in a laboratory.
- The female is given hormones to halt her menstrual cycle prior to being given hormones to induce superovulation/production of many eggs.
- The female is given hormones/FSH to stimulate the production of many mature eggs via superovulation.
- Eggs are watched to determine when they reach maturation/blastocyst stage of development.
- Eggs are extracted.
- Eggs are put into nutrient medium with sperm.
- Usually at least 75,000 sperm per egg
- Fertilised eggs left to develop into embryos/blastocyst
- Embryos inserted into uterus
- Usually no more than 3 embryos inserted
- Can lead to multiple births
- Ethical issues exist with IVF/many feel IVF goes against natural processes
- IVF is used to treat infertility.
- Embryonic research using unused blastocyst/embryonic stem cells is an ethical concern.

21. (a) One point is awarded for each of the following if they are drawn and labelled correctly:
- Dendrites
- Axon
- Myelination/Schwann cells
- Terminal buttons
- Terminal branches

   (b)
- Action potential arrives at terminal button
- Calcium-gated ion channels open
- Calcium diffuses into the terminal button

- Vesicles containing neurotransmitter undergo exocytosis
- Neurotransmitter released into synapse
- Neurotransmitter binds to receptors on sodium channels
- Sodium channels open, causing sodium to enter the presynaptic cell/dendrite of next neuron/muscle cell
- Postsynaptic membrane reaches threshold/–50 mV (+/– 5)
- Action potential travels through postsynaptic cell
- Neurotransmitter undergoes reuptake/enzyme degradation
- One point is awarded for any named neurotransmitter (examples: dopamine, acetylcholine, GABA, serotonin)

(c)

- Hypothalamus controls body temperature/thermostat of the body is the hypothalamus
- Body temperature is detected to be too low, so the body will begin to shiver in order to produce heat to warm the body tissues. (You must explain why to score the point)
- Arterioles/blood vessels will constrict and send blood to the internal vital organs to keep them warm
- Body may curl up to prevent heat loss from exposed areas
- Body temperature is detected to be too low, so the body will begin to shiver in order to produce heat to warm the body tissues. Arterioles to the skin's surface will dilate, bringing more blood to the surface for cooling
- Body may spread out to release more heat because increased surface area is exposed

ADDITIONAL
HL MATERIAL

Nucleic Acids

CENTRAL IDEAS

- DNA structure
- DNA replication
- Protein synthesis (transcription and translation)
- Protein structure and function

TERMS

- Antiparallel
- Exons
- Introns
- Nucleosome
- Polysome

INTRODUCTION

The discovery of deoxyribonucleic acid (DNA) revolutionized the molecular world of biology. Since its discovery, scientists have been investigating and manipulating genes coded for in DNA. The impacts of DNA technologies are just beginning to be seen. The future of biotechnology developments based on DNA may help in the discovery, treatment, cure and prevention of many human diseases.

Deoxyribonucleic acid (DNA) contains all the molecular information needed for a cell to carry out life functions. Proteins are coded for in DNA and transcribed into RNA in order to carry out translation at the ribosomes. Transcription and translation are both needed for protein synthesis.

NOS Connections

- Careful observation and analysis of research lead to scientific understanding (discovery of DNA structure, protein synthesis).
- Cooperation and collaboration (local, national and international) are important in scientific endeavours (DNA discoveries).
- Technology aids in scientific research (mapping genomes, visualizing processes, comparing gene sequences).
- Patterns, trends and discrepancies exist in nature (nucleic acid structure, protein structure and organization).

NUCLEIC ACIDS

TOPIC CONNECTIONS

- 2.6 Structure of DNA and RNA

DNA Structure

Nucleic acids are all built from nucleotides that consist of a sugar, a phosphate and a base. The nucleotides that make up a strand of DNA are held together by phosphodiester bonds between the sugar of one nucleotide and the phosphate of another. The sugar in DNA is a 5-carbon (pentose) sugar. Each carbon is identified as 1 prime (1′) through 5 prime (5′). The first carbon in the central carbon ring of deoxyribose is identified as 1′, the second as 2′, the third as 3′, the fourth as 4′ and the carbon on the side chain of the ring above the 4′ carbon as 5′. Individual nucleotides are joined together to form a chain of alternating deoxyribose sugars and phosphates that make up the backbone of DNA. Phosphodiester bonds form between the 3′ carbon of one deoxyribose sugar and a central phosphate that is linked to the 5′ carbon of another deoxyribose sugar. The two strands of alternating deoxyribose sugars and phosphates that make up the backbone of DNA are held together by hydrogen bonds between the nitrogenous bases of the two strands. (See Figure 7.1).

On one side of a DNA strand, the 3′ carbon is at the top of the structure. On the other side, the same carbon is at the bottom of the structure. Based on the bonding of the top and bottom prime carbons, DNA is considered to run in a 3′ to 5′ direction on one strand and in a 5′ to 3′ direction on the opposite strand. Since the two strands run opposite of each other, DNA is referred to as being antiparallel, as shown in Figure 7.1.

Collaboration among scientists led to the discovery of the structure of DNA in the 1950s.

- James Watson and Francis Crick: Modelled the proposed shape of DNA with sticks and balls.
- Rosalind Franklin and Maurice Wilkins: Used X-ray diffraction to show the twisted nature of DNA.

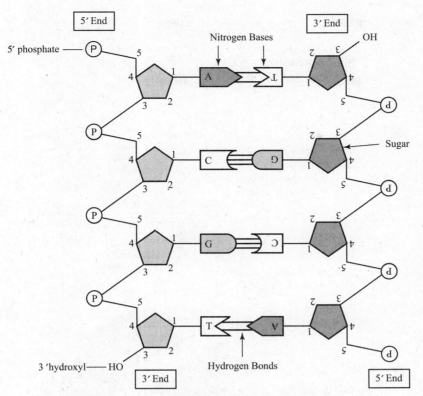

Figure 7.1. DNA structure showing antiparallel nature

NUCLEOSOMES

Since DNA is over three meters long it must be packaged into nucleosomes in order to fit into the nucleus of eukaryotic cells. Prokaryotic cells do not package their DNA since it is much shorter and easily fits inside the cytoplasm of the bacterial cell.

Each nucleosome consists of 8 core histone proteins stacked in a 4 by 2 arrangement with DNA wrapped around the core histones twice. In addition, a stabilizing histone is located outside the core structure and holds the DNA in place. Many nucleosomes occur on each strand of DNA, making DNA appear as "beads on a string" when viewed with an electron microscope. Nucleosomes further fold over onto each other as DNA becomes supercoiled. (See Figure 7.2.)

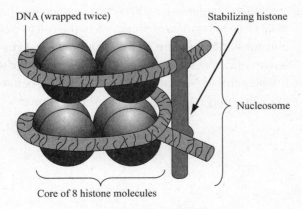

Figure 7.2. (a) Nucleosome

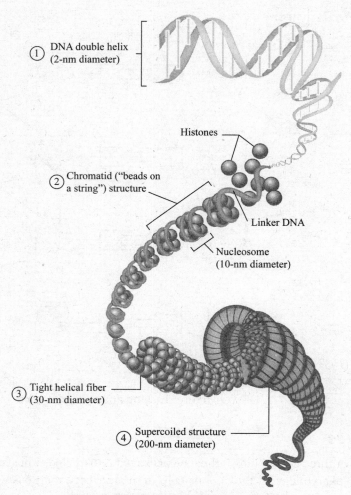

① DNA double helix
(2-nm diameter)

Histones

② Chromatid ("beads on a string") structure

Linker DNA

Nucleosome
(10-nm diameter)

③ Tight helical fiber
(30-nm diameter)

④ Supercoiled structure
(200-nm diameter)

Figure 7.2. (b) Supercoiling of DNA

Highly Repetitive Sequences (Tandem Repeats)

The base arrangements found in DNA are either single-copy genes (which code for proteins) or are noncoding, highly repetitive sequences (which do not code for proteins). The highly repetitive sequences are often referred to as satellite DNA or as tandem repeats. Most of the DNA of eukaryotic organisms consists of highly repetitive, noncoding regions. Although the function of these highly repetitive sequences is still being investigated, they are speculated to play a role in regulatory processes in the cell as well as in maintaining chromosome structure. Since each individual shows variation in the length of DNA tandem repeats, these repeats are used in DNA profiling. Repetitive sequences typically contain between 5 and 300 base pairs that may repeat up to half a million times!

REMEMBER

Regions of DNA that Do Not Code for Proteins:

- Regulators of gene expression
- Introns
- Telomeres (the ends of chromosomes)
- Genes that code for tRNA

EXONS AND INTRONS

Eukaryotic genes contain both coding and noncoding regions. Both regions undergo transcription. This results in a piece of pre-mRNA that must undergo splicing before leaving the nucleus in order to remove the introns. Mature mRNA (with introns removed) exits the nucleus to bind with a ribosome and carry out the process of translation in order to produce a protein. Since prokaryotic cells lack introns, they do not require splicing.

REMEMBER

Exons are **ex**pressed and **ex**it the nucleus.

Introns are not expressed and remain **in** the nucleus. (Note that introns are removed from mature mRNA by splicing).

DNA Replication

During DNA replication, the 5′ end of a free nucleotide is always added to the 3′ end of an existing DNA strand. Since the DNA strands run in opposite directions (see Figure 7.3) and DNA replication must occur in a 5′ to 3′ direction, replication occurs in opposite directions on each strand of DNA.

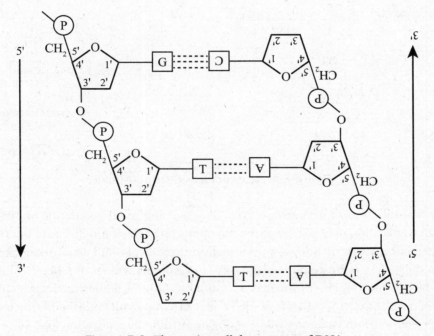

Figure 7.3. The antiparallel structure of DNA

DNA replication begins when the enzyme helicase unwinds the DNA strands and breaks the hydrogen bonds between base pairs, exposing the complementary bases of DNA. Then the enzyme DNA gyrase relieves strain on the DNA strand so helicase can more easily unwind and open the DNA strand. Once the strands are exposed, RNA primase lays down a primer so that DNA polymerase III can begin adding nucleotides. Each nucleotide added to the original strand of DNA begins as a deoxynucleoside triphosphate as shown in Figure 7.4. Each deoxynucleoside triphosphate loses 2 phosphates in order to become a nucleotide. The energy released from breaking the 2 phosphate bonds is used to form a new phosphodiester bond that binds the nucleotide to the growing strand.

NUCLEIC ACIDS

REMEMBER

Helicase hacks and
ligase links!

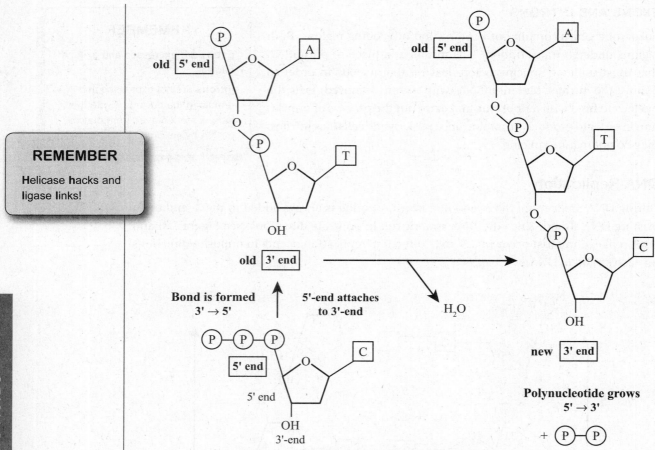

Figure 7.4. Deoxynucleoside triphosphate

Since replication must occur in a 5′ to 3′ direction and the DNA strands run in opposite directions, replication occurs in opposite directions on each strand. Because the DNA polymerase III that is acting on both strands is really one molecule with two working parts, it can function only when both strands are in fairly close contact with each other. On the leading strand, the DNA polymerase III works in a continuous fashion, adding nucleotides in a 5′ to 3′ direction towards the replication fork. However, as DNA polymerase III adds nucleotides on the lagging strand, it must do so while working away from the replication fork. Once it moves a certain distance, the DNA polymerase III will be "pulled" back to the replication fork and has to start over. This occurs due to the pull of the DNA polymerase III moving forward on the leading strand. Each time DNA polymerase III starts over on the lagging strand, a new primer must be laid down by RNA primase. The result of this discontinuous replication is the formation of small sections of replicated DNA nucleotides known as Okazaki fragments. Okazaki fragments must be joined together by phosphodiester bonds. DNA ligase forms phosphodiester bonds between the adjacent nucleotides of the Okazaki fragments, thereby creating a continuous strand of DNA.

DNA polymerase I replaces the RNA primers that were laid down by RNA primase on both DNA stands with DNA nucleotides. This creates two new strands of DNA, each containing half of the original strand. This is called semiconservative replication. (Refer to Figure 7.5 and Table 7.1.)

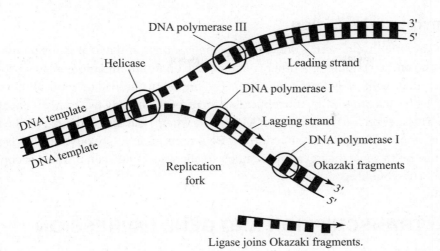

Figure 7.5. DNA replication fork

Table 7.1: Enzymes of Replication

Enzyme	Function
Helicase	Unwinds DNA strand and breaks hydrogen bonds between complementary bases
Gyrase (DNA gyrase)	Relieves strain on the DNA strand to facilitate the unwinding and separation of DNA strands by helicase
RNA primase	Lays down the RNA primers on the leading and lagging strands
DNA polymerase III	Adds nucleotides to the growing DNA strands
DNA polymerase I	Excises the RNA primers and replaces them with DNA nucleotides
DNA Ligase	Forms phosphodiester bonds between the nucleotides of the Okazaki fragments

FEATURED QUESTION

May 2008, Question #6

Which events take place in DNA replication?

  I. Formation of messenger RNA
 II. Unwinding of the DNA double helix
III. Formation of complementary strands by DNA polymerase

A. I and II only
B. I and III only
C. II and III only
D. I, II and III

(Answer is on page 554.)

EUKARYOTIC REPLICATION

Since eukaryotic chromosomes are very long, replication is initiated at many points in order to copy the entire genome in a timely manner. Each point of initiation creates two replication forks that work away from each other. The numerous individual replication "bubbles" meet up with each other when the replication forks meet, creating two newly replicated DNA strands. The sites on the replicating DNA molecule where helicase is opening up the DNA are known as origins of replication. In contrast, DNA replication in prokaryotic cells is initiated at only one point since their genome is much smaller in size and can be quickly copied with one replication bubble.

7.2 TRANSCRIPTION AND GENE EXPRESSION

TOPIC CONNECTIONS

- 2.7 DNA replication, transcription and translation

Transcription is the process of encoding the information stored in DNA into a strand of mRNA. The process of transcription occurs inside the nucleus of eukaryotic cells and in the cytosol of prokaryotic cells. During transcription, new nucleotides are always added to the 3′ end of the existing DNA strand, making the mRNA transcript run in a 5′ to 3′ direction.

Sense and Antisense Strands

During the process of transcription, only one strand of DNA serves as the mRNA template. The strand that is transcribed is referred to as the antisense strand. The strand that is not transcribed is referred to as the sense strand. (Refer to Figure 7.6.) The DNA strand that runs 3′ to 5′ is the antisense strand since during transcription, nucleotides must be added to it in order to obey the 5′ to 3′ directional rule. Due to complementary base pairing, the newly copied strand actually matches the nontranscribed strand. This is why the nontranscribed strand is referred to as the sense strand.

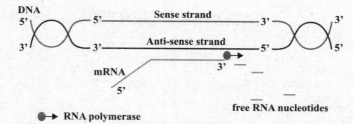

Figure 7.6. DNA sense and antisense strands

Cells must be able to identify the location of each specific gene within the entire genome in order to locate and transcribe a particular gene when needed. This is accomplished by incorporating specific arrangements of nucleotides (bases) that identify where a gene transcript starts and where it ends. The nucleotide arrangement that identifies the location for the transcript of a gene is referred to as the promoter region. The nucleotide arrangement that signals the end of the gene transcript is referred to as the terminator.

Once the location of a desired gene is identified, RNA polymerase attaches to the promoter region and begins to unwind the section of DNA that contains the transcript for the gene. Helicase is not needed for unwinding DNA and breaking the hydrogen bonds for transcription.

The mRNA transcript is synthesized with RNA polymerase using nucleoside triphosphates. Each nucleoside triphosphate loses 2 phosphates, releasing energy that is used to form new phosphodiester bonds in the growing strand of mRNA. After the 2 phosphates have been removed, a nucleoside triphosphate becomes the nucleotide that will be added to the elongating mRNA molecule. When RNA polymerase reaches the terminator, transcription is complete and the newly formed mRNA molecule is released from the antisense strand. Hydrogen bonds reform between the complementary bases of DNA, leaving DNA unchanged after the process of transcription is complete. Figure 7.7 illustrates the process of transcription.

REMEMBER

Promoters and terminators are examples of noncoding regions of DNA that do have a function.

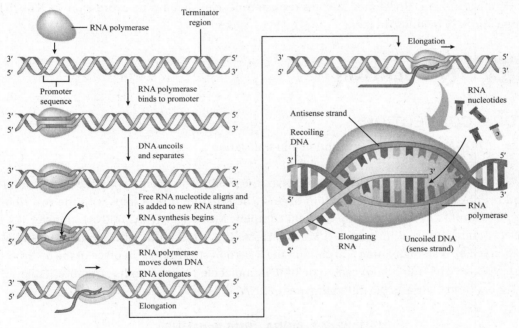

Figure 7.7. Transcription

The process of transcription is nearly identical in prokaryotic and eukaryotic cells. One main difference is that transcription occurs in the cytoplasm of prokaryotic cells since these cells do not have a nucleus. Another unique feature is that translation of the newly synthesized mRNA transcript can begin in a prokaryotic cell before transcription is complete since the two processes both occur in the cytoplasm. In addition, since prokaryotic cells lack introns, there is no need for splicing (removal of introns) prior to beginning translation. Since eukaryotic DNA contains introns that are copied during transcription, these introns must be removed (spliced) before translation.

Nucleosomes and Transcription

As described earlier, nucleosomes act as the packaging unit of DNA by helping to supercoil chromosomes. However, nucleosomes also play a central role in regulating transcription.

NUCLEIC ACIDS

Nucleosomes do not coil those areas of DNA that need to be transcribed. Instead, they leave these areas exposed, allowing transcription to be carried out more easily.

If a gene is located in a part of the DNA wrapped around the core histones, it must first be exposed (uncoiled from the nucleosome) before transcription can occur. Hence, selective packaging of DNA in nucleosomes serves as a method of transcription regulation. Nucleosomes can form, open, and then re-form on new sections of DNA in order to control which genes are exposed for transcription. The particular genes exposed vary based on the specific needs of the cell.

Gene expression can be controlled by the presence of proteins that bind to specific sequences of nucleotides in DNA. These proteins are known as transcription factors. They impact the workings of RNA polymerase, the formation of nucleosomes and other processes involved in transcriptional processes. In addition, both the internal and the external environment of a cell or an organism impact the rate of gene expression. Some mammal species, such as the Artic Fox, express different fur colour during their lifetime based on the temperature of their environment. Hormones, such as testosterone, can signal gene expression such as the appearance of baldness in men.

7.3 TRANSLATION

TOPIC CONNECTIONS

- 2.7 DNA replication, transcription and translation

Translation follows transcription and involves the production of a polypeptide based on the codons present on mRNA. Translation begins when mature mRNA joins with the two ribosomal subunits to form a transcriptional complex. Messenger RNA (mRNA), transfer RNA (tRNA) and ribosomal RNA (rRNA) are all involved in translation. All RNA molecules contain uracil instead of thymine and contain the 5-carbon sugar ribose in place of the 5-carbon deoxyribose. The function of each type of RNA molecule is based on its structure. Table 7.2 compares and contrasts the different types of RNA.

Table 7.2: mRNA, tRNA and rRNA

Type of RNA	Process Involved In	Function
Messenger RNA (mRNA)	Transcription	Carries the genetic code for protein synthesis (copies the code from DNA) Each triplet base pair (codon) codes for one amino acid
Transfer RNA (tRNA)	Translation	Brings the appropriate amino acid to the ribosome based on matching the tRNA anticodon to the mRNA codon
Ribosomal RNA (rRNA)	Translation	Along with proteins, rRNA makes up ribosomes and catalyses the process of protein synthesis at the ribosome

Each transfer RNA molecule is folded into an upside-down cloverleaf shape with an amino acid attached to the 3' end. Transfer RNA (tRNA) activating enzyme adds the amino acids to the 3' end of each tRNA using ATP as energy to form the bond holding the amino acid in place. The base triplet CCA is located at the terminal end of each tRNA, allowing the enzyme to recognize the bonding site. The central fold in the tRNA contains the anticodon that will match up with its complementary mRNA codon during translation. This ensures that the correct amino acid (out of the 20 different amino acids) is brought to the ribosome at the right location in order to form a specific polypeptide. (Refer to Figure 7.8.)

REMEMBER

The tRNA molecule is another example of how structure relates to function!

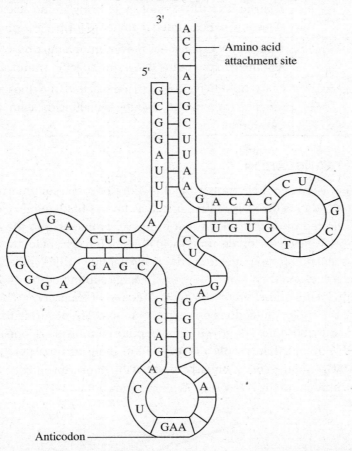

Figure 7.8. Transfer RNA (tRNA). ATP is the energy used to add the amino acid. The enzyme catalysing this addition is the tRNA activating enzyme.

Ribosome Structure

Ribosomes consist of two subunits (large and small) that are each composed of ribosomal RNA (rRNA) and protein. The two ribosomal subunits exist as separate units in the cytoplasm until they join with mRNA in order to carry out translation. A molecule of mRNA joins with a ribosome at a binding site on the ribosome, thereby forming the perfect structure for protein synthesis.

The large ribosomal subunit has three binding sites for tRNA molecules to bind to as they deliver and drop off amino acids. The binding sites are the E (exit) site, the P (peptidyl) site and the A (aminoacyl) site. The tRNA moves from the A to the P to the E site. From the E site, it is then released back into the cytoplasm to pick up a new amino acid.

Steps of Translation

Translation is divided into four steps—initiation, elongation, translocation and termination. Each step of translation is marked by specific processes that occur during the different phases of protein synthesis. Table 7.3 describes each step and the processes involved.

Table 7.3: Steps of Translation

Steps of Translation	Processes Involved
Initiation	Ribosomal subunits join with mRNA in the cytoplasm.
Elongation	Transfer RNA (tRNA) molecules bring amino acids to the ribosome to be bonded together to form a growing polypeptide.
Translocation	Transfer RNA molecules move through the ribosome from the A to the P to the E site, delivering specific amino acids.
Termination	Once the stop codon has been reached, the ribosomal subunits separate from mRNA, thereby signalling the completion of translation.

PROCESS OF TRANSLATION

Translation begins when an mRNA transcript joins with ribosomal subunits in the cytoplasm of a cell. The first codon is the start codon. It signals for the first amino acid to be brought into the P site followed by another tRNA that brings an amino acid to the A site. The start codon is always AUG and codes for the amino acid methionine. Except for the first tRNA arriving at the ribosome and binding directly to the P site, all other tRNA molecules enter the A site, move to the P site, and then to the E site on the large subunit of the ribosome. This A to P to E movement occurs in the required 5′ to 3′ direction of translation. The triplet codons on mRNA match the triplet anticodons on each tRNA molecule, ensuring the correct amino acid is brought to the ribosome. The growing polypeptide continues to elongate until the stop codon is reached. Although there is only one start codon, there are three possible stop codons: UAA, UGA and UAG. The stop codon signals the end of the transcript and does not code for an amino acid to be brought to the ribosome. (Refer to Figure 7.9.)

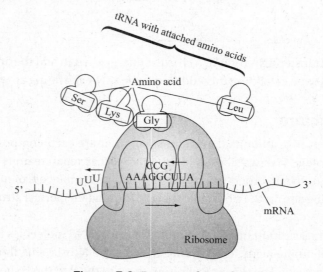

Figure 7.9. Process of translation

NUCLEIC ACIDS

> **Ribosome Binding Sites**
>
> **A site** (aminoacyl-tRNA accepting site): A tRNA molecule carrying a single amino acid binds to the site and accepts the growing polypeptide chain from the tRNA in the P site before moving from the A site to the P site.
>
> **P site** (peptidyl-tRNA accepting site): Holds a tRNA carrying the growing polypeptide chain before the polypeptide is passed to the tRNA in the A site. Once the polypeptide chain is passed to the tRNA in the A site, the now-empty tRNA moves from the P site to the E site.
>
> **E site** (exit site): Holds the empty tRNA until the tRNA is released from the ribosomal complex.

DIPEPTIDE FORMATION

Condensation reactions occur between amino acids as they are joined by peptide bonds at the ribosome. The amine group of one amino acid loses hydrogen (H^+), and the carboxyl group of another amino acid loses a hydroxyl group (OH^-). A peptide bond forms between the available nitrogen and the available carbon atom of each of the amino acids. Water (H_2O) forms as a by-product.

POLYSOMES

When a cell needs to make a lot of one single type of protein (polypeptide), polysomes form. These polysomes make many polypeptides from one mRNA transcript. Once the first ribosome has moved far enough away from the start codon along the mRNA, another ribosome attaches to the transcript and begins translation as well. Ribosomes keep attaching to the mRNA and keep producing more polypeptides as long as the signal for producing more of the same protein is being received. Pancreatic cells form many polysomes in order to produce the hormones insulin and glucagon and to produce many of the digestive enzymes needed by the body. Figure 7.10 illustrates the function of a polysome.

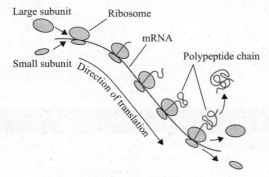

Figure 7.10. Polysome

Free and Bound Ribosomes

In eukaryotic cells, translation can occur on a free ribosome or on a bound ribosome. A free ribosome is not attached to the rough endoplasmic reticulum (rough ER). A bound ribosome is attached to the rough ER. If a ribosome is attached to the rough ER, the proteins it synthesizes are sent into the interior of the rough ER, where they are modified and packaged into a vesicle for export out of the cell (exocytosis). The rough ER also produces lysosomes. Lysosomes package and hold digestive enzymes that remain in the cytoplasm. Packaging digestive enzymes into these vesicles ensures that the enzymes will not damage cellular components and will be ready for use when they are needed. However, if the cell is producing proteins for immediate use within the cell, protein synthesis occurs on a free ribosome.

Protein Structure

Following protein synthesis at the ribosome, the protein must undergo structural changes in order for it to carry out its function. There are four levels of protein structure, each involving unique bonding patterns. How a protein folds is based on the composition of the R-groups of the amino acids in the primary chain. Each level of protein folding is outlined in Table 2.8.

GLOBULAR AND FIBROUS PROTEINS

Depending on their function, proteins may exist as either globular or fibrous structures. Globular proteins are spherical in shape, have a compact structure, usually carry out metabolic processes, have a complex tertiary structure, often have a quaternary structure and are soluble in water. Fibrous proteins are linear in shape, have an extended structure, rarely have a complex tertiary structure and usually perform structural roles in the body. Table 7.5 lists the functions of several globular and of several fibrous proteins.

Table 7.5: Globular and Fibrous Proteins

Globular Proteins	
Example	Function
Haemoglobin	Transports oxygen
Enzymes (amylase, helicase, etc.)	Catalyse reactions
Antibodies	Provide protection by targeting antigens
Fibrous Proteins	
Example	Function
Keratin	Gives strength to hair and nails
Collagen	Acts as connective tissue
Thrombin	Involved in blood clotting

Polar and Nonpolar Amino Acids

The R-groups of amino acids vary in structure, allowing proteins to express both polar (hydrophilic) and nonpolar (hydrophobic) properties. Channel proteins must consist of both polar and nonpolar amino acids in order to span the nonpolar interior of the plasma membrane and to exist within the polar cytoplasm. In addition, enzymes must be able to catalyse reactions that involve both polar and nonpolar substrates. An enzyme that catalyses a reaction involving a polar substrate must have polar amino acids in the structure of the active site in order to catalyse the reaction. An enzyme that catalyses reactions involving a nonpolar substrate must have nonpolar amino acids in the structure of the active site in order to catalyse the reaction. The structure of the amino acids at the active site of the enzyme ensures an attraction between the enzyme and substrate in order to catalyse a specific reaction efficiently.

Protein Functions

- **Catalysing reactions:** For example, enzymes such as amylase
- **Transport:** For example, haemoglobin transports oxygen
- **Communication (hormones):** For example, insulin
- **Movement:** For example, actin and myosin in muscle cells
- **Protection:** For example, antibodies (immunoglobins)
- **Blood clotting:** For example, thrombin and fibrin

1. Which of the following are functions of nucleosomes?

   I. Supercoil chromosomes
   II. Regulate transcription
   III. Store ribosomes

   A. I only
   B. I and II only
   C. II and III only
   D. I , II and III

2. Which of the following shows the correct function of each enzyme?

	Helicase	DNA Polymerase I	DNA Ligase
A.	Unwinds DNA and breaks hydrogen bonds between complementary bases	Adds nucleotides to the DNA strand	Links Okazaki fragments
B.	Adds nucleotides to the DNA strand	Unwinds DNA and breaks hydrogen bonds	Replaces the RNA primers
C.	Unwinds DNA and breaks hydrogen bonds between complementary bases	Replaces the RNA primers with DNA	Links Okazaki fragments
D.	Adds nucleotides to the DNA strand	Replaces the RNA primers with DNA	Unwinds DNA and breaks hydrogen bonds between complementary bases

3. During transcription, the nontranscribed strand is known as the:

   A. Sense strand
   B. Antisense strand
   C. Nonsense strand
   D. Activating strand

4. Splicing involves the removal of:

   A. Introns
   B. Exons
   C. Genes
   D. Nucleosomes

NUCLEIC ACIDS

5. Which of the following shows the correct direction of each process?

	Replication	Transcription	Translation
A.	5′ to 3′	5′ to 3′	3′ to 5′
B.	3′ to 5′	3′ to 5′	3′ to 5′
C.	5′ to 3′	3′ to 5′	5′ to 3′
D.	5′ to 3′	5′ to 3′	5′ to 3′

6. Which of the following is/are a function(s) of ribosomes attached to the rough endoplasmic reticulum?

   I. Synthesizing proteins for export
   II. Synthesizing proteins for use within the cell
   III. Production of lysosomes

   A. I only
   B. I and II only
   C. I and III only
   D. I, II, and III

7. Which of the following help in regulating gene expression in eukaryotic cells?

   A. DNA ligase
   B. Transcription factors
   C. Plasmids
   D. Nucleotides

8. Which enzyme relieves tension on the DNA strand to allow helicase to work more efficiently?

   A. DNA polymerase I
   B. DNA ligase
   C. DNA gyrase
   D. DNA polymerase II

9. The process of translation occurs in the:

   A. Cytosol
   B. Nucleus
   C. Cell membrane
   D. Nucleolus

10. The term polysome refers to:

    A. Many tRNA molecules attaching to a single ribosome.
    B. Many mRNA molecules attaching to a single ribosome.
    C. Many ribosomes attaching to a single mRNA molecule.
    D. Many amino acids attaching to a single tRNA molecule.

11. Which of the following enzymes are needed for transcription of DNA?

    I. DNA polymerase
    II. Helicase
    III. RNA polymerase

    A. I only
    B. I and II only
    C. II and III only
    D. III only

12. Posttranscriptional processing of eukaryotic RNA includes:

    A. Addition of introns
    B. Removal of introns
    C. Addition of exons
    D. Removal of exons

13. Which of the following statements is correct concerning prokaryotic and eukaryotic DNA replication?

    A. DNA replication in prokaryotic cells is initiated at many sites, and eukaryotic replication is initiated at only one site.
    B. DNA replication in eukaryotic cells is initiated at many sites, and prokaryotic replication is initiated at only one site.
    C. Both prokaryotic and eukaryotic DNA replication occur in the nucleus.
    D. Both prokaryotic and eukaryotic DNA replication occur in the cytosol.

14. The average population sizes of periwinkles inhabiting the marsh grass at a state park were calculated in 5-meter quadrants. Six population estimates were calculated at distances of 5, 10, 15, 15, 25, and 30 meters from the boat landing dock. The graph below shows the results of the population sampling.

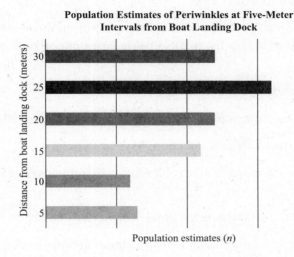

Population Estimates of Periwinkles at Five-Meter Intervals from Boat Landing Dock

Population estimates (*n*)

(a) State the distance at which the greatest number of periwinkles was found. (1)

(b) Calculate the range for the mean population estimate of the periwinkles. (2)

(c) Suggest reasons for the distribution of periwinkles at the different intervals from the boat landing dock. (2)

15. Compare the roles of DNA polymerase I and DNA polymerase III. (2)

16. State 4 functions of proteins. Name an example of each. (4)

17. Outline splicing. (2)

18. (a) Draw tRNA. (4)

(b) Compare DNA and RNA. (4)

(c) Outline protein structure. (6)

ANSWERS EXPLAINED

1. **(B)** Nucleosomes supercoil chromosomes and help regulate transcription. Genes located in areas between nucleosomes are more easily transcribed than those wound around the nucleosomes (histones).

2. **(C)** Helicase unwinds DNA and breaks hydrogen bonds between the bases; DNA polymerase I replaces the RNA primers. DNA ligase forms phosphodiester bonds between Okazaki fragments.

3. **(A)** The sense strand is the nontranscribed strand.

4. **(A)** Splicing removes introns (non-coding regions) in eukaryotic cells.

5. **(D)** All processes run in a 5′ to 3′ direction.

6. **(C)** Ribosomes attached to the rough ER synthesize proteins for export and produce lysosomes.

7. **(B)** Transcription factors regulate the process of transcription.

8. **(C)** DNA gyrase relieves tension on the DNA strand in order to allow helicase to work more efficiently.

9. **(A)** Translation occurs in the cytosol of the cell.

10. **(C)** A polysome consists of many ribosomes translating off of a single mRNA strand.

11. **(D)** Only RNA polymerase is needed for transcription.

12. **(B)** Posttranscriptional processing of mRNA includes splicing, which is the removal of introns.

13. **(B)** DNA replication in eukaryotic cells is initiated at many sites, and prokaryotic replication is initiated at only one site.

14. (a) 25 meters (+/− 1) units required.
    (b) 32 − 12 = 20 range of periwinkles (+/− 2). The calculation must be shown.
    (c)
    - Less human intervention away from boat landing dock
    - Less pollution away from boat landing dock
    - Less disturbance from waves produced by boats
    - At the greatest distance, the population drop may be due to less turbulence in water, leading to less nutrient movement/change in soil and therefore resulting in lower levels of grass marsh

15.
    - DNA polymerase I replaces RNA primers with DNA.
    - DNA polymerase III lays down nucleotides, producing complementary strands.
    - DNA polymerase I replaces many more RNA primers on the lagging strand than on the leading strand.
    - Both enzymes are used during replication.

16.

Function of Protein	Example
Transport	Haemoglobin transports oxygen around the body.
Protection	Antibodies help destroy free antigens in the body.
Communication	Some hormones, such as insulin, help tissues communicate.
Movement	Actin and myosin in muscle cells function for movement.
Structural support	Keratin in skin and nails adds support and strength. Collagen helps support tissues.

17.

- Splicing involves the removal of introns from pre-mRNA to make mature mRNA.
- Splicing occurs in eukaryotic cells only/prokaryotic cells do not have introns.
- Splicing occurs after transcription but before mRNA leaves the nucleus.
- Introns are noncoding regions.

18. (a)

- One point is awarded for the correct cloverleaf shape
- 3′ end holding amino acid
- ATP energy for adding amino acid to 3′ end
- Anticodon on middle of cloverleaf as a triplet base
- 5′ and 3′ ends labelled correctly
- CCA as last triplet base pairs on 3′ end

(b)

- DNA is double stranded. RNA is single stranded.
- DNA contains the sugar deoxyribose. RNA contains the sugar ribose.
- DNA contains the bases adenine, thymine, cytosine and guanine. RNA contains the bases adenine, uracil, cytosine, and guanine.
- DNA codes for all the genes. RNA codes for one gene at a time.

(c)

- Proteins are composed of amino acids.
- Amino acids are joined together by peptide bonds.
- Chains of amino acids (polypeptides) are synthesized at the ribosome.
- Primary structure is a chain of amino acids held together by peptide bonds.
- Secondary structure involves the polypeptide forming alpha helixes or beta pleats.
- Secondary structure is due to hydrogen bonding.
- Tertiary structure forms when the secondary structure folds over on itself.
- Disulphide bonds are responsible for tertiary structure.

- R-groups determine the folding of proteins.
- Quaternary structure involves two or more polypeptides joining together due to many types of bonds.
- Quaternary structure may involve the addition of prosthetic groups.
- Award one point for naming an example of a protein with a quaternary structure (for example, haemoglobin).

Metabolism, Cellular Respiration and Photosynthesis

CENTRAL IDEAS
- Metabolic processes
- Cellular respiration
- Photosynthesis

TERMS
- Activation energy
- Carboxylation
- Chemiosmosis
- Decarboxylation
- Oxidation
- Reduction

INTRODUCTION

Enzymes are biological catalysts that speed up the rate of reactions in living organisms. Photosynthesis relies on enzymes to catalyse carbon fixation and to produce organic compounds. Cellular respiration relies on enzymes to facilitate the release and transfer of energy stored in organic molecules. Without enzymes, processes such as photosynthesis and respiration would not be able to efficiently carry out energy transfer and transformation. A cell must be able to regulate enzyme activity in order to control metabolic pathways such as those involved in photosynthesis and respiration.

NOS Connections
- Technology allows scientists to model scientific investigations and also enhances scientific research (bioinformatics, modelling and investigating metabolic pathways).
- Collaboration among scientists leads to greater understanding of the scientific world (metabolic pathways involved in photosynthesis and respiration, such as chemiosmosis).

8.1 METABOLISM

TOPIC CONNECTIONS

- 2.7 DNA replication, transcription and translation

Metabolic Pathways

In order for metabolic processes to occur fast enough to support homeostasis, enzymes are needed. Enzymes catalyse metabolic pathways. These pathways consist of chains and cycles of reactions. Although enzymes bind to a specific substrate in order to catalyse a reaction, some fluctuation at the active site allows enzymes to bind to more than one substrate. The original thought was that each enzyme catalysed a reaction with only one substrate. This is known as the lock-and-key model since it is analogous to how one key opens only one lock. The discovery that some enzymes can act on more than one substrate led to the induced-fit model of enzyme-substrate specificity. Both the enzyme and the substrate can slightly modify their shapes in order to catalyse multiple reactions as long as the substrates are similar in shape.

Enzyme Action

Enzymes lower the activation energy of cells, which is the energy needed to facilitate the start of a reaction. Enzymes accomplish this by stressing and bending the bonds involved in the reaction to make them easier to break or to form as the reaction proceeds. Regardless of whether the reaction is exothermic (releases energy) or endothermic (absorbs energy), enzymes always lower the activation energy. The graph in Figure 8.1 shows the same reaction both with and without a catalyst. Notice the difference in the activation energy. Without an enzyme present, reactions would not occur fast enough to support homeostasis in the body.

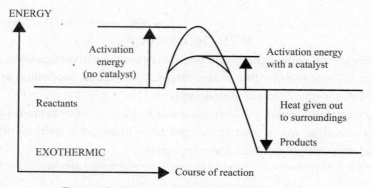

Figure 8.1. Enzymes and activation energy

COMPETITIVE AND NONCOMPETITIVE INHIBITION

The rate that enzyme-catalysed reactions occur can be regulated through both competitive and noncompetitive inhibition. Competitive inhibition occurs when a molecule structurally similar to the substrate competes with and binds to the active site of the enzyme. This temporarily inhibits the substrate from binding and slows the rate of the reaction. Noncompetitive inhibition occurs when a molecule that is not structurally similar to the substrate binds to another site on the enzyme (not the active site), changing the shape of the active site. This other site is referred to as the allosteric site. The binding of the molecule to the allosteric site on the enzyme alters the entire three-dimensional structure of the enzyme, including the shape of the active site. The new shape of the active site prevents substrate binding and slows the reaction rate. (Refer to Figure 8.2 and Table 8.1.)

Substrate binding and inhibitor binding (both competitive and noncompetitive) are not permanent. The substrate or the inhibitor attach to and are released from the active site. This allows the site to alternate between being available for catalysing a reaction and being inhibited from catalysing a reaction. The ratio of inhibitor to substrate concentration determines the rate at which the substrate can bind. The presence of more substrate helps to overcome the effect of both competitive and noncompetitive inhibition by ensuring the substrate has a higher probability of binding to the active site.

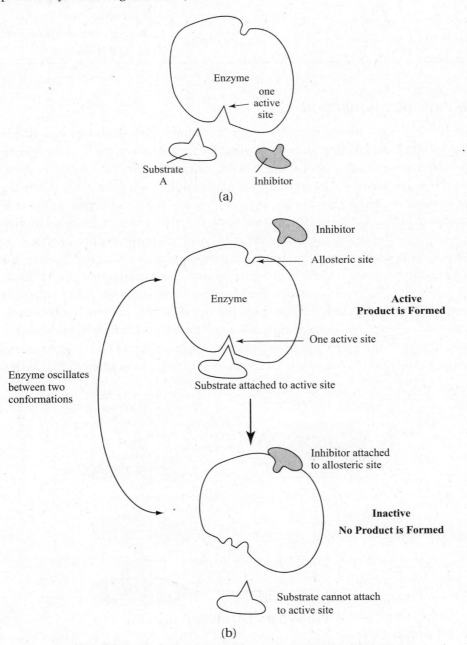

Figure 8.2. (a) Competitive and (b) noncompetitive inhibition.

Table 8.1: Examples of Competitive and Noncompetitive Inhibition

Type of Inhibition	Example	Effect
Competitive	Malonate competes with succinate for the active site on a dehydrogenase enzyme.	Since succinate is an intermediate in the Krebs cycle, malonate competitively slows the process of cellular respiration.
Noncompetitive	ATP binds to the enzyme phosphofructokinase.	Phosphofructokinase is an enzyme that catalyses the phosphorylation of glucose in order for it to enter glycolysis. The presence of ATP stops glycolysis from continuing until more ATP is needed.

END-PRODUCT INHIBITION

Allosteric sites are sites on enzymes where noncompetitive inhibitors can attach. When inhibitors attach to allosteric sites, their presence puts stress on the enzyme and alters the shape of the active site. The altered shape of the active site inhibits the substrate from attaching to the active site of the enzyme. End-product inhibition occurs when the final product in an enzyme-catalysed pathway allosterically inhibits the first enzyme in the reaction as illustrated in Figure 8.3. Regardless of how many enzymes and substrates are involved in an enzyme-catalysed pathway, the final product always inhibits the very first enzyme involved in the pathway in end-product inhibition. A great example of allosteric end-product inhibition is ATP noncompetitively inhibiting phosphofructokinase and stopping glucose from undergoing glycolysis. This process ensures that glycolysis occurs only when ATP is needed. When the cell is not using ATP, the ATP builds up and is available to inhibit glycolysis from occurring allosterically and noncompetitively. Allosteric inhibition of phosphofructokinase by ATP inhibits glycolysis and keeps cells from using glucose to produce ATP. The glucose is stored as glycogen and released by hydrolysis for use in cellular respiration when ATP is needed again.

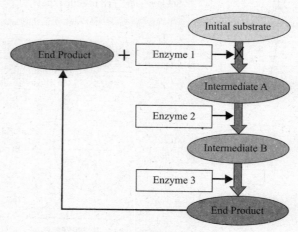

Figure 8.3. End-product inhibition

Another self-regulatory process of end-product inhibition involves the production of the amino acid isoleucine. Isoleucine can be produced through a series of enzymatic reactions that begin with the amino acid threonine. Threonine enters the metabolic pathway to produce isoleucine when isoleucine levels are low. When isoleucine levels rise, enough of the molecule is available to bind to threonine deanimase and noncompetitively inhibit the production of more isoleucine. This end-product (isoleucine) inhibition is reversed when the isoleucine levels drop due to a cell's use of isoleucine in metabolic reactions.

8.2 CELLULAR RESPIRATION

TOPIC CONNECTIONS

- 2.8 Cellular respiration

Oxidation and Reduction

Oxidation and reduction are processes that frequently occur simultaneously during metabolic pathways. Oxidation involves the gain of oxygen during a chemical reaction and reduction involves the loss of oxygen during a chemical reaction. In addition, oxidation involves the loss of elections while reduction involves the gain of electrons. During oxidation hydrogen is lost and during reduction hydrogen is gained. (See Table 8.2)

> Use OIL RIG to remember what is occurring with elections and hydrogen during chemical pathways involving oxidation and reduction.
>
> **O**xidation **I**s a **L**oss (OIL)
>
> **R**eduction **I**s a **G**ain (RIG)

Table 8.2: Oxidation and Reduction

Particle Lost or Gained	Oxidation	Reduction
Electron	Lost	Gained
Hydrogen	Lost	Gained
Oxygen	Gain	Lost

Glycolysis

Glycolysis is an anaerobic process that occurs in the cytoplasm of cells. (Refer to Figure 8.4.) It involves the conversion of one hexose (6-carbon) sugar, usually glucose, into two 3-carbon compounds (pyruvate). The lysis of glucose releases energy that produces a net gain of 2 ATP molecules and 2 NADH (reduced nicotinamide adenine dinucleotide) molecules. In order for glucose to enter into glycolysis, the glucose molecule must first be phosphorylated. This simply means that phosphates are added to the glucose molecule. Phosphorylation makes glucose less stable and facilitates glycolysis. The addition of phosphates to glucose requires energy supplied by 2 ATP. After phosphorylation, glucose is split into 2 pyruvate molecules, resulting in the production of 4 ATP. Although 4 ATP are produced, the net gain is only 2 ATP per glucose molecule due to the investment of 2 ATP molecules for the phosphorylation of glucose. After glucose is split, the resulting two 3-carbon molecules are oxidized, releasing hydrogen atoms that are used to reduce 2 NAD^+ (nicotinamide adenine dinucleotide) molecules to NADH.

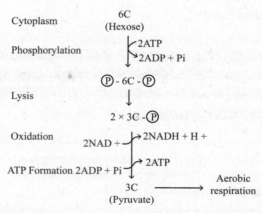

Figure 8.4. Glycolysis: phosphorylation, lysis and oxidation

Aerobic Respiration—Krebs Cycle

If oxygen is available, the 2 pyruvate molecules produced in glycolysis enter the matrix of the mitochondria to be broken down further, yielding more ATP, NADH and $FADH_2$. The first process that occurs in the matrix of a mitochondrion is known as the link reaction. In the link reaction, pyruvate is decarboxylated, forming an acetyl molecule and releasing a molecule of CO_2 and yielding 1 NADH molecule.

The resulting 2-carbon acetyl molecule reacts with coenzyme A and forms the 2-carbon molecule acetyl-CoA. Acetyl-CoA enters the Krebs cycle when it bonds with a 4-carbon compound to form a 6-carbon compound. The 6-carbon compound is decarboxylated into a 5-carbon compound that is then decarboxylated back into a 4-carbon compound. This 4-carbon compound is the same molecule that originally joined with acetyl-CoA. The energy molecules resulting from each acetyl-CoA that enters the Krebs cycle include 1 ATP, 1 $FADH_2$, and 3 NADH. $FADH_2$ is another energy-carrying coenzyme, just like NADH. Remember that 2 acetyl-CoA molecules are produced from each glucose molecule, so the net gain in energy is doubled. The total net gain from 1 glucose molecule (2 acetyl-CoA molecules) in the Krebs cycle is 2 ATP, 6 NADH and 2 $FADH_2$. (Refer to Figure 8.5.)

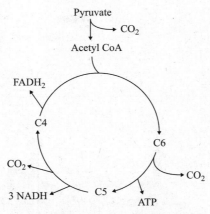

Figure 8.5. Krebs cycle

Energy Molecules Gained During the Krebs Cycle

When 2 acetyl-CoA molecules from the 2 pyruvate molecules produced during glycolysis enter the link reactions, the following molecules are produced:

- 2 ATP
- 6 NADH
- 2 FADH$_2$

Aerobic Respiration—Electron Transport Chain

The coenzymes NADH and FADH$_2$ that were reduced in the Krebs cycle will be oxidized in order to release hydrogen ions (protons) and electrons. The electrons will travel through the intermembrane of the mitochondria through the electron transport chain (ETC). As the electrons travel through the ETC, they move from a high-energy state to a low-energy state. Simultaneously, the electrons release energy as they travel down the electron transport chain from protein to protein. These proteins are called cytochromes. The released energy is used to power proton pumps in the intermembrane of a mitochondrion. Proton pumps move the hydrogen ions from the matrix into the intermembrane space from an area of low to an area of high concentration. This creates a concentration gradient between the matrix and the intermembrane space. The hydrogen ions pass through the integral protein ATP synthase as they diffuse back into the matrix from high to low concentrations. This is an excellent example of facilitated diffusion. The flow of hydrogen ions through ATP synthase creates energy that is used to phosphorylate ADP and create anywhere from 32 to 34 ATP. (Refer to Figure 8.6.)

The electrons resulting from the oxidation of NADH and FADH$_2$ that have travelled through the electron transport chain join with oxygen and 2 hydrogen ions (protons). For this reason, oxygen is known as the final electron acceptor. Oxygen helps maintain the proton concentration gradient required for chemiosmosis by accepting 2 hydrogen ions (protons) and forming water. The resulting water molecule (oxygen + 2 protons + 2 electrons) is known as metabolic water since it is produced during metabolic processes. (Refer to Figure 8.7.)

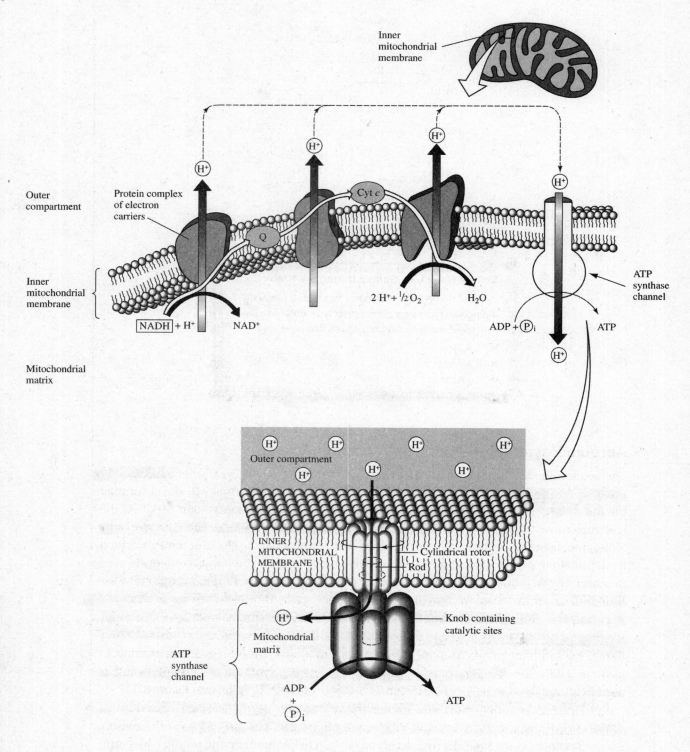

Figure 8.6. The electron transport chain, proton gradient, and ATP synthase

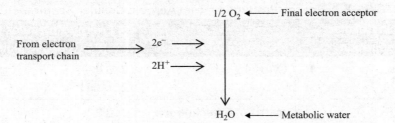

Figure 8.7. Acceptance of electrons and protons by water

CHEMIOSMOSIS

The oxidation (loss of electrons and protons) of NADH and $FADH_2$ provides both the electrons that will travel through the ETC and the protons that will be pumped into the intermembrane space. The electrons travel from a high-energy level to a low-energy level as they move from cytochrome to cytochrome (from protein to protein) in the ETC. Energy released as the electrons travel is used to produce ATP by phosphorylating ADP. The resulting ATP provides the energy for the hydrogen pumps in the inner mitochondrial membrane. The hydrogen pumps move hydrogen protons (H^+) from an area of low concentration in the matrix into an area of high concentration in the intermembrane space. This creates a concentration gradient that leads to the movement of the protons (H^+) back into the matrix (from a high to a low concentration) though ATP synthase as they undergo facilitated diffusion. The facilitated diffusion of the protons through ATP synthase is referred to as chemiosmosis. Chemiosmosis produces 32 to 34 ATP per original glucose molecule. The production of the 32 to 34 ATP is known as oxidative phosphorylation since the oxidation of NADH and $FADH_2$ provides the energy (electrons and protons) that are ultimately used to produce ATP. (Refer to Figure 8.8.)

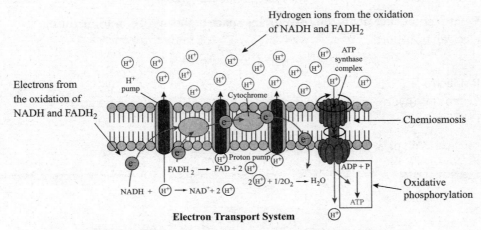

Figure 8.8. Oxidative phosphorylation and chemiosmosis

Each glucose molecule that enters into aerobic cellular respiration results in the production of an estimated 36 to 38 ATP. Table 8.3 summarizes cellular respiration and ATP formation.

Table 8.3: Energy Production from Aerobic Cellular Respiration

Step in Cellular Respiration	Location of Process	Energy Molecules Gained (Per Glucose Molecule)
Glycolysis	Cytosol	2 ATP 2 NADH
Krebs cycle	Matrix of mitochondria	2 ATP 6 NADH 2 $FADH_2$ *(1 ATP, 3 NADH and 1 $FADH_2$ per acetyl-CoA)
Electron transport chain (ETC)	Intermembrane of mitochondria	32 to 34 ATP

Total ATP gained per glucose molecule entering aerobic cellular respiration (see Figure 8.9):

Glycolysis (2) + Krebs cycle (2) + ETC (32 to 34) = 36 to 38 ATP/glucose

FEATURED QUESTION

May 2004, Question #28

What accumulates in the intermembrane space of the mitochondrion during electron transport?

A. ATP
B. Electrons
C. Protons (hydrogen ions)
D. Oxygen

(Answer is on page 554.)

METABOLISM/CELL RESPIRATION

Two concepts will help you understand the process of energy conversion in cellular respiration. When a chemical bond is formed, energy is stored. When a chemical bond is broken, energy is released. For example, the formation of ATP and of ADP stores energy. The lysis of glucose and of pyruvate releases energy.

A charge moving in only one direction in a contained area releases energy. The moving charge can be positive (protons, H^+) or negative (electrons). For example, electrons moving through the ETC release energy. Protons moving through ATP synthase also release energy.

This concept is readily seen with electricity. Electrons moving in one direction through insulated wires release energy, which is called electricity. In biological systems, the type of charge does not matter. Both moving electrons and moving protons release energy. This movement of charged particles is the key to energy transfer in cellular respiration.

Phosphorylation: The addition of a phosphate—the phosphorylation of ADP to produce ATP.

- *Substrate-level phosphorylation:* Phosphorylation using energy provided by the lysis of a molecule (substrate)

  □ **Glycolysis:** Lysis of glucose provides the energy for phosphorylation of ADP.
  □ **Krebs cycle:** Decarboxylation of molecules provides energy for the phosphorylation of ADP.

- *Oxidative phosphorylation:* Phosphorylation using energy provided by the loss of electrons and hydrogen ions (oxidation) from a molecule

  □ **ETC:** Oxidation of NADH and $FADH_2$ provides the energy to phosphorylate ADP during chemiosmosis.

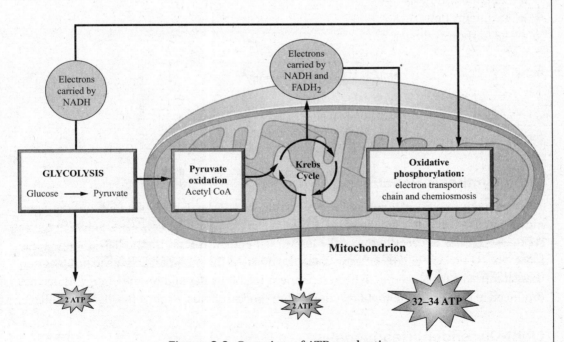

Figure 8.9. Overview of ATP production

Mitochondrion Structure and Function

Since structure always relates to function, the structure of a mitochondrion is perfectly designed to carry out the organelle's related functions. (Refer to Table 8.4.) The presence of two membranes forms a space in which a concentration gradient can be built with protons. This space in the mitochondria allows the process of chemiosmosis to occur (through ATP synthase) as the protons move from the intermembrane space into the matrix. The folds in the intermembrane (cristae) greatly increase the surface area, allowing space for the electron transport chain. The fluid in the matrix provides a medium for all the enzymes needed for the Krebs cycle.

Table 8.4: Mitochondrion Structure and Function

Structure	Function
Presence of 2 membranes	Separation of space to allow proton accumulation and chemiosmosis
Presence of folds in intermembrane (cristae)	Increased surface area for electron transport chain (ETC)
Presence of fluid inside of the matrix	Contains enzymes for the Krebs cycle

8.3 PHOTOSYNTHESIS

TOPIC CONNECTIONS

- 2.9 Photosynthesis
- 4.2 Energy flow
- 4.3 Carbon cycling

> **Photon**: packet of light energy
>
> **Synthesis**: to build
>
> **Photosynthesis**: to build organic molecules using energy from the sun (photons)

Photosynthesis Reactions

The two main processes involved in photosynthesis are the light-dependent and the light-independent (Calvin cycle) reactions. The light-dependent reactions use light energy to produce ATP and NADPH that fuel the light-independent reactions. The light-independent reactions use the ATP and NADPH produced by the light-dependent reactions to produce carbohydrates (sugar). Since the light-independent reactions depend on the energy molecules produced in the light-dependent reactions, both processes must occur when light is available.

Light-Dependent Reactions

The light-dependent reactions of photosynthesis occur in the thylakoid membrane of the chloroplast. (Refer to Figure 8.10.) Two photosystems (collections of pigments) are located in the thylakoid membrane. The photosystems contain pigments that can absorb radiant

L▷ Chorophyll

energy (light energy) and convert it into chemical energy. The main photosynthetic pigment is chlorophyll. However, many accessory pigments are located in the photosystems as well. Accessory pigments absorb light energy at slightly different wavelengths than chlorophyll, so their presence enhances the amount of radiant energy that can be absorbed. The accessory pigments pass the energy they absorb to chlorophyll. Chlorophyll is the only pigment that can absorb enough energy to donate electrons to the electron transport chain (ETC) in order to create a moving charge. In the fall, when chlorophyll breaks down and is no longer produced by the plant, the leaves "change colours". The leaves are not really changing colour. The loss of chlorophyll allows you to see the accessory pigments since chlorophyll is no longer the dominant pigment. Since the accessory pigments cannot lose electrons to fuel the ETC, the coloured leaves will not last long. Plants must have chlorophyll in order to donate the electrons that fuel ATP and NADPH production.

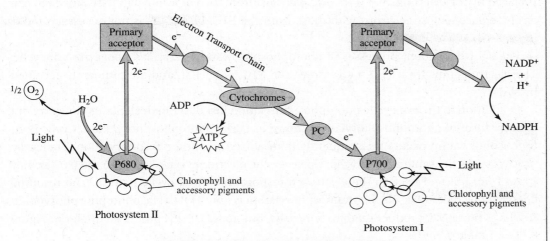

Figure 8.10. Light-dependent reactions

Photosystem II (PSII) absorbs wavelengths most efficiently in the 680 nm range and is often referred to as P680. Photosystem I absorbs wavelengths most efficiently in the 700 nm range and is often referred to as P700. PSII is the first photosystem involved in the light-dependent reactions, followed by photosystem I. Although this doesn't seem to make sense, the photosystems are named for the order in which they were discovered. The term photoactivation is used to describe the process of radiant energy being absorbed by the photosystems, activating chlorophyll and ultimately causing that chlorophyll to lose electrons.

Photosystem II absorbs radiant (light) energy that strikes Earth in packets of radiant (light) energy known as photons. The energy absorbed by chlorophyll causes the molecule to lose 2 electrons. The 2 electrons lost by chlorophyll are sent to a primary electron acceptor (membrane protein) and down the electron transport chain.

The electrons lost from PSII travel through the ETC from cytochrome to cytochrome (protein to protein). As the electrons move from a high-energy level to a low-energy level, energy is released. The energy released by the travelling electrons is used to phosphorylate ADP and produce ATP in the process of photophosphorylation.

> ***Photophosphorylation***: The production of ATP from ADP using radiant energy from *pho*tons.

METABOLISM, CELLULAR RESPIRATION AND PHOTOSYNTHESIS 267

The electrons lost from chlorophyll are replaced by the photolysis of a water molecule. Recall that water is polar, possessing both a negative and a positive region (pole) within the molecule.. Since chlorophyll has lost electrons, it has taken on a positive charge that pulls on the negative pole of the water molecule. The "pull" of chlorophyll on water causes the water molecule to split into ½ of an O_2 molecule (since oxygen is always diatomic), 2 electrons, and 2 protons (H^+). The oxygen is released as a by-product. The electrons released from the water are transferred to the chlorophyll, replacing the 2 that were lost.

Meanwhile, photons of light are also striking photosystem I (PSI). The chlorophyll in PSI loses 2 electrons to a second electron transport chain. The electrons from PSI are ultimately accepted by nicotinamide adenine dinucleotide phosphate ($NADP^+$). Nicotinamide adenine dinucleotide phosphate ($NADP^+$) accepts 2 electrons and a proton (H^+) to form NADPH. The NADPH has been reduced because it gained a hydrogen ion. The electrons lost by PSI are replaced by the electrons that were originally lost from the chlorophyll in PSII. Since PSI can easily replace its lost electrons with those from the ETC, photolysis is not necessary and is associated only with PSII.

The light-dependent processes of photophosphorylation described above are noncyclic. The electrons flow from PSII, move down the ETC, enter PSI, and are ultimately accepted by $NADP^+$.

In addition to the noncyclic photophosphorylation just described, all plants can carry out another form of photophosphorylation known as cyclic photophosphorylation. Cyclic photophosphorylation occurs less frequently than noncyclic photophosphorylation. The cyclic process starts the same as noncyclic. However, the electrons "cycle" through the ETC as they travel from PSI and back to the electron transport chain in a cyclic pattern. The resulting energy molecule from cyclic photophosphorylation is only ATP. Cyclic photophosphorylation results in more ATP production than noncyclic, but no NADPH is produced. (Refer to Figure 8.11 and Table 8.5.)

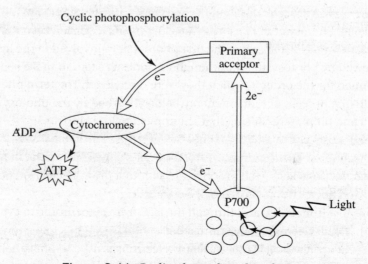

Figure 8.11. Cyclic photophosphorylation

Table 8.5: Photophosphorylation

Noncyclic Photophosphorylation	Cyclic Photophosphorylation
Normal pathway	Rare pathway
Involves PSII and PSI	Involves only PSI
Produces both ATP and NADPH	Produces more ATP than does noncyclic and produces NO NADPH

Both ATP and NADPH are needed by the light-independent reactions. However, more ATP than NADPH is used in the light-independent reactions. It is believed that cyclic photophosphorylation begins when the plant has enough NADPH but needs more ATP. The cyclic reactions keep the ratios of these energy molecules at the right level for cellular processes.

Phosphorylation And Chemiosmosis

The ATP produced by the ETC is actively used to move hydrogen ions from the stroma into the thylakoid lumen though proton pumps. In other words, energy is needed to accomplish this. The result is a proton gradient between the stroma and the inner thylakoid space (lumen). The protons flow back into the stroma through the process of chemiosmosis facilitated by the integral protein ATP synthase.

Although it seems like a waste of energy to use ATP to pump hydrogens into the thylakoid space in order to drive ATP formation with chemiosmosis, the enzyme ATP synthase can produce much more ATP than was invested in fuelling the proton pump. Except for the location where chemiosmosis occurs, the process is identical in both photosynthesis and cellular respiration.

Light-Independent Reactions

The ATP and NADPH produced in the light-dependent reactions are the sources of energy for the light-independent reactions. The phosphate bond in ATP and the hydrogen bond in NADPH are broken. The energy released is used to fuel the light-independent reactions. The light-independent reactions use that energy to form carbon bonds while producing carbohydrates.

The light-independent reactions (often referred to as the Calvin cycle) occur in the stroma of the chloroplast. A 5-carbon molecule of ribulose bisphosphate (RuBP) joins with a 1-carbon molecule of carbon dioxide. This reaction is catalysed by the enzyme ribulose bisphosphate (RuBP) carboxylase (commonly called RuBisCO). The process is known as carbon fixation. The resulting 6-carbon compound (5-C RuBP + 1-C CO_2) quickly splits into two 3-carbon molecules of glycerate-3-phosphate (GP).

> **Carbon fixation**: The process of converting inorganic (atmospheric) carbon into organic compounds.

Glycerate-3-phosphate (GP) is rearranged into two 3-carbon molecules of triose phosphate (TP). Rearranging molecules takes energy. The energy for this process comes from ATP and NADPH supplied by the light-dependent reactions. Three carbon dioxide (CO_2) molecules are needed to produce 1 TP molecule. This should make sense since there are 3 carbons in 1 molecule of TP. Most of the TP will go on to regenerate RuBP. Only 1 out of every 6 molecules of TP produced will be used for carbohydrate production.

As shown in Figure 8.12, the three phases of the light-independent reaction include:

1. Carbon fixation
2. Reduction of GP to TP
3. Regeneration of RuBP

Insert similar figure with noted changes. Label the first molecule on the right (3-phosphoglycerate) as glycerate 3-phosphate and omit the second one (1, 3 bisphosphoglycerate) and label the glyceraldehydes-3-phosphate as triose phosphate to match IB terms. Leave the ATP and NADPH use. Replace Calvin cycle with light-dependent reactions.

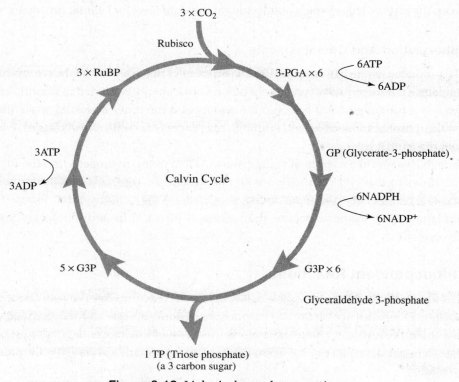

Figure 8.12. Light-independent reactions

Chloroplast Structure and Function

Since structure always relates to function, the structure of the chloroplast is perfectly designed to carry out the related functions. The presence of three membranes results in space where a concentration gradient can be built with protons. The space in the chloroplast also allows chemiosmosis to occur (through ATP synthase) as the protons move from the inner thylakoid space into the stroma. The stacks of thylakoid greatly increase the surface area allowing space for the photosystems and the ETC. The fluid in the stroma provides a medium for all the enzymes needed for the light-independent reactions. Table 8.6 summarizes this information.

Table 8.6: Chloroplast Structure and Function

Structure	Function
Presence of 3 membranes	Separation of space to allow proton accumulation and chemiosmosis
Presence of multiple stacked membranes (thylakoids)	Increased surface area for photosystems and electron transport chain (ETC)
Presence of fluid in the stroma	Contains enzymes for the light-independent reactions

Factors Limiting Photosynthesis

Light intensity, temperature and the concentration of carbon dioxide in the environment all limit the rate of photosynthesis. As light intensity increases, the rate of photosynthesis also increases. More available light means more absorption of photons that leads to more chlorophyll molecules being photoactivated. This increases the amount of ATP and NADPH available for the light-independent reactions. However, at high light intensities, other factors begin to affect the rate of photosynthesis. The rate of the photosynthesis reactions plateau (level off) as shown in Figure 8.13.

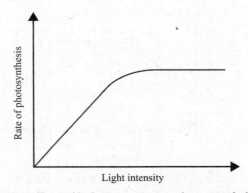

Figure 8.13. The effect of light intensity on the rate of photosynthesis

At very low temperatures, molecular reactions are too slow to support photosynthesis. This occurs because molecules move very slowly at very low temperatures and therefore rarely collide. As the temperature increases, more reactions occur because molecules move faster and collide more often. The rate of photosynthesis increases. At very high temperatures, though, enzymes that catalyse photosynthetic processes begin to denature. This causes photosynthesis to slow and eventually stop. (Refer to Figure 8.14.)

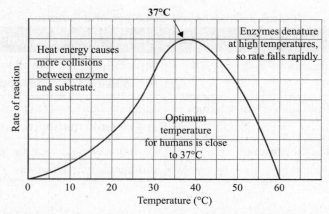

Figure 8.14. The effect of temperature on the rate of photosynthesis

The rate of photosynthesis increases with increasing levels of carbon dioxide. Higher levels of carbon dioxide allow the enzyme RuBP carboxylase (RuBisCO) to catalyse carbon fixation more rapidly. However, the enzyme eventually reaches a maximum rate (where all of the active sites are maximally occupied) of catalytic ability, causing the rate to level off as shown in Figure 8.15.

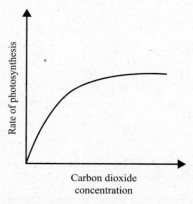

Figure 8.15. The effect of carbon dioxide concentration on the rate of photosynthesis

1. What is the correct order of events in the process of glycolysis?

    A. Lysis, oxidation, phosphorylation
    B. Oxidation, lysis, phosphorylation
    C. Phosphorylation, lysis, oxidation
    D. Phosphorylation, oxidation, lysis

2. Which of the following are products of glycolysis?

    I. ATP
    II. Pyruvate
    III. Acetyl-CoA

    A. I only
    B. I and II only
    C. II and III only
    D. I, II and III

3. What processes are carried out in BOTH respiration and photosynthesis?

    I. Phosphorylation
    II. Chemiosmosis
    III. Photoactivation

    A. I only
    B. I and II only
    C. II and III only
    D. I, II and III

4. Which graph represents the effect of light intensity on the rate of photosynthesis?

    A.

    C.
    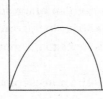

    B.

    D.
    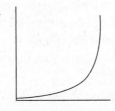

5. Which structure of the mitochondrion allows for the establishment of a proton gradient?

    A. The presence of a fluid in the matrix.
    B. The presence of multiple membranes.
    C. The presence of ribosomes.
    D. The presence of the exterior cytosol.

6. The light-independent reactions occur in the:

    A. Inner membrane of the chloroplast
    B. The stroma of the chloroplast
    C. Cytosol of the cell
    D. Thylakoid space of the chloroplast

7. How many ATP are produced per glucose molecule as a direct result of the Krebs cycle?

    A. 2
    B. 4
    C. 6
    D. 32

8. The final electron acceptor in cellular respiration is:

    A. Hydrogen
    B. Pyruvate
    C. Oxygen
    D. Carbon dioxide

9. During cellular respiration, protons (H^+) diffuse from the:

    A. Matrix to the intermembrane space
    B. Cytosol to the intermembrane space
    C. Intermembrane space to the cytosol
    D. Intermembrane space to the matrix

10. What best describes the process of oxidation?

    A. NADPH losing a hydrogen ion to form $NADP^+$
    B. $NADP^+$ accepting a hydrogen ion to form NADPH
    C. Oxygen accepting electrons
    D. Carbon dioxide joining with RuBP

11. The following diagram shows end-product inhibition. Which enzyme-catalysed step will be inhibited by the final product?

    A -------- Product B ------- Product C ------- Product D ------ Final Product E
        ↑   ↑   ↑   ↑
        A.   B.   C.   D.

12. Which of the following are products of the light-dependent reactions?

    I. Carbon dioxide
   II. ATP
 III. NADPH

   A. I only
   B. I and II only
   C. II and III only
   D. I, II and III

13. Which of the following are correctly matched for oxidation and reduction?

	Oxidation	Reduction
A.	Loss of oxygen	Gain of oxygen
B.	Gain of oxygen	Gain of hydrogen
C.	Loss of oxygen	Loss of hydrogen
D.	Gain of hydrogen	Loss of hydrogen

14. What is the name of the enzyme that catalyses carbon fixation?

   A. ATP synthase
   B. ADP synthase
   C. RuBP carboxylase
   D. RuBP decarboxylase

15. Which locations for the processes of carbon fixation and the ETC are correctly matched?

	Carbon Fixation	ETC
A.	Stroma	Stroma
B.	Thylakoid membrane	Stroma
C.	Stroma	Thylakoid membrane
D.	Thylakoid membrane	Thylakoid membrane

16. The graph below shows the amount of oxygen consumption by germinating and non-germinating seeds at 12 degrees Celsius and at 22 degrees Celsius.

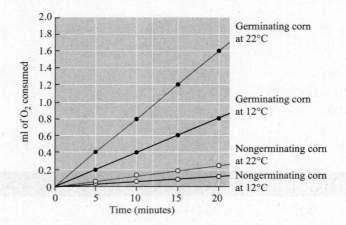

(a) Describe the relationship between temperature and oxygen consumption. (2)

(b) Calculate the difference between the rate of oxygen consumption in germinating and in nongerminating seeds at 20 minutes and 12 degrees Celsius. (2)

(c) Deduce why germinating seeds may use more oxygen than nongerminating seeds. (2)

(d) Predict the rate of oxygen consumption at 22 degrees Celsius at 25 minutes. (1)

17. Explain chemiosmosis in terms of ATP production. (3)

18. State two methods of inhibiting the rate of enzyme-catalysed reactions. (2)

19. Explain why light intensity is a limiting factor for the rate of photosynthesis. (4)

20.

(a) Draw the structure of the mitochondrion. (4)

(b) Explain the processes involved in the Krebs cycle. (6)

(c) Explain the concept of limiting factors in relation to photosynthesis. (8)

ANSWERS EXPLAINED

1. **(C)** Phosphorylation occurs first, followed by lysis and finally oxidation.

2. **(B)** Two ATP and pyruvate are both products of glycolysis. Acetyl-CoA is produced during the link reaction (after glycolysis) and enters into the Krebs cycle.

3. **(B)** Phosphorylation and chemiosmosis both occur in photosynthesis as well as in cellular respiration. Photoactivation refers to the photosystems absorbing photons during photosynthesis.

4. **(A)** As light intensity increases, so will the rate of photosynthesis. However, the reaction rate will reach a plateau due to other limiting sources, such as the amount of CO_2 and the temperature.

5. **(B)** The presence of two membranes allows for the separation of space in order to develop a concentration gradient for protons.

6. **(B)** The light-independent reactions occur in the stroma of the chloroplast.

7. **(A)** Two ATP are produced in the Krebs cycle (1 ATP per pyruvate with 2 pyruvate per glucose).

8. **(C)** Oxygen is the final electron acceptor in cellular respiration. Oxygen accepts 2 electrons and 2 protons, forming metabolic water.

9. **(D)** Protons diffuse (via facilitated diffusion) from the intermembrane space into the matrix through ATP synthase.

10. **(A)** Oxidation involves a loss of either protons or electrons. Oxidation can instead involve a gain of oxygen. NADPH losing a hydrogen ion (H^+) is an example of oxidation.

11. **(A)** In end-product inhibition, the final product always inhibits the action of the first enzyme in the metabolic pathway.

12. **(C)** ATP and NADPH are products of the light-dependent reactions.

13. **(B)** Oxidation involves a gain of oxygen, and reduction involves a gain of hydrogen.

14. **(C)** RuBP carboxylase catalyses carbon fixation.

15. **(C)** Carbon fixation (light-independent reactions) occurs in the stroma, and the ETC is located in the thylakoid membrane.

16. (a)
    - As temperature increases, the rate of oxygen consumption increases.
    - At higher temperatures/22 degrees Celsius, the rate is much higher than at lower temperatures/12 degrees Celsius.
    - One point is awarded for calculating the difference at a set time.
    - The relationship is linear.

    (b) $0.8 - 0.1 = 0.7$ mL difference in the production of oxygen (units required)

    (c)
    - Germinating seeds are carrying out cellular respiration at rates high enough to support the processes of germination.

- Nongerminating seeds are carrying out photosynthesis as well as respiration but at lower levels of respiration than when germinating.
- Germinating seeds must carry out cellular respiration in order to make energy for use in producing early plant structures.

(d) Near 1.8 mL of oxygen (+/− 0.1)

17. One point is awarded for any of the following statements:
    - Chemiosmosis occurs in both the mitochondria and the chloroplast.
    - Chemiosmosis is a type of facilitated diffusion.
    - Chemiosmosis involves protons moving from the intermembrane space to the matrix in cellular respiration.
    - In photosynthesis, chemiosmosis involves protons moving from the intermembrane space to the stroma.
    - Chemiosmosis is passive.
    - Chemiosmosis occurs through the protein ATP synthase.
    - Energy from the moving protons is used to phosphorylate ADP, producing ATP.

18.
    - Competitive inhibition
    - Noncompetitive inhibition
    - End-product inhibition

19.
    - As light intensity (and therefore temperature) increases, the rate of photosynthesis increases as well.
    - The rate levels off/plateaus.
    - Other limiting factors (CO_2 availability, temperature) cause the plateau in the rate.
    - Plants need light for the light-dependent reactions.
    - Light allows for the production of ATP and NADPH.
    - One point is awarded for a correctly drawn graph (only the shape is needed).

20. (a) One point is awarded for any of the following if drawn and correctly labelled:
    - Outer membrane
    - Inner membrane
    - Matrix
    - Cristae
    - Intermembrane space
    - Circular DNA in matrix
    - Ribosomes (70S)

(b)
    - Krebs cycle occurs in the matrix
    - Oxygen must be available in order for the Krebs cycle to occur/aerobic process
    - Krebs cycle uses acetyl-CoA
    - Acetyl-CoA results from the link reaction
    - Decarboxylation of pyruvate produces acetyl-CoA
    - Acetyl-CoA is a 2-carbon compound
    - Acetyl-CoA joins with a 4-carbon compound (oxaloacetate), resulting in a 6-carbon intermediate
    - 6-C intermediate is decarboxylated, resulting in a 5-C compound

- 5-C compound is decarboxylated back into a 4-C compound (oxaloacetate)
- 4-C product starts the cycle over by accepting another acetyl-CoA
- Net yield in ATP is 2 (1 per pyruvate)
- 6 NADH produced
- 2 $FADH_2$ produced
- One point is awarded for an annotated diagram

(c)

- Limiting factors influence the rate at which photosynthesis can occur.
- Temperature is a limiting factor.
- Light intensity is a limiting factor.
- Carbon dioxide concentration is a limiting factor.
- At low temperatures, photosynthesis is slow/molecules collide too slowly at low temperatures to allow for efficient photosynthetic processes.
- An optimal temperature exists for photosynthesis.
- At higher temperatures, rate decreases.
- Enzymes denature at higher temperatures, decreasing the rate.
- If the temperature is too high, photosynthesis will stop.
- As light intensity increases, so does the rate.
- Rate levels off at high light intensities.
- Levelling off occurs due to other limiting factors (carbon dioxide concentration/temperature).
- As carbon dioxide concentrations increase, the rate increases.
- The rate levels off at a maximum.
- All enzyme active sites are maximally occupied.
- CO_2 is used in carbon fixation.
- One point is awarded for any correctly drawn graph.

Plant Biology

CENTRAL IDEAS
- Plant structure
- Plant modifications
- Meristem tissue
- Plant hormones
- Water absorption and transport
- Xerophytes and hydrophytes
- Seed dispersal
- Seed germination
- Flowering in plants

TERMS
- Apical meristem
- Dicotyledonous
- Fertilization
- Germination
- Lateral meristem
- Monocotyledonous
- Phloem
- Phytochrome
- Seed dispersal
- Transpiration
- Xylem

INTRODUCTION

Plants must adapt to the environment in which they live and be able to respond to changing environmental conditions. Plant structures have evolved that enable plants to exist in specific environments. Xerophytes, which are plants that live in arid environments, possess physiological and structural adaptions for water absorption and conservation. Hydrophytes, which are plants that live in aquatic environments, possess physiological and structural adaptations to prevent too much water from entering their tissues.

NOS Connections
- Models can enhance scientific knowledge and display scientific processes (plant structure, transpiration).
- Technology allows scientists to visualize, analyse and thoroughly investigate scientific phenomenon (plant hormones, plant structure, water transport).
- Collaboration and cooperation among scientists leads to the development of scientific understanding (plant anatomy and physiology).

9.1 TRANSPORT IN THE XYLEM OF PLANTS

TOPIC CONNECTIONS

- 2.2 Water
- 2.9 and 8.3 Photosynthesis

Plant Leaves

Plant leaves are designed to absorb radiant energy and optimize the rates of photosynthesis. Figure 9.1 is a cross section of a plant leaf. Table 9.1 lists the structures found in a plant leaf and their functions. The cuticle is a waxy layer located on the top and bottom of the leaf. The cuticle helps retain water in the leaves (water conservation) of terrestrial plants. The stomata (singular: stoma) are openings in the plant leaf that allow gas exchange and facilitate water movement out of the plant (transpiration). The stomata can open and close as needed by the regulation of the guard cells. The veins of the leaf contain the vessels for transporting water and nutrients throughout the leaf as well as throughout the entire plant. Xylem (the top layer of cells in the vascular bundles) passively transports water and minerals from the roots to the top of the plant. The phloem (the bottom layer of cells in the vascular bundles) actively transports the products of photosynthesis throughout the plant. Collenchyma tissue is found above the bundle sheaths and provides support for the leaf. The mesophyll (*meso-* means "middle") of the plant is the main site of photosynthesis. However, any green part of the plant can carry out photosynthesis. The palisade and spongy mesophyll cells are packed with chloroplast-containing chlorophyll in order to absorb radiant energy. The spongy layer consists of air spaces between the cells in order to facilitate gas exchange (carbon dioxide and oxygen).

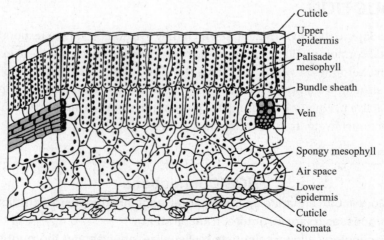

Figure 9.1. Plant leaf (cross section)

Table 9.1: Leaf Structures and Functions

Structure	Function
Waxy cuticle	Water retention
Chloroplast (mainly in mesophyll—spongy and palisade layer)	Carry out photosynthesis—contain photosynthetic pigments (chlorophyll and accessory pigments)
Stomata with guard cells	Guard cells regulate the opening and closing of the stomata to facilitate gas exchange and transpiration (water movement)
Collenchyma cells	Support vascular bundles (xylem and phloem) and give leaf structure
Vascular tissue	<ul><li>Xylem: transports water and minerals from the roots to the shoot (stems, flowers, buds, leaves) by passive transport</li><li>Phloem: transports products of photosynthesis throughout the plant by active transport</li></ul>

Transparation

> **Transpiration:** The passive movement and evaporation of water from the stems and leaves of a plant.

In vascular plants, water travels through xylem vessels. Xylem is dead tissue when it is functionally mature. This allows the cells of xylem to exist as hollow structures that facilitate maximum movement of water from the roots to the shoot (stems, flowers, buds, leaves). Tracheid cells and vessel elements make up xylem tissue. Vessel elements are shorter and wider than tracheid cells. Both cell types are tubelike in structure and contain primary and secondary cell walls. Pits in the sides of the cells allow water to flow between the cells of the xylem. In woody plants, secondary xylem (additional layers) develop and are often referred to as wood. Thick layers of secondary xylem are the main component of wood building materials due to the strength of the thickened cell walls.

Cohesive forces (hydrogen bonding) hold the water molecules together in a streamlike fashion as they travel up the xylem vessels. Evaporation of water molecules creates a "pull" on the stream of molecules, causing them to move up the xylem tissue. Water is also entering into the roots of the plant, creating a "push" of water molecules up the xylem as well. This push and pull on the water molecules creates tension that is continually forcing water to move up the xylem vessels. (Refer to Figure 9.2.) Ultimately, this is all driven by the evaporation of water due to the radiant energy from the sun.

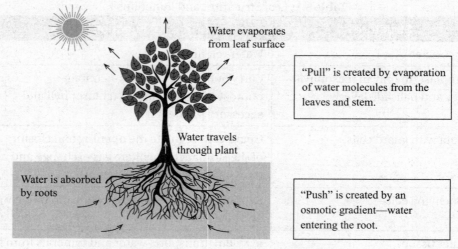

"Pull" is created by evaporation of water molecules from the leaves and stem.

"Push" is created by an osmotic gradient—water entering the root.

Figure 9.2. Transpiration (push + pull = tension)

The adhesive properties of water contribute to its ability to travel in a transpiration stream as well. Due to adhesion, water is attracted to the sides of the xylem vessels. The water molecules adhere to the sides of the vessels and resist the gravitational pull downwards.

STOMATA REGULATION

Guard cells regulate transpiration by opening and closing the stomata. When the stomata are open, the evaporation of water proceeds. When the stomata are closed, water remains in the plant. Stomata open when the guard cells are turgid (central vacuole is full of water) and close when the guard cells are flaccid (central vacuole lacks water). This is illustrated in Figure 9.3. Plants control the opening and closing of their stomata by changing the osmotic pressure in the guard cells. In order for the guard cells to open (become turgid), solutes (mainly potassium K^+) are pumped into the guard cells. This creates a hypotonic environment, and water enters the guard cells. In order for guard cells to close (become flaccid), solutes are pumped out of the guard cells. This creates a hypertonic environment, and water exits the guard cells.

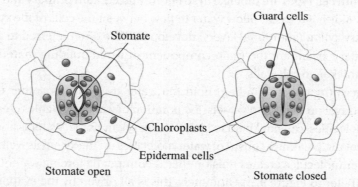

Figure 9.3. Regulation of guard cells (a) turgid and open, (b) flaccid and closed

ABSCISIC ACID

When plants detect a lack of available water, they trigger the closing of stomata in order to prevent water loss through transpiration. Since water enters plants through the roots, the roots are the first plant structure to sense low water availability. When the roots detect low water availability, root cells release the hormone (chemical messenger) abscisic acid (ABA). The ABA travels up the xylem to the stomata, causing the stomata to close.

ABIOTIC FACTORS AFFECTING TRANSPIRATION

Abiotic factors affect the rate of transpiration in terrestrial plants, as shown in Table 9.2. Increased light, temperature (heat) and wind intensity all increase the rate of transpiration. Humidity (moisture in the air) decreases the rate of transpiration. The amount that each abiotic factor affects transpiration varies based on the effects of other abiotic factors. For example, added light and temperature both increase the rate of transpiration but only to a certain point.

Table 9.2: Abiotic Factors and Transpiration Rates

Abiotic Factor	Effect on Transpiration	Explanation
Light	Increases transpiration up to a certain level; the level then plateaus	As light intensity increases, the rate of the light-dependent reactions increases (more light energy is available). The light-independent reactions increase when the light-dependent reactions increase. The light-independent reactions require carbon dioxide that enters the plant through open stomata. Open stomata allow more water to exit the plant by transpiration. Increased light also increases the temperature of the air surrounding the plant, which increases transpiration as well.
Temperature	Increases transpiration up to a certain level (maximum temperature) at which the transpiration rate rapidly drops	As temperature increases, water molecules evaporate more rapidly. Since evaporation drives transpiration, increased temperature causes increased rates of transpiration. High temperatures cause the denaturation of enzymes required for photosynthetic processes to continue.
Wind	Increases transpiration up to a point; then the wind is too powerful to allow the plant to support itself	Wind carries humidity in the air away from the stomata. Since water evaporates more rapidly from areas of higher concentration (within the plant) to areas of lower concentration (air surrounding the plant), the water evaporates more rapidly from the stomata.
Humidity	Decreases transpiration at increasing levels of humidity	Increased moisture (water) in the air decreases the concentration gradient of water between the stomata opening and the atmosphere. Since water moves from higher (in the atmosphere) to lower concentrations (in the plant), added moisture in the air surrounding the stomata slows transpiration.

Root System and Water Uptake

The root system is vital to plants. It facilitates the absorption of water and minerals from the soil. Terrestrial plants rely on the soil for obtaining water and minerals. These plants possess adaptations that enable them to do so efficiently. The root system consists of branching roots that grow down and spread out into the soil, creating a large surface area for water and mineral absorption. Root hairs further increase the surface area of the roots and are the site of water and mineral absorption. Root hairs are only 1-cell thick. So water and minerals can quickly enter into the tissues. The cell wall of root hairs is thicker than that of other plant cells. The additional cellulose in the cell wall helps to attract and absorb more water molecules. An example of the absorbent properties of cellulose can be seen in cotton balls. Cotton balls are made of cellulose. When placed into water, cotton balls quickly absorb the water. This demonstrates the role of the thickened cellulose in root hairs.

Uptake of Mineral Ions

Mineral ions in the soil enter the root passively through diffusion, via mass flow with water and by way of mutualistic relationships with fungi.

- **Diffusion**: The passive movement of molecules from high to low concentration
- **Mass flow with water**: The passive process by which mineral ions enter the root cells through protein channels (via facilitated diffusion) in the root hair cells.
- **Fungal hyphae** (rootlike structures): Mineral ions move into the branching cells of fungal hyphae, which have formed a close association with the roots of the plant. This is a mutualistic relationship in which the fungi facilitate increased mineral ion uptake for the plant and the plant provides nutrients for the fungi.

Active Transport of Minerals

Mineral ions are usually found at a much lower concentration in the soil than in the root cells. In order to facilitate the movement of the mineral ions from an area of low concentration (from the soil) to an area of high concentration (into the root), the cells must use active transport. The root cells possess protein pumps that use ATP to pump the mineral ions into the root cells. Mitochondria in the root hair cells produce the energy (ATP) that is used to pump the mineral ions against the concentration gradient. Roots must have oxygen available to them in order to carry out cellular respiration and produce the ATP needed to transport mineral ions into the root hair cells actively. This is the reason soil must be aerated (tilled—turned over to ensure air spaces are present) prior to planting crops. Aerated soil contains available oxygen. It is also why plants can die from too much water. The excess water saturates the soil and limits oxygen available to the plant.

Adaptations of Xerophytes

Xerophytes are plants that live in arid (dry) environments. Xerophytes constantly battle the need to open stomata for gas exchange and the need to close stomata for water conservation. For this reason, xerophytes have developed adaptations that allow them to keep their stomata open long enough to support the need for gas exchange while, at the same time, limit the rate of transpiration. (Refer to Table 9.3.)

Table 9.3: Adaptations of Xerophytes

Adaptation (Structure)	Function
Reduced leaves (sometimes to spines)	Less surface area for water loss. For example, cactus spines.
Rolled leaves	Rolled leaves block the wind from reaching the stomata. Since wind increases transpiration, rolled leaves decrease transpiration rates.
Deep roots	Deep, branching roots cover more surface area in the soil for water absorption and allow for absorption of water deep below the surface.
Thickened, waxy cuticle	The waxy, hydrophobic cuticle repels water and helps keep it in the tissues of the leaves.
Reduced number of stomata	Since water is lost through the stomata, having fewer stomata decreases the number of sites available for water loss.
Stomata in pits (sunken)	By keeping the stomata in pits up in the leaf tissue, the wind cannot come in contact with the water molecules at the surface of the stomata, which would increase the rate of transpiration.
Hairs	Hairs on leaves help to trap water, thereby increasing the humidity near the stomata and decreasing transpiration.
Water storage organ	Some xerophytes, like cacti, can store water in their tissues for times of drought.
Low growth form	By reducing the size of the plant and the number of stems the plant possesses, the overall surface area for water loss is decreased.
CAM (crassulacean acid metabolism) and C_4 metabolism	Both CAM and C_4 plants possess the enzyme PEP carboxylase, which can fix CO_2 at very low concentrations. As a result, the plants open their stomata for gas exchange less frequently. This decreases the amount of time available for transpiration but meets the CO_2 needs of the plant. In addition, CAM plants open their stomata only at night, reducing transpiration since the light intensity and temperature are low.

FEATURED QUESTION

May 2008, Question #40

A plant has a waxy cuticle, reduced leaves, reduced number of stomata, and CAM physiology. What type of plant could this be?

A. A hydrophyte
B. A filicinophyte
C. A bryophyte
D. A xerophyte

(Answer is on page 554.)

Adaptations of Halophytes

Halophytes are plants that live in areas with high saline concentrations. The increased saline in the environment makes it difficult for the plants to maintain proper water balance (carry out osmoregulation). Halophytes possess several adaptations in order to survive in saline conditions. These include salt glands that pump excess salt out of plant tissues, vacuoles that store excess salt and hardened tissues that prevent the entry of salt into plant tissues.

9.2 TRANSPORT IN THE PHLOEM OF PLANTS

TOPIC CONNECTIONS

- 1.4 Membrane transport

The vascular tissue that transports sugars and amino acids throughout the plant is phloem. The process that moves organic material through phloem is known as translocation. During translocation, phloem uses energy to carry organic material from the source (location of synthesis) to the sink (location of use). Because energy is used, this is a form of active transport. Since phloem must use energy to transport organic nutrients, the phloem must be alive at functional maturity. However, it needs to be hollow in order to carry the large amount of organic material that needs to be transported throughout the plant.

The cells that make up phloem include hollow sieve tubes associated with companion cells. The cytoplasm of sieve tubes and companion cells are connected by plasmodesmata. Plasmodesmata are channels that allow for some of the cytoplasmic material to be shared among the cells. Companion cells house the organelles (mitochondria and nucleus) that regulate and produce products for both cell types. This allows the sieve tubes to exist as living cells without having a crowded cytoplasm that would block translocation of organic molecules.

Molecules can travel within plant tissues by two distinct routes. Apoplastic movement involves material moving through the cell walls within plant tissue. Symplastic movement involves material moving though the plasmodesmata after entering the cytoplasm of a plant cell. Translocation moves organic molecules symplastically through the plasmodesmata. (Refer to Figure 9.4.)

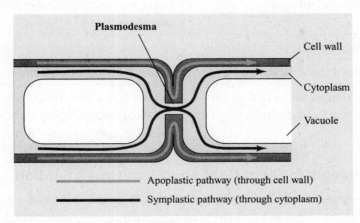

Figure 9.4. Plasmodesmata

The source for the organic molecules produced during photosynthesis is most often the leaves of the plant. However, plants store excess sugars (carbohydrates) as amylase in storage organs such as underground roots and stems. These underground storage organs (bulbs, tubers, storage roots) act as the source of organic molecules when a new growing season arrives. The sink may be a developing bud or fruit on the stem of a plant. It could also be one of the storage organs. For this reason, translocation can occur in many directions, always moving from a source to a sink.

Phloem and xylem are located adjacent to each other in order for water to move easily into the phloem through osmosis. The high solute concentration near the sink creates an osmotic gradient that causes water to move into the phloem. Remember that water moves from high to low concentration. So the water will diffuse into the phloem from the xylem. Water, the universal solvent, dissolves sucrose and other organic material that is transported throughout plant tissues in phloem. Water also aids in the movement of organic material throughout the plant. Organic material moves from the source to the sink, from areas of high concentration to areas of low concentration during translocation within the phloem.

The solute concentration in the phloem cells near the source is high since the products of photosynthesis (sucrose and amino acids) are being pumped into the phloem cells. The solute concentration in the sink is lower due to the cells in the sink taking up the products of photosynthesis. Higher solute levels in the cells near the phloem cause water to enter the phloem cells and produce hydrostatic pressure (turgor pressure) in the phloem cells near the source. Lower solute levels in phloem cells near the sink cause water to leave the cells and lower hydrostatic pressure (turgor pressure) in the phloem cells causes water to enter the cells near the sink. Table 9.4 compares and contrasts transpiration and translocation.

Table 9.4: Transpiration and Translocation

Transpiration	Translocation
Passive—water diffuses into the xylem and evaporates from the leaves	Active—sucrose and amino acids are pumped into the phloem
Transports water and minerals	Transports sucrose, amino acids and hormones
Consists of vessel elements and tracheids	Consists of sieve tubes and companion cells
Dead at functional maturity	Alive at functional maturity
Moves water and minerals from the roots to the shoot (one direction of travel)	Moves organic molecules from the source (leaves or storage organ) to the sink (growing plant structure: fruits, seeds, roots)

Xylem and Phloem in Dicot Root and Stem Cross Section

Xylem and phloem are found in vascular bundles within plant tissue. Vascular cambium, which is located between the xylem and phloem of each vascular bundle, is made of undifferentiated tissue that can become either xylem or phloem. The close association of xylem next to phloem is important because xylem supplies the water needed as a solvent for the organic molecules travelling in phloem. In dicot plants the vascular bundles are arranged in a ring within the stem. The phloem is on the outside of the vascular bundle with xylem facing the inside of the stem. In roots the xylem is located in the central region of the root with the phloem surrounding it facing the outside of the root. The structures of the stem and root are shown in Figure 9.5.

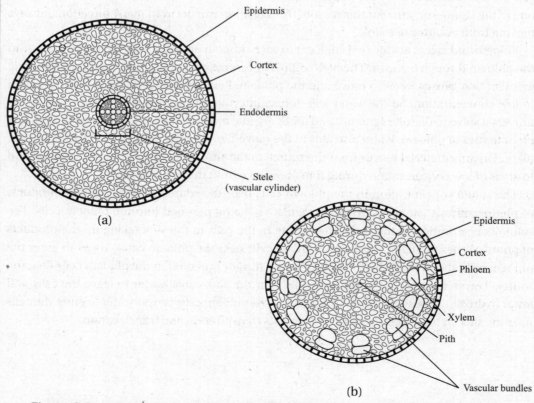

Figure 9.5. (a) Dicot root and (b) dicot stem showing location of xylem and phloem

9.3 GROWTH IN PLANTS

TOPIC CONNECTIONS

- 3.5 Genetic modification and biotechnology

Plants possess hormones that allow them to respond to changes in the environment. Plant hormones also help to regulate growth, reproduction (flowering) and water loss from a plant due to transpiration. They regulate plant growth by controlling the rate of mitosis and cell division in the meristem tissue found in the stems and roots of plants.

Meristem Tissue

Meristem tissue contains undifferentiated tissue that can eventually become many different tissue types. It is analogous to stem cells in animals. Meristem tissue is located in places where the plant exhibits major growth. Dicotyledonous plants possess apical and lateral meristem tissue. Plant hormones control growth due to cellular division in meristem tissue.

APICAL AND LATERAL MERISTEM TISSUE

Apical meristems are responsible for the primary growth of a dicotyledonous plant. They are located at the top of the plant (shoot). Apical meristems allow plants to grow and produce new leaves and flowers. Apical meristem tissue located in the roots allows the roots to grow down into the soil.

Lateral meristems are responsible for secondary growth in woody plants. Lateral meristem tissue allows woody plants to increase stem and root thickness. Herbaceous (nonwoody) plants do not exhibit secondary growth. In these plants, lateral growth occurs due to the presence of a vascular cambium that produces new xylem and phloem each year and cork cambium that produces the bark seen on woody plants. For this reason, lateral meristem tissue is often referred to as cambium.

Tropisms

Auxin is a type of plant hormone that is important in regulating plant growth as well as the process of phototropism.

Directional movement of plants in response to a stimulus is referred to as tropism. Plant shoots respond to environmental cues by trophic movements.

> **REMEMBER**
>
> The name *auxin* is derived from the Greek work "auxein," which means *to grow*.

> **Tropic Movement**
>
> Tropic movement refers to the directional growth or movement of plant tissue in response to an environmental cue.
>
> - **Phototropic movement:** Movement of plant structures based on the presence of light. For example, plant stems bend toward the light.
> - **Gravitropic movement:** Movement of plant structures based on the presence of gravitational pull. For example, roots grow downward in the soil.

Although plants cannot uproot and move to new locations on their own, they can change the direction in which their tissues are growing. When plants move toward light the process is referred to as *phototropism*.

AUXIN

The hormone responsible for phototropism is auxin. Auxin affects cells on the "dark side" (side of the stem not facing the light) of the plant, causing the cells to elongate. The elongated cells are not as strong as the cells on the "light side" of the stem. As a result of the elongated cells, the stem begins to bend toward the light as shown in Figure 9.6.

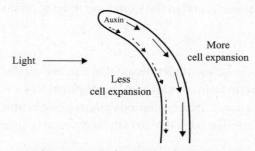

Figure 9.6. Auxin and phototropism

When light shines directly down on a plant shoot, the levels of auxin are evenly distributed within the tissue of the shoot. This even distribution signals the plant to grow straight. When the light source is projected at an angle on the plant shoot, auxin levels change within the shoot tissue. The shoot bends towards the light source. Auxin efflux pumps lower the concentration of auxin inside of cells by moving auxin out of the cells. Concentration gradients of auxin play a role in determining whether or not a cell will elongate. Auxin influences gene expression, causing genetic changes in cells and resulting in uneven localized tissue growth. In phototropism, cells on the light side of the plant stem (the side facing the sun) pump auxin out of the intracellular fluid. This causes auxin to accumulate in cells on the dark side of the stem. Higher levels of intracellular auxin on the dark side stimulate gene expression and lead to cell growth (elongation) on the dark side. The tissue on the dark side now consists of elongated cells, which bend and facilitate the tropic movement of the stem towards light.

9.4   REPRODUCTION IN PLANTS

Angiospermophytes

Angiospermophytes (angiosperms) are the most successful plants on Earth. Angiosperms have evolved to rely on insects for pollination. The flowers produced by angiosperms, as shown in Figure 9.7, are designed to attract pollinators and facilitate pollination. The structure of a dicotyledonous flower is shown in Figure 9.7.

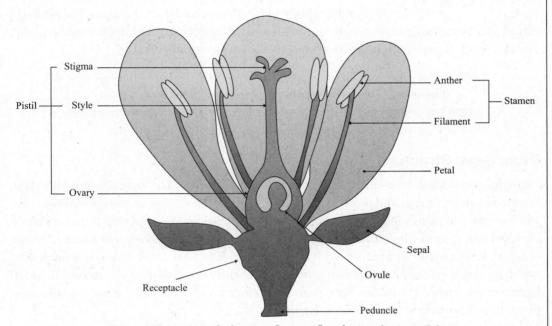

Figure 9.7. Dicotyledonous flower (floral parts in 4s or 5s)

- **Sepals**: Modified leaves that protected the flower when it was a bud. As a group, all the sepals are referred to as the calyx.
- **Petals**: Coloured leaves that protect the reproductive structures and attract pollinators.
- **Anther**: A male part of the flower that produces pollen (male gamete for sexual reproduction).
- **Filament**: A male part of the flower that holds up the anther.
- **Stigma**: A female part of the flower that contains a sticky substance to which pollen adheres.
- **Style**: A female part of the flower that holds up the stigma and transfers sperm from the pollen grain to the ovary.
- **Ovary**: A female part of the flower that holds the ovules (eggs) that will be fertilized by the sperm and that will develop into seeds encased in fruit.
- **Stamen**: The collective male parts of the flower: anther and filament.
- **Pistil**: The collective female parts of the flower: stigma, style and ovary.

REMEMBER

Plants have developed mutualistic relationships with pollinators in order to facilitate sexual reproduction. For example, hummingbirds and honeybees (and many other insects) receive nectar (food) from the flowers they pollinate.

Pollination

Pollination is the term used to describe pollen landing on the stigma. After the pollen lands on the stigma, a pollen tube grows from the pollen down the style and delivers the sperm to the ovary. *Fertilization* describes the process of the sperm joining with the ovules (eggs) in the ovary of the flower.

The fertilized ovules in the ovary undergo mitosis and develop into mature seeds. The ovary then swells into fruit that surrounds the seeds. Animals are attracted to the fruit and consume the fruit, which contain the seeds. The seeds go through an animal's digestive system and are deposited in a location far from the parent plant. Plants rely on animals for this process, which is known as *seed dispersal*.

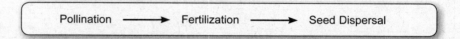

Pollination ⟶ Fertilization ⟶ Seed Dispersal

Dicot Seed Structure

Dicotyledonous seeds contain two cotyledons for food storage. The *testa* (commonly called the seed coat) surrounds and protects the seed. The embryo contains an *embryonic shoot* that will become the plant's first leaves and an *embryonic root* that will develop into the plant's complex root system. The *hilum* is the scar left on the seed from where it was attached to the ovary of the plant and through which it received nutrients. Next to the hilum is a small scar called the *micropyle* where the pollen tube grew into the ovule (before it was a seed) in order to deliver the sperm. The kidney bean (*Phaseolus vulgaris*) is an example of a dicotyledonous plant. Its seed structure can be seen in Figure 9.8.

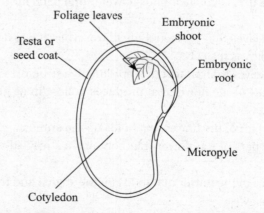

Figure 9.8. Dicotyledonous kidney bean seed (*Phaseolus vulgaris*)

Seed Germination

Although some seeds rely on light to stimulate germination, most seeds do not need light for germination. Unlike mature plants that rely on light for energy to carry out photosynthesis, seeds have all the energy they need for germination in stored amylose (starch). The amylose is stored in the cotyledons of the seeds and is used when the seed is ready to germinate.

Every seed must have water, oxygen and the right atmospheric temperature available in order to undergo germination. The embryo cannot break out of the protective testa (seed coat) by itself. The absorption of water (imbibition) breaks the testa and allows water to enter the seed. Once inside the seed, the water supports metabolic processes.

Seeds cannot carry out photosynthesis because they have no photosynthetic structures. Therefore, seeds must use the energy stored in the cotyledons for cellular processes. In order to carry out cellular respiration to produce energy (ATP) from the stored amylose (starch), the seed must have oxygen available. The oxygen is used for aerobic respiration and for breaking down energy stored in the cotyledon (amylose) into ATP. In addition, seeds require enough warmth to facilitate molecular reactions. In fact, some seeds require extreme temperatures, like wildfires, in order to begin germination.

After the seed absorbs water, the cotyledon produces the hormone gibberellin, which is also known as gibberellic acid or GA. Gibberellin triggers the aleurone layer (a layer of cells directly inside of the testa) to produce the enzyme amylase (alpha amylase). This enzyme digests the stored amylose and breaks it into maltose. Maltose is then further broken down into glucose, which is used for cellular respiration by the embryo to produce energy for germination.

Phytochrome and Flowering

Plants must regulate when they produce flowers in order to ensure there is sufficient light, enough water and the appropriate environmental temperature to support the production of seeds and fruit. The reaction of plants to the length of daylight is known as *photoperiodism*. The ratio of daylight to darkness induces many plants to flower. Long-day plants flower when the ratio of daylight to darkness is high. In contrast, short-day plants flower when the ratio of daylight to darkness is low. Some plants, though, don't respond to the ratio of daylight to darkness to regulate flowering. These plants, which are called day-neutral plants, use other environmental cues, such as temperature, to trigger flowering.

Plants that flower based on photoperiodism use the pigment phytochrome in order to tell if the days are long or short. Phytochrome exists in two forms—phytochrome red (P_r) and phytochrome far-red (P_{fr}). Phytochrome far-red is the active form of phytochrome used to regulate biological activities such as flowering. Phytochrome is produced from the direct conversion of P_r and can be converted back into P_r. During daylight hours, P_r is converted to P_{fr}.

At night, P_{fr} is converted back to P_r. The nighttime conversion of P_{fr} to P_r is much slower than the daylight conversion of P_r to P_{fr}. Plants use the amount of P_{fr} to determine the length of the daylight. Long days facilitate the conversion of high levels of P_r to P_{fr}, and short nights restrict the time available for the slower conversion of P_{fr} back to P_r. For this reason, long-day plants are induced to flower when they detect high levels of P_{fr}, which signal long days. Short-day plants have less time (light) to convert P_r to P_{fr} and more time in darkness to convert P_{fr} back to P_r. For this reason, short-day plants are induced to flower when they detect low levels of P_{fr}, which signal short days. (Refer to Figure 9.9.)

The chrysanthemum is a short-day plant since it flowers when the nights are long. Short-day plants can be induced to produce flowers during periods of short nights by artificially controlling the length of the day within greenhouses. The same method can be used to stimulate flowering artificially in long-day plants and to induce long-day plants to flower during short days.

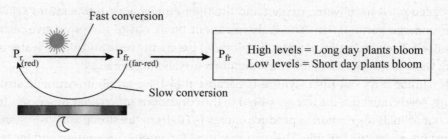

Figure 9.9. Phytochrome and flowering. The pigment phytochrome exists as two incontrovertible forms, P_r and P_{fr}. The level of P_{fr} determines flowering.

MONOCOTS AND DICOTS

Flowering plants are divided into two main groups based on the basic structure of their stems, roots, leaves and floral parts. These two groups are the monocotyledonous plants (monocots) and the dicotyledonous plants(dicots). Table 9.5 presents the main differences between the monocots and dicots.

Table 9.5: Monocots and Dicots

Monocotyledonous Structures	Dicotyledonous Structures
One cotyledon	Two cotyledons (embryonic leaves)
Parallel veins in leaves	Netlike veins in leaves
Floral organs in multiples of 3	Floral organs in multiples of 4 or 5
Vascular tissue in stems scattered	Vascular tissue in stems distributed in rings
Fibrous roots	Taproots with lateral branches

PRACTICE QUESTIONS

1. Adaptations of xerophytes include:

    I. Reduced roots
    II. Rolled leaves
    III. Hairy leaves

    A. I only
    B. I and II only
    C. II and III only
    D. I, II and III

2. The hormone abscisic acid is responsible for:

    A. Closing of the stomata
    B. Phototropism
    C. Seed germination
    D. Photoperiodism

3. In order for organic molecules to travel from the leaves of a plant to the roots of a plant, what tissue must transport them?

    A. Xylem, actively
    B. Xylem, passively
    C. Phloem, actively
    D. Phloem, passively

4. What structure produces pollen?

    A. Anther
    B. Filament
    C. Stigma
    D. Ovary

5. Which of the following shows the processes in the correct order?

    A. Seed dispersal, fertilization, pollination
    B. Pollination, seed dispersal, fertilization
    C. Pollination, fertilization, seed dispersal
    D. Fertilization, seed dispersal, pollination

6. Which of the following are required by ALL seeds in order to germinate?

 I. Light
 II. Oxygen
 III. Water

 A. I only
 B. I and II only
 C. II and III only
 D. I, II and III

7. Which of the following are features of dicotyledonous plants?

 I. Taproot with lateral branches
 II. Flowers in 3s
 III. One cotyledon

 A. I only
 B. I and II only
 C. II and III only
 D. I, II and III

8. Which of the following shows correct statements about xylem and phloem?

	Tissue	Method of Transport	Material Transported	Name of Process
A.	Xylem	Active	Water and minerals	Transpiration
B.	Phloem	Active	Sucrose and amino acids	Translocation
C.	Xylem	Passive	Sucrose and amino acids	Transpiration
D.	Phloem	Passive	Sucrose and amino acids	Translocation

9. Which of the following aid in the process of transpiration?

 I. Cohesion
 II. Adhesion
 III. Polarity of water

 A. I only
 B. I and II only
 C. II and III only
 D. I, II and III

10. Plants will die with too much water due to:

    A. Water competing with minerals to enter the cell
    B. Water keeping photosynthetic products from entering the roots
    C. Water limiting the amount of carbon dioxide in the soil
    D. Water limiting the amount of oxygen in the soil

11. Identify the floral structures in the diagram below.

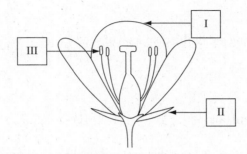

	I	II	III
A.	Sepal	Petal	Anther
B.	Petal	Sepal	Stigma
C.	Stigma	Sepal	Anther
D.	Petal	Sepal	Anther

12. Which of the following increases the rate of transpiration the most?

    A. High humidity, light wind
    B. Low humidity, high temperatures
    C. Low humidity, low temperatures
    D. High humidity, no wind

13. Translocation refers to:

    A. Water loss from plant tissue
    B. The active transport of sucrose within plants
    C. Water uptake by the roots of plants
    D. The passive transport of sucrose within plants

14. Water uptake by roots is assisted by the presence of:

    I. Branching roots
    II. Root hairs
    III. Phloem

    A. I only
    B. I and II only
    C. II and III only
    D. I, II and III

15. Which of the following aid in the dispersal of seeds?

    A. Mammals
    B. Bees
    C. Bacteria
    D. Microscopic organisms

16. The rate of transpiration and the absorption of light were measured over a 24-hour period. The results are shown in the graph below.

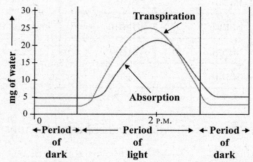

    A. At what time of day is the rate of transpiration the highest? (1)
    B. Suggest reasons for the absorption rate following the rate of transpiration. (2)
    C. Calculate the difference in absorption levels between 6 A.M. AND 2 P.M. (1)

17. List three adaptations of xerophytes that help prevent water loss. (3)

18. Explain how abiotic factors can affect the rate of transpiration. (4)

19. Label the parts of the seed and give the function of each labelled structure. (4)

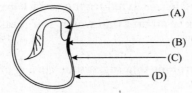

20.

    A. Draw, label and name a dicotyledonous flower. (4)

    B. Compare monocotyledonous and dicotyledonous plants. (6)

    C. Outline the process of mineral ion uptake in roots. (8)

ANSWERS EXPLAINED

1. **(C)** Xerophytes have rolled leaves to reduce the surface area exposed to wind. They also have hairy leaves to block wind and trap water. The roots of xerophytes are deep and branching.

2. **(A)** Abscisic acid is the hormone that triggers the closing of the stomata.

3. **(C)** Phloem is the only vascular tissue that can transport material from the leaves to the roots. Materials in xylem always travel from the roots to the shoot. Phloem is living tissue that actively pumps organic material from the leaves to the roots of a plant.

4. **(A)** The anther produces the male gametophyte—pollen.

5. **(C)** Pollination must occur before fertilization. Then the seed will develop and be dispersed.

6. **(C)** All seeds require water and oxygen in order to germinate. Only some seeds also require light.

7. **(A)** Dicotyledonous plants possess one main taproot with lateral branches. Monocotyledonous plants have one cotyledon and flower in 3s.

8. **(B)** Phloem uses active transport. The phloem transports sucrose and amino acids, thereby carrying out the process of translocation.

9. **(D)** Cohesion and adhesion both contribute to the process of transpiration. Hydrogen bonds that form between water molecules due to the molecule's polarity cause water to be cohesive.

10. **(D)** Too much water in the soil limits the amount of oxygen available to the roots. Roots require oxygen to carry out cellular respiration to produce ATP.

11. **(D)** I—petal; II—sepal; III—anther

12. **(B)** Low humidity and high temperatures increase the rate of transpiration the most.

13. **(B)** Translocation is the active movement of sucrose, amino acids and hormones through plant tissues.

14. **(B)** Branching roots (as well as deep roots) and root hairs increase the surface area, allowing more water to be absorbed by the plant.

15. **(A)** Mammals assist plants in seed dispersal. This is why plants produce fruit with fructose to encourage mammals to consume the fruit (which contains seeds).

16. (a) 2 P.M. (+/− 1hour)

    (b)
    - Light increases the rate of evaporation of water from the leaves due to the increase in temperature caused by the light energy.
    - Light increases the rate of the light-dependent reactions, which use water/photolysis and thereby cause more water to enter the xylem.

    (c) 25 mg (high) – 4 mg (low) = 21 mg (+/− 2)

17.
- Reduced leaves/spines for leaves
- Rolled leaves
- Deep/branching roots
- Thick cuticle
- Reduced number of/few stomata
- Sunken stomata/stomata in pits
- Vertical growth/low growth form; hairy leaves
- CAM/C4 photosynthesis
- One point is awarded for including an example (cactus, and so on)

18. Temperature:
- As temperature increases, so does the rate of transpiration.
- Temperature increases the evaporation of water.
- Very high/high temperatures limit transpiration due to plant metabolic processes not functioning/denaturation of enzymes.
- Very cold temperatures limit metabolic processes and slow transpiration/water molecules don't collide and move as fast.
- One point is awarded for a correctly drawn graph.

Light:
- As light intensity increases, so does transpiration.
- The rate of transpiration will level off as other factors (besides light intensity) become limiting
- Higher light intensities give off more energy to heat the plant surface, increasing the evaporation of water.
- Higher light intensity increases the light-dependent reactions, which use water/ cause water to undergo photolysis and cause more water to be drawn into the plant.
- One point is awarded for a correctly drawn graph.

Wind:
- As wind increases, so does the rate of transpiration.
- Wind decreases the humidity near the surface of the stomata/aids in water evaporating from the leaves due to changing the gradient of water concentration between the atmosphere and the surface of the stomata. Since water diffuses from areas of high concentration to areas of low concentration, and the decreased water vapour in the atmosphere increases the water gradient between the two locations, water moves from the stomata into the atmosphere, increasing the rate of water loss from the plant.

Humidity:
- Humidity decreases the rate of transpiration.
- Since water diffuses from areas of high concentration to areas of low concentration, and the increased water vapour in the atmosphere decreases the water gradient between the two locations, water moves more slowly from the stomata into the atmosphere, decreasing the rate of water loss from the plant/amount of water in air is higher relative to low humidity, and water will not evaporate as rapidly.

19. (a)
- Embryo
- The baby plant
- Will grow into the mature plant

(b)
- Hilum
- Location where the seed was attached to the ovary in order to receive nutrients

(c)
- Micropyle
- Site of entry of the pollen tube
- Site of entry of sperm into the ovule in order for fertilization to occur

(d)
- Testa/seed coat
- Protects the seed until the seed is ready to germinate

20. (a) One point is awarded for including each of the following if they are correctly labelled, including their function:
- Petal: Protect reproductive organs/attract pollinators
- Sepal/calyx (collectively): Originally protected flower as a bud
- Anther: produces the pollen/male gametophyte
- Filament: holds up the anther to aid in pollen distribution
- Stigma: contains a sticky substance for pollen/pollen lands here
- Style: holds up stigma to facilitate pollination/pollen tube travels down style/connects stigma to ovary
- Ovary: contains ovules/eggs for fertilization/contains eggs that will be fertilized to become seeds

(b)
- Monocotyledonous plants/monocots have one cotyledon, and dicotyledonous plants/dicots have two cotyledons.
- Monocots possess floral organs/parts in 3s, and dicots possess floral organs/parts in 4s or 5s.
- Monocots have fibrous adventitious roots, and dicots have a taproot with lateral branches.
- Monocots have parallel veins in their leaves, and dicots have netlike veins.
- Monocots have vascular bundles scattered in their stems, and dicots have vascular bundles in rings.
- One point is awarded for naming an example for both monocots and dicots.

(c)
- Mineral ion uptake can occur by mass flow with the passive entry of water into the root cell by osmosis.
- Mineral ions can enter through mutualistic relationships with fungi/mycorrhiza.
- Fungi absorb minerals for the plant/increase surface area for absorption of minerals and plants give organic nutrients to the fungi.
- Minerals can diffuse into the roots if the concentration of minerals is higher in the soil.
- Typically more minerals are in the root cells than in the soil/higher concentration in root cells.
- Root cells must use energy to pump in mineral ions/ATP is used to pump in mineral ions.
- Oxygen must be available in the soil in order to allow aerobic cellular respiration to produce enough ATP to pump in the mineral ions.
- Soil lacking mineral ions will not support the plant's metabolic processes.

Genetics and Evolution

CENTRAL IDEAS
- Gene recombination
- Dihybrid crosses and gene linkage
- Changes in allele frequencies
- The process of speciation

TERMS
- Autosomes
- Chiasmata
- Linkage group
- Polygenic inheritance
- Sex chromosomes
- Speciation

INTRODUCTION

Variation within populations is vital in order for populations to maintain the capacity to adapt to changing environmental conditions. Sexual reproduction contributes to variation by producing offspring with unique combinations of genes. Meiosis allows the formation of new combinations of genes that can result in new phenotypes in populations of sexually reproducing organisms. Geneticists evaluate the probable genetic combinations of offspring by looking at the genetic makeup of the parents. Changes in the genetic makeup of organisms (allele frequencies) can lead to speciation.

NOS Connections
- Collaboration and cooperation among scientists results in greater scientific understanding (about patterns of inheritance and processes of evolution).
- Patterns, trends and discrepancies exist in nature (genetic inheritance patterns, mutations, speciation).
- Careful observations and data collection are keys to scientific understanding (genetic inheritance patterns, speciation).
- Skepticism exists within the scientific community (evolutionary processes).

GENETICS AND EVOLUTION

10.1 MEIOSIS

Meiosis is a special type of cell division that produces reproductive cells—the sperm and the ova (eggs). Meiosis produces four new haploid daughter cells from one original diploid parent cell. The cells produced by meiosis are all genetically different from each other as well as from the parent cell from which they originated.

Meiosis I reduces the chromosomal number from diploid to haploid. The homologous chromosomes line up as tetrads at the metaphase plate and are pulled towards opposite poles of the cell. Tetrads are made up to two homologous chromosomes that code for the same traits. Chromatids on the same chromosome are referred to as sister chromatids. Chromatids located on different homologs are referred to as nonsister chromatids. (Refer to Figure 10.1.) The centromere is in the same location on both chromosomes, and the length of each chromosome is the same.

Autosomes and Sex Chromosomes

Autosomes are nonsex chromosomes. Humans have 22 pairs of autosomes.

Sex chromosomes are the X and Y chromosomes that determine the gender of an individual. Humans possess one pair of sex chromosomes (XX = female; XY = male).

Assortment: Autosomes and sex chromosomes assort independently during meiosis.

Although homologous chromosomes code for the same traits, they may or may not have different alleles (forms of the gene) for each trait. Each homolog (member of a homologous pair) consists of two sister chromatids that contain exact copies of the genes at the same loci (location) on the chromatids. In other words, homologs have the same alleles. The chromatids are produced during the S-phase of the cell cycle through DNA replication to ensure that each new cell receives an exact copy of the parent cell DNA. The kinetochore is the region around the centromere where the spindle fibres attach to and pull on to separate the two chromatids into individual chromosomes.

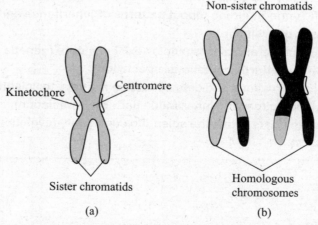

Figure 10.1. (a) Chromosome with sister chromatids and (b) tetrad with homologous chromosomes

The replicated chromosomes become visible (condense) and form tetrads with their homologs during prophase I. In metaphase I, the tetrads line up at the metaphase plate (centre of the cell). During anaphase I, spindle fibres attach to the kinetochores and pull the chromosomes to opposite poles (ends) of the cell. In telophase I, the nuclear membranes reform around the chromosomes. Telophase I is followed by cytokinesis to produce two new, genetically different daughter cells. Crossing-over during prophase I followed by independent assortment of the chromosomes during metaphase produces the variation in genetic makeup between the new cells and the parent cell from which they originated. (Refer to Figure 10.2.)

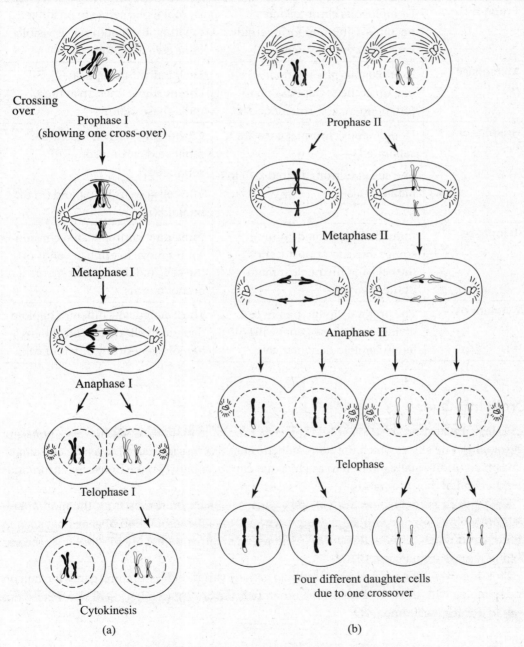

Figure 10.2. Comparison of (a) meiosis I and (b) meiosis II

Meiosis II follows meiosis I. Meiosis II is similar to mitosis since it is responsible for separating the sister chromatids into individual chromosomes. No tetrads are present since each set of homologous (chromosomes that code for the same trait) were separated into individual chromosomes occupying a newly created haploid daughter cell in meiosis I. Table 10.1 compares meiosis I and meiosis II.

Table 10.1: Meiosis I and Meiosis II

Process	Meiosis I	Meiosis II
Prophase	Homologous chromosomes become visible and form tetrads. Crossing-over may occur.	Homologous pairs are no longer together. Each homolog is visible in its own daughter cell.
Metaphase	Homologous pairs (tetrads) line up at the metaphase plate. Independent assortment occurs.	Each individual replicated chromosome lines up at the metaphase plate.
Anaphase	Homologous chromosomes are separated. The cell goes from diploid ($2n$) to haploid (n).	Chromatids of each individual replicated chromosome are separated. The cell goes from haploid (n) to haploid (n).
Telophase	Individual replicated chromosomes move to opposite ends of the cell, and the nuclear membrane reforms.	Individual nonreplicated chromosomes move to opposite ends of the cell, and the nuclear membrane reforms.
Cytokinesis	The original diploid parent cell divides into 2 new genetically different haploid daughter cells.	The 2 genetically different haploid daughter cells divide into 4 new haploid genetically different cells.

Crossing-Over

Crossing over in prophase I involves the switching of genetic material between nonsister chromatids. The site at which the nonsister chromatids cross is referred to as the chiasmata. Chiasmata (individually referred to as chiasma) can form between the nonsister chromatids at either end of the chromatids.

Each pair of homologous chromosomes consists of a chromosome from the mother and a chromosome from the father. Crossing-over leads to a recombination of genes not seen in either parent. New combinations of genes not seen in either parent are referred to as recombinants, as shown in Figure 10.3.

Crossing-over in prophase I contributes to genetic variation by switching genes located on one homolog with those located on the other. This creates a combination of new alleles not seen in the original homologs.

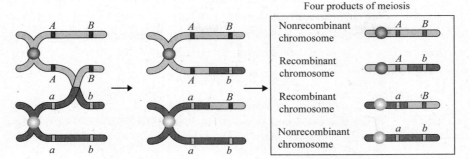

Figure 10.3. Recombinants formed from crossing-over

Figure 10.3 shows how crossing-over can result in the formation of new combinations of genes in the daughter chromosomes. The original parental chromosomes carried the allele combinations of *AB* and *ab*. Crossing-over produced new combinations (recombinants) of alleles not seen in either parent (*Ab* and *aB*) as well as two parental type chromosomes (*AB* and *ab*).

Independent assortment of the chromosomes as they line up at the metaphase plate further increases the genetic variation in the chromosomes of the offspring. Each homologous pair (tetrad) lines up independently of the others and can have either homolog on either side of the metaphase plate. This random orientation of each homologous pair leads to the offspring receiving different percentages of the original maternal or paternal chromosomes. For each homologous pair, there exists a 50 percent chance that the daughter cell will receive a paternal chromosome and a 50 percent chance it will receive a maternal chromosome. Since there are 23 homologous pairs of chromosome in the human genome, the number of possible resulting combinations of the maternal and paternal chromosomes is close to 8 million (2^{23})! This is why no two offspring look exactly like a parent or exactly like each other (unless they are identical twins).

In meiosis, genetic variation results from:

- Crossing-over during prophase I
- Independent assortment (random orientation) during metaphase I

Mendel's Law of Independent Assortment

Gregor Mendel was a monk who lived in the 1800s and studied heredity in pea plants. His observations led to the development of two laws of heredity. Mendel's first law is the law of segregation, as illustrated in Figure 10.4. It states that the alleles of homologous chromosomes will separate during the formation of gametes (meiosis).

Mendel's second law is the law of independent assortment. This law states when alleles separate, they separate independently of each other. In other words, the selection of an allele for one gene in a gamete has no bearing on the selection of other genes for the gamete. The law of independent assortment applies to all genes unless they are located on the same chromosome (linked genes).

> **REMEMBER**
>
> Mendel Developed Two Laws
> 1. The Law of Segregation
> 2. The Law of Independent Assortment

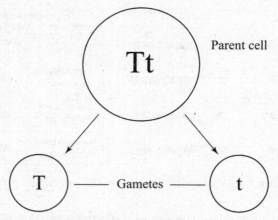

Figure 10.4. Law of segregation

During metaphase I of meiosis, the chromosomes line up at the metaphase plate. Each homologous pair lines up independently of the other pairs. Since each homolog carries its own alleles, when the homologs separate, each new daughter cell receives one of the two possible alleles for each gene. Mendel's law of independent assortment is directly related to meiosis since random assortment occurs during metaphase I when the chromosomes line up independently of each other and the alleles are separated from each other during anaphase I.

Mendel was fortunate that the traits he investigated were located on separate chromosomes because they were unlinked and exhibited independent assortment. If he had investigated genes on the same chromosomes (linked genes), he would not have gathered the same data. The phenotypic ratios would have been very different. Note that Mendel had no idea about genes, chromosomes, or assortment. He just happened to choose unlinked genes.

Thomas Morgan carried out investigations on fruit flies that showed different patterns of inheritance than those proposed by Mendel. Morgan's genetic testing on fruit flies (*Drosophila melanogaster*) led to the discovery of sex-linked traits (traits carried on the X or Y chromosome) and genes coding for traits carried on the same chromosome (linked genes).

10.2 INHERITANCE

Dihybrid Crosses

Dihybrid crosses involve genetic recombination between two genes. During meiosis, the two alleles of a gene separate from each other (segregate) into individual haploid sex cells (sperm and egg). The alleles present in the parent cells represent the possible alleles that can be assorted into gametes during meiosis (see Figure 10.5). At fertilization, the haploid alleles in the sperm and ovum (egg) join with each other and produce a diploid zygote. The possible genotypes (allele combinations) in the resulting offspring are determined by conducting a dihybrid cross.

GENETICS AND EVOLUTION

In order for the alleles to separate independently of each other, the genes must be on separate autosomal (nonsex) chromosomes. Dihybrid crosses involve parents who are both heterozygous for the two traits. Dihybrid crosses always result in a 9:3:3:1 phenotypic ratio.

	TY	Ty	tY	ty
TY	TTYY	TTYy	TtYY	TtYy
Ty	TTYy	TTyy	TtYy	Ttyy
tY	TtYY	TtYy	ttYY	ttYy
ty	TtYy	Ttyy	ttYy	ttyy

Figure 10.5. Dihybrid cross involving parent genotypes TtYy

Figure 10.5 shows the possible allele combinations in the F_1 gametes that were determined by using FOIL. Using FOIL allows you to determine the possible allele combinations for the gametes quickly. Figure 10.6 shows how independent assortment results in different allele combinations simply based on how the homologous chromosomes line up during metaphase I.

> ### REMEMBER
>
> FOIL stands for "first, outer, inner, last". For example, with the dihybrid TtYy combine the *first* alleles of each gene (TY), combine the *outer* alleles of each gene (Ty), combine the *inner* alleles of each gene (tY) and combine the *last* alleles of each gene (ty).

> The number of different possible allele combinations that can result from recombination in meiosis (possible allele combinations found in newly created sperm or egg) can be calculated using the formula 2^n, where n equals the number of heterozygous gene combinations present in the parental genotype. For example, the allele combinations of *AaBBCcDd* with 2 heterozygous combinations (*AaCc*) would result in a possibility of 2^2 or 4 different allele combinations in the resulting gametes. If there are no heterozygous combinations (*AABBCCDD*), the probability is 2^0 or only 1 gene combination in the gametes.

VARIATION DUE TO CROSSING-OVER

Crossing-over in prophase I can result in an exchange of parts of a chromatid that produces a new combination of alleles in the resulting gametes. Figure 10.6 shows how the formation of the chiasmata and the resulting crossing-over produce new combinations of alleles in the chromosomes. Crossing-over is the only method of recombination of alleles in meiosis for linked genes.

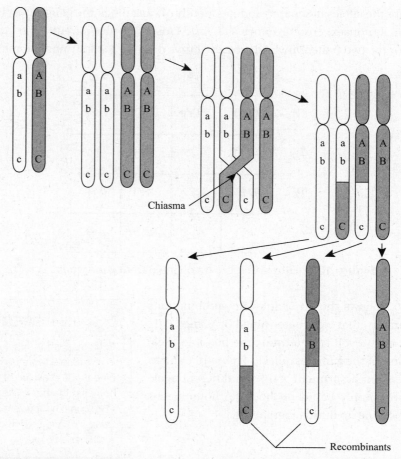

Figure 10.6. Crossing-over produces new allele combinations (recombinants).

Linked Genes

> **Linkage group:** A group of genes (gene loci) located on the same chromosome.

Figure 10.6 shows linkage groups. The alleles *a*, *b* and *c* located on the first chromosome (white) represent a linkage group. The linkage group on the homolog (yellow chromosome) consists of genes A, B and C. Remember that homologous chromosomes code for the same genes but may or may not have the same forms of the genes (alleles). So both homologs will code for genes A, B and C but can have either the dominant or the recessive form of the gene. The genes shown on the homologous chromosomes in Figure 10.6 all code for the same traits, but they possess different alleles for the linkage group.

FEATURED QUESTION

May 2003, Question #26

What constitutes a linkage group?

A. Genes carried on the same chromosome
B. Genes whose loci are on different autosomes
C. Genes controlling a polygenic trait
D. Alleles for the inheritance of ABO blood groups

(Answer is on page 554.)

CROSSES INVOLVING LINKED GENES

Linked genes do not show independent assortment and can show variation only due to cross-ing-over. Linked genes will be identified as being on the same chromosome either by words or diagrams. On the IB Biology Exam, unlinked genes will be shown side by side (*AaBb*) and linked genes will be shown as vertical pairs, as shown in Figure 10.7.

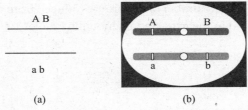

(a) (b)

Figure 10.7. (a) Linked genes as shown on the IB exam and
(b) representative chromosomes showing linkage

When the chromosomes separate during anaphase, the dominant alleles *A* and *B* shown on one homolog will always assort together. The recessive alleles *a* and *b* shown on the matching homolog will always stay together. However, crossing-over between the nonsister chromatids can result in new combinations of alleles in offspring. If crossing-over occurs (and it does not always occur), the resulting recombinants for the alleles shown in Figure 10.7 would be *aB* and *Ab* (see below).

Crossing-over between the dominant allele (*A*) and the recessive allele (*a*) for gene A results in the recombinants *Ab* and *aB*. Since crossing-over does not always occur, most of the gam-etes will be like the parents (*AB* and *ab*). If any of the recombinants appear (*Ab* and *aB*), they are due to crossing-over in prophase I of meiosis. The number of recombinants produced in a two-point testcross can be used to calculate how often crossing-over occurred. A testcross always involves the use of an individual showing the dominant trait and a homozygous recessive individual. Two-point testcrosses involving a heterozygous parent and a homozygous parent always results in a 1:1:1:1 (25% for each) ratio of phenotypes in the offspring. This type of cross is always used when asked to identify recombinants in linked genes or calculate percent recombination. (Refer to Figure 10.8.)

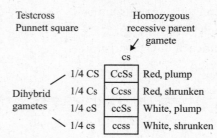

Parent genotypes: CcSc × ccss

Testcross Punnett square

Homozygous recessive parent gamete

cs

Dihybrid gametes

1/4 CS	CcSs	Red, plump
1/4 Cs	Ccss	Red, shrunken
1/4 cS	ccSs	White, plump
1/4 cs	ccss	White, shrunken

Figure 10.8. Two-point (two genes) testcross showing 1:1:1:1 expected phenotypic and genotypic ratio

A two-point test cross will result in the expected 1:1:1:1 phenotypic ratio as long as the genes are located on separate chromosomes (unlinked). If the genes are located on the same chromosome (linked), the ratio of recombinants will be much lower than that of the parent type. The example in Figure 10.9 shows the results of a two-point testcross with linked and with unlinked genes. The total number of offspring produced from each cross was 100. Notice that the first cross resulted in the expected phenotypic ratio of 1:1:1:1 (25% of each), but the second cross did not.

AaBb × aabb

	ab				ab	
AB	AaBb	25		AB	AaBb	40
Ab	Aabb	25		Ab	Aabb	10
aB	aaBb	25		aB	aaBb	12
ab	Aabb	25		ab	aabb	38
	(a)				(b)	

Figure 10.9. Testcross with (a) unlinked genes and with (b) linked genes

$$\text{Percent recombination} = \frac{\text{\# of recombinants}}{\text{Total offspring}} \times 100$$

Example, using numbers from Figure 10.9(b):

$$\frac{22}{100} \times 100 = 22\% \text{ recombination}$$

Recombinants

Recombinants are any new genetic combination of alleles seen in offspring that are not seen in either parent. In the cross shown in Figure 10.9, the recombinants are *Aabb* and *aaBb* since they are different from either parent. These new gene combinations are the result of independent assortment in nonlinked genes and of crossing-over in linked genes. Recombinants do not occur as often in linked genes. The genes cannot assort independently of each other because they are on the same chromosome. In addition, recombinants can result only from crossing-over. Since crossing-over does not always occur, the recombinants will always occur with less frequency than the parental genotypes.

Polygenic Inheritance

> **Polygenic inheritance:** The inheritance of traits that are coded for by many genes. The combined influence of the genes determines the phenotype.

REMEMBER

The chi-square test is the appropriate statistical analysis tool to use when evaluating the results of a genetic cross to determine if the observed phenotypic results are statistically significantly different than the expected phenotypic ratio.

Polygenic traits exhibit greater variation than monogenic traits. The variation seen in polygenic inheritance tends to show a continuous pattern. Most monogenic traits are expressed as one of two forms—either the dominant or the recessive form. When many genes play a role in determining the expression of a trait, the phenotypes of the trait exhibit a bell curve pattern. Figure 10.10 shows the typical bell curve pattern for the polygenic trait of skin colour. Other traits, such as height and hair colour, are also polygenic.

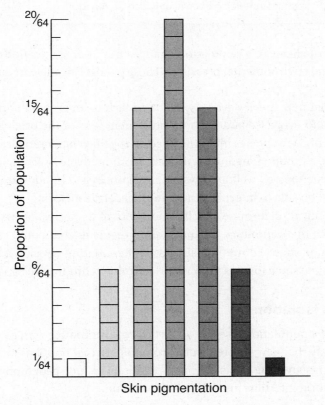

Figure 10.10. Polygenic inheritance of skin colour

GENETICS AND EVOLUTION

10.3 GENE POOLS AND SPECIATION

TOPIC CONNECTIONS

- 5.1 Evidence for evolution

Species and Speciation

The genetic makeup of reproducing organisms in a population determines the genetic makeup of future populations. Each offspring will receive one form of a gene (allele) from each of the two parents. If there are more dominant alleles in the population, future offspring will inherit a greater percentage of dominant than recessive alleles. Random mating encourages natural changes in the allele frequency of a population.

As the environment changes over time, the percentages of alleles in the population can change as well. If the environment favours a recessive trait over a dominant trait, more individuals possessing the recessive trait will survive to reproduce. This selection for the recessive allele will cause the recessive trait to be the more prevalent form of the gene in the population.

> **Allele frequency:** The percentage of various forms of a gene (alleles) that appear in a population (usually *A* and *a*).
>
> **Gene pool:** All the forms of a gene (alleles) that are found in an interbreeding population (such as *A* and *a*).

The frequency of alleles in a population changes from one generation to the next due to random mating and environmental pressures. These changes in allele frequency are the basis of evolution.

It is believed that new species evolve when members of an original population or species become isolated and evolve separately from other members of the original population. The unique evolution of each species gives the species specific characteristics. The term *species* is used to describe a group of organisms that are reproductively isolated and share common morphological structures as well as physiological processes. In addition, members of the same species must be able to interbreed and produce fertile offspring.

It is often difficult to determine specific members of a species. Issues arise when dealing with organisms that reproduce asexually and organisms that could possibly reproduce together but do not share common habitats. Differentiating species based on fossilized remains is impossible since there is no way to determine reproductive capabilities.

Reproductive Isolation

When members of a population lose the opportunity to reproduce with each other, they may evolve in such a way that new species evolve from the original population. As shown in Table 10.2, barriers that isolate gene pools within populations include geographical isolation, temporal isolation, hybrid infertility and behaviour isolation.

Table 10.2: Barriers Between Gene Pools

Type of Barrier	Description	Example
Geographic isolation (location isolation)	Land or water forms separate existing populations	Darwin's finches in the Galapagos Islands—each species of finch evolved from an original population that became separated by water due to island formation.
Temporal isolation (time isolation)	Timing of reproduction (availability of sperm and egg)	American toad and Fowler's toad (closely related species)—American toads mate in early summer and Fowler's toads mate in early fall.
Behavioural isolation	Behaviours that encourage re-production—many courtship behaviours must be exhibited before mating is allowed	Different species of crickets look very similar but do not mate with each other due to different "songs" they each create. Therefore, they do not interbreed.
Hybrid infertility	Members of different species mate and produce offspring, but the offspring are sterile.	The donkey and horse can mate and produce offspring (a mule). The hybrid mule is sterile. Hence the two populations cannot be considered the same species.

Polyploidy

Polyploidy occurs when a cell receives extra sets of chromosomes. Normal diploid ($2n$) cells contain two sets of chromosomes (one from each parent). Although all cells with extra sets of chromosomes are referred to as polyploid, a cell with three sets of chromosomes is more specifically referred to as triploid ($3n$). A cell with four sets is referred to as tetraploid ($4n$).

Polyploidy is rare in animal cells. In most cases, a polyploid animal zygote will not survive. In plants, polyploidy is not only common, it is one of the main sources of plant speciation. Polyploidy can occur with mitotic or meiotic failures of chromosome sets to separate fully, with mutations that lead to extra sets of chromosomes, or when more than two gametes undergo fertilization ($n + n + n$ (extra gamete) $= 3n$). (Refer to Figure 10.11.) The newly created polyploid organism can no longer reproduce with the population it originated from due to the differences in chromosomal numbers. Since plants can reproduce asexually (without union of gametes) and many species produce both male and female flowers that allow for self-fertilization, polyploidy is not an issue in plants. The new population of plants with polyploid chromosomal numbers is considered to be a new species since it can no longer reproduce with the population from which it originated. Most angiosperm species are thought to have evolved due to polyploidy.

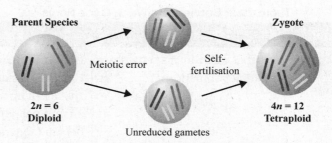

Figure 10.11. Meiotic error leading to polyploidy

Allopatric and Sympatric Speciation

Speciation refers to an event that separates an existing species into two or more new species. Speciation occurs when members of an existing species become reproductively isolated from each other. This isolation results in the development of separate gene pools for each of the separated populations. Over time, the separate gene pools evolve in such a way that they are no longer reproductively compatible. There are two basic types of speciation: allopatric speciation and sympatric speciation. (Refer to Table 10.3 and Figure 10.12.)

Table 10.3: Allopatric and Sympatric Speciation

Speciation Type	Isolation	Example
Allopatric (geographic isolation)	Speciation is due to members of the population living in *different geographic areas*. Members of an existing population become physically separated from each other.	Darwin's finches on the Galapagos Islands.
Sympatric	Speciation due to different behaviours of members of the population. Members of an existing population in *the same geographic area* become reproductively isolated.	Behavioural or temporal isolation. Members of an existing species begin reproducing at different times or exhibit different behaviours that must be present in order to allow reproduction.

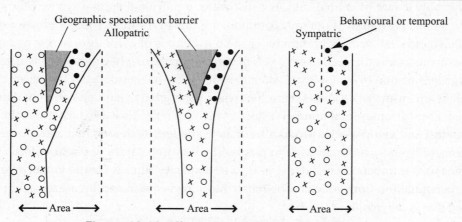

Figure 10.12. Allopatric and sympatric speciation

Adaptive Radiation

Many species evolved from one common ancestor. The evolution of new species occurs when variations among members of an original population allow some of the members to inhabit new niches and evolve for optimal survival in the new habitat. Natural selection for survival in the new niche drives the evolution of the population members in such a way that they are no longer able to reproduce with the original population. The separate niches of the population members could be the result of either allopatric or sympatric speciation.

An excellent example of adaptive radiation can be seen in the finches of the Galapagos Islands. Each species evolved from an original species when they became separated into new environments and evolved in such a way as to take advantage of each habitat's food source. Each new species of the Galapagos finches evolved from a common species with different beak adaptations that allowed them to feed on the food source available in the varied niches.

Convergent and Divergent Evolution

Convergent evolution and divergent evolution describe the development of similar structures that are present in different species. Divergent evolution occurs when species evolve from a common ancestor in such a way that their features become less alike. Homologous structures (same basic structures used for different functions) can be seen in divergent evolution. Although the monkey, whale, pig and bird are all descendants of mammals, they have evolved to inhabit specific niches. The basic structure of their forelimbs is the same, but the function of each of the structures is different. The limbs all contain the same number of bones with variations from the original pentadactyl (5-digit) limb.

Convergent evolution occurs when species evolve from different ancestor species in such a way that their features become more alike. These organisms originated from different ancestral lineages but developed adaptions for similar environments. Convergent evolution is associated with the development of analogous structures (different structural adaptions used for the same function). The bat, insect, and bird wing all evolved in order to allow flight. Each wing serves the same function (flight) but exhibits a different structure. (Refer to Figure 10.13.)

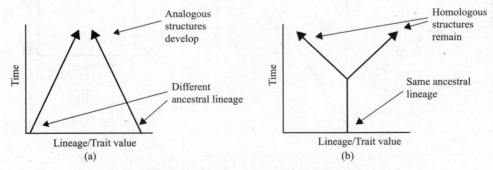

Figure 10.13. Convergent and divergent evolution

The pace of evolution seems to follow that of either gradualism or punctuated evolution. Gradualism involves the slow progression of the evolution of a new species. Punctuated equilibrium involves a more rapid and sudden development of a new species. As shown in Table 10.4 and Figure 10.14, species can evolve by either or by both forms of speciation.

Table 10.4: Gradualism and Punctuated Evolution

Gradualism	Punctuated Evolution
Slow process of speciation	Long process of speciation
Small variations seen during evolution of new species	Large variations seen suddenly during speciation
Constant process	Sudden process with long periods of stability
Due to natural selection favouring small variations	Due to sudden mutations favouring a new form of the species or due to a sudden drastic change in the environment favouring members of the species best adapted to the changed environment
Supported by fossil record showing slow changes as speciation occurs (for example, evolution of the modern horse)	Supported by fossil evidence showing stable species with the sudden appearance of new species without evidence of intermediate species (for example, mass extinctions followed by new species)

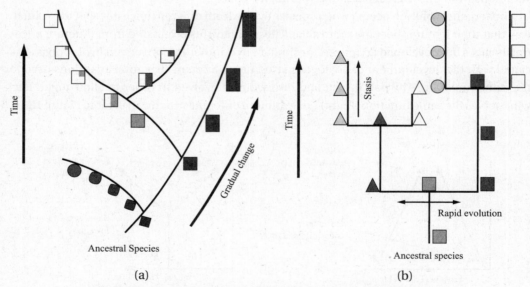

Figure 10.14. (a) Gradualism and (b) punctuated equilibrium

Transient Polymorphism

Polymorphism refers to the existence of more than one form of an organism within a population. Most species exhibit polymorphism in order to ensure variation within the population. Without variation, species may not be able to adapt to changes within their environment. Over time, as the environment changes, one form will gradually become more prevalent than the other. The selection for either form can change as the environment changes and favours one form over the other.

The best example of polymorphism exists within the populations of *Bison betularia*, the peppered moth. Two distinct variations, one grey (peppered) and one black (melanic), of the moth exist within the population. Previous to the British Industrial Revolution, the grey form was favoured in the environment and most of the population exhibited grey colouration. This was due to the fact that peppered moths rest on tree trunks that are light in colouration and the lighter form of the moth was better camouflaged from predators. The dark form was visible to predators. During the Industrial Revolution, great amounts of soot and pollution were released into the environment and settled on the trees, making their bark darker in colouration. This change in the environment favoured those few dark moths. Their dark colour now exhibited the best camouflage for the environment. The peppered moths were quickly seen by predators and consumed at greater rates then were the black (melanic) forms. Over a relatively short period of time, the peppered moth populations evolved to have mainly black forms with only a few of the light peppered moths present. Currently, due to the institution of clean air acts and less pollution being emitted into the environment, the bark of the trees has returned to a lighter shade, which naturally selects for the peppered moth over the black form. The peppered moth is currently the dominant form of the moth population, with few of the dark forms present.

> **REMEMBER**
>
> Within populations living in industrial regions, changes in colouration that favour darker forms over lighter forms due to the presence of industrial pollution are referred to as *industrial melanism*.

BALANCED POLYMORPHISM AND SICKLE CELL

Balanced polymorphism exists when more than one form (allele) of a trait exists within a population and there is no selection for one trait over the others. An excellent example is seen in the gene that codes for sickle cell. Two forms of the gene exist. One form of the gene codes for normal haemoglobin and results in normal red blood cells. The other form codes for abnormally shaped haemoglobin and results in sickle-shaped red blood cells.

Individuals who are heterozygous (contain one of each form of the gene) have an advantage over either homologous combination. Although possessing two of the recessive alleles results in sickle-shaped cells that prevent the development of malaria, these cells cannot carry oxygen efficiently and are prone to clotting. Possessing two of the dominant alleles results in normal red blood cells, but does not protect the individual from developing malaria. Individuals who are heterozygous for the sickle cell trait do not have problems carrying oxygen or have issues with blood clotting. In addition, they are protected from developing malaria due to the slightly curved shape of the red blood cells that prevents malaria from developing. Both alleles are selected for since the dominant form allows normal red blood cells to be produced and the recessive form offers protection from malaria. (Refer to Table 10.5.)

Table 10.5: Sickle Cell Anaemia Trait

Heterozygous for Trait (Hb^{A}Hb^S)*	Homozygous Dominant for Trait (Hb^{A}Hb^A)	Homozygous Recessive for Trait (Hb^{S}Hb^S)
SCT	Normal	SCA
Protected from malaria	No protection from malaria	Protection from malaria
Slightly curved red blood cells	Normal shape of red blood cells	Sickle shape of red blood cells
Can carry oxygen efficiently	Can carry oxygen efficiently	Cannot carry oxygen efficiently—leads to anaemia and blood clotting

*Hb^A = normal allele; Hb^S = sickle cell trait

Directional, Stabilizing and Disruptive Natural Selection

The three possible modes of natural selection for the form of a trait best adapted for the environment include directional, stabilizing and disruptive selection. (Refer to Figure 10.15.) Directional selection favours the phenotype at one of the ends of the ranges of observed traits. Stabilizing selection favours the intermediate phenotype between the two ends of the range of observed traits. Disruptive selection favours the phenotypes found at both ends of the ranges of observed traits. The type of selection seen in a population for a specific trait can change as the environment changes.

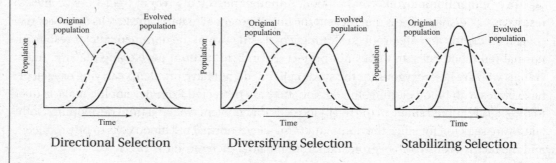

Directional Selection   Diversifying Selection   Stabilizing Selection

Figure 10.15 Modes of natural selection

1. Crossing-over most frequently occurs during:

   A. Prophase I of meiosis
   B. Prophase of mitosis
   C. Prophase II of meiosis
   D. Telophase I of meiosis

2. Which of the following genetic crosses will result in a 3:1 phenotypic ratio?

   A. Dihybrid cross
   B. Monohybrid cross
   C. One-point testcross
   D. Two-point testcross

3. When more than one gene plays a role in determining the phenotype of an individual, the trait is called:

   A. Monogenic
   B. Autosomal
   C. Polygenic
   D. Multiple alleles

4. The site at which nonsister chromatids undergo the process of crossing-over is referred to as the:

   A. Centromere
   B. Chromatid
   C. Chiasmata
   D. Kinetochore

5. The following cross was performed: *AaBb* × *aabb*. Which of the offspring produced from this cross are recombinants?

   A. *AaBb* and *aaBb*
   B. *Aabb* and *aabb*
   C. *AaBb* and *aabb*
   D. *Aabb* and *aaBb*

6. Which of the following processes contribute to genetic variation?

   I. Mitosis
   II. Crossing-over
   III. Independent assortment

   A. I only
   B. I and II only
   C. II and III only
   D. I, II and III

7. Which phase of cell division is shown below?

A. Metaphase I of meiosis
B. Anaphase I of meiosis
C. Metaphase II of meiosis
D. Anaphase II of meiosis

8. Which of the following traits are polygenic?

I. Skin colour
II. Height
III. Colour blindness

A. I only
B. I and II only
C. II and III only
D. I, II and III

9. Autosomes are:

A. Nonsex chromosomes
B. Sex chromosomes
C. Both nonsex and sex chromosomes
D. Only the X chromosome

10. When does recombination occur in linked genes?

A. During independent assortment
B. During crossing-over
C. During independent assortment and crossing-over
D. No recombination can occur in linked genes

11. What mode of natural selection favours both extreme phenotypes?

A. Stabilizing
B. Disruptive
C. Directional
D. Destabilizing

12. Which of the following can lead to speciation?

   I. Geographic isolation
   II. Temporal isolation
   III. Behavioural isolation

   A. I only
   B. I and II only
   C. II and III only
   D. I, II and III

13. Define gene pool. (1)

14. Perform the following cross, and identify the phenotypic ratios that will be seen in the offspring: *AaBb* × *AaBb*. (4)

15. A two-point testcross (involving two genes) was carried out on an individual heterozygous for both traits. The following numbers and genotypes of offspring were produced:

   AaBb: 27   *aaBb*: 5   *Aabb*: 6   *aabb*: 32

   a. Are the genes linked?
   b. Which offspring are recombinants?
   c. How can recombinants occur in linked genes?
   d. How can recombinants occur in unlinked genes? (4)

16. (a) Draw a tetrad. (4)
    (b) Explain how crossing-over leads to an exchange of alleles. (6)
    (c) Outline the major events involved in meiosis. (8)

ANSWERS EXPLAINED

1. **(A)** Crossing-over occurs during prophase I of meiosis.

2. **(B)** Monohybrid crosses result in 3:1 phenotypic ratios.

3. **(C)** Polygenic genes are coded for by more than one gene. They will exhibit a bell-curve pattern and include traits such as skin colour, height and hair colour.

4. **(C)** The chiasmata (singular: chiasma) is the site at which nonsister chromatids cross over during prophase I of meiosis.

5. **(D)** *Aabb* and *aaBb* are recombinants because they are different from either parent (a new combination of alleles not seen in either parent).

6. **(C)** Crossing-over and independent assortment contribute to genetic variation. Mitosis creates identical daughter cells.

7. **(A)** Metaphase I of meiosis is shown since homologous chromosomes are present and the chromosomes are lined up at the metaphase plate.

8. **(B)** Skin colour and height are polygenic traits. Colour blindness is a sex-linked trait (carried on the X chromosome).

9. **(A)** Autosomes are nonsex chromosomes.

10. **(B)** Linked genes can exhibit variation only due to the process of crossing-over. Linked genes do not exhibit independent assortment.

11. **(B)** Both extremes (both ends of the range of variation in the phenotype) are favoured in disruptive selection.

12. **(D)** Geographic, temporal and behavioural isolation can all lead to speciation.

13. A gene pool consists of all the possible alleles of all of the genes present in an interbreeding population.

14. The phenotypic ratio for a dihybrid cross is always 9:3:3:1.

	AB	*Ab*	*aB*	*aB*
AB	*AABB*	*AABb*	*AaBB*	*AaBb*
Ab	*AABb*	*AAbb*	*AaBb*	*Aabb*
aB	*AaBB*	*AaBb*	*aaBB*	*aaBb*
ab	*AaBb*	*Aabb*	*aaBb*	*aabb*

15. (a) Yes, they are linked.

    (b) The allele combinations of *aaBb* and *Aabb* are recombinants.

(c) Recombinants result from crossing-over in prophase I, leading to an exchange of alleles in the tips of the nonsister chromosomes.

(d) Independent assortment and crossing-over lead to genetic variation in unlinked genes.

16. (a) One point is awarded for a properly drawn and labelled diagram of a tetrad.
    - Sister chromatid
    - Nonsister chromatids
    - Centromere
    - Homologous chromosomes
    - Chiasmata

(b)
    - Crossing-over occurs during prophase I/meiosis
    - Involves nonsister chromatids
    - Occurs between homologous chromosomes
    - Occurs at the chiasmata
    - Alleles are switched between the nonsister chromatids
    - Crossing-over does not always occur
    - Leads to greater genetic diversity

(c)
    - Meiosis consists of meiosis I and meiosis II
    - In meiosis I, the chromosome number is reduced from diploid to haploid
    - Meiosis I produces two genetically different daughter cells
    - Chromosomes become visible in prophase I
    - Crossing-over can occur in prophase I
    - Homologous pairs/tetrads line up at the metaphase plate in metaphase I
    - Anaphase I separates homologous chromosomes
    - In telophase I, the nuclear membranes reform
    - Cytokinesis/interkinesis occurs after telophase I
    - Prophase II has visible haploid chromosomes/one set of the homologs
    - In metaphase II, the individual chromosomes line up at the metaphase plate/centre of the cell
    - Anaphase II separates the replicated chromosomes into individual chromosomes/chromatids separate
    - Meiosis II begins with 2 haploid cells and creates sperm and egg
    - Meiosis is the main source of genetic variation
    - Variation in meiosis is due to crossing-over and independent assortment

Animal Physiology

CENTRAL IDEAS
- Excretory processes
- Immunological processes
- Muscles and movement
- Reproductive processes

TERMS
- Active immunity
- Antibody
- Antigen
- Excretion
- Ligament
- Osmoregulation
- Passive immunity
- Sarcomere
- Tendon

INTRODUCTION

The human body is composed of multiple systems that assist in maintaining the body at ideal homeostatic levels. The excretory system expels waste from the body and aids in osmoregulation. The immune system helps prevent and eliminate the invasion of pathogenic agents. Muscles allow for movement of organisms and for the movement of internal substances (peristalsis and contractions of the heart). The process of reproduction enables populations to continue to exist and to pass on traits most beneficial for survival.

NOS Connections
- Ethical issues exist in scientific exploration (vaccine studies and early testing—Jenner and smallpox, infertility treatments).
- Curiosity leads to scientific discoveries (animal systems and physiology).
- Advances in technology contribute to scientific discoveries (animal systems and physiology).
- Causation and correlation need to be scientifically established (hormone use in food products and side effects—infertility, early puberty, risk of vaccinations—possible links to autism).

ANIMAL PHYSIOLOGY

TOPIC CONNECTIONS

- 6.3 Defence against infectious disease
- 11.4 Sexual reproduction

> **Innate immunity:** Immunity present at birth
>
> **Adaptive immunity:** Immunity developed during life

Once a pathogen has breached the first- and second-line (nonspecific) defence, the specific immune system goes to work. (Refer to Table 11.1.) The immune system must be able to identify body cells (self) from invading pathogens (nonself) and identify infected cells from non-infected cells. The immune system uses a group of cell surface molecular protein markers known as the major histocompatibility complex (MHC) to determine this. These molecular markers are found on the surface of all nucleated cells. (Note that red blood cells lack a nucleus and hence do not have MHC markers.) With the exception of identical twins, MHC markers are unique to each individual. There are two different classes of MHC markers—those located on immune cells (lymphocytes and macrophages) and those located on nucleated body cells. When body cells are infected with pathogens, the cells begin breaking down the antigen. The cells then attach parts of the antigen to an MHC marker and transport this to the surface of the cell membrane. This combination of self (MHC marker) and nonself (fragment of pathogen) alerts the immune system of the cellular invasion of the pathogen. The MHC markers are constantly being produced and displayed on the cell surface to ensure that any invading pathogens can be quickly identified and targeted for removal. The two classes (forms) of MHC markers allow the immune system to distinguish between infected body cells (class I) and antigen-presenting immune cells (class II). This way, the immune system can target infected body cells for destruction and target immune cells to proliferate.

REMEMBER

Pathogens can be species specific or can cross barriers and infect more than one species.

The immune system consists of different types of lymphocytes: B-cells and T-cells. These cells recognize and destroy specific pathogens. Both cell types are produced in the bone marrow and have distinct functions in the immune system. B-cells function to destroy any free (not inside of a cell) pathogen. T-cells function to destroy any body cells infected with pathogens, to destroy multicellular pathogens, and to destroy cancerous cells.

T-cells need to be carefully regulated since they destroy cells and the body is made of cells. For this reason, T-cells leave the bone marrow and migrate to the thymus (a gland located right above the heart) to complete their development and to be screened to make sure they destroy only foreign cells or infected body cells. Unlike the T-cells, B-cells are fully functional when they leave the bone marrow.

- **B-cells** originate and fully mature in the bone marrow (centre of long bones). B-cells are named for their location of maturation—the **b**one marrow.
- **T-cells** originate in the bone marrow but must migrate to the thymus to mature fully. T-cells are named based on their location of maturation—the **t**hymus.

ANIMAL PHYSIOLOGY

Table 11.1: Nonspecific Innate Defence Mechanisms

First Line Defence	Structure	Function
	Skin—slightly acidic, keratinized, stratified (layered)	Low pH inhibits many pathogens; keratin (a fibrous protein) adds structure to outer layer to help prevent entry of pathogens; multiple layers make skin impermeable to pathogens
	Mucous membranes	Trap pathogens and contain antimicrobial agents
	Stomach acid	Acidic environment destroys many pathogens

Second Line Defence	Structure	Function
	Phagocytes/macrophages	Engulf foreign material and alert the specific immune system of the invader
	Inflammatory response	Increases blood supply to the affected area to bring more white blood cells and antimicrobial agents to the site of infection
	Interferon	A chemical released by cells infected with a virus blocking cell-to-cell infection
	Complement system	A group of proteins that can lyse invading cells

> The development of T-cells that recognize the body's own cells (self cells) leads to autoimmune diseases in which the immune system destroys healthy cells. Examples of autoimmune diseases include lupus, rheumatoid arthritis and multiple sclerosis.

Cell-Mediated and Humoral Immunity

The immune system is divided into two types of responses based on the function of the B-cells and T-cells. B-cells are involved in the humoral immune response. T-cells are involved in the cell-mediated immune response. (Refer to Table 11.2.) Both responses are controlled by T-cells known as helper T-cells. Helper T-cells activate both humoral and cell-mediated responses when presented with an invading pathogen. Helper T-cells cannot recognize a pathogen unless it is presented with the antigen by an antigen-presenting cell (APC). Macrophages and B-cells can both present antigens to helper-T cells.

Table 11.2: Humoral and Cell-Mediated Immunity

Feature	Humoral Immunity	Cell-Mediated Immunity
Lymphocytes involved	B-cells	T-cells
Effector cells	Plasma cells	Cytotoxic T-cells
Target/function	Destroys all free antigens (those not inside of cells)	Destroys all infected cells, multicellular pathogens, and cancerous cells
Mode of action	Plasma cells produce antibodies that surround and neutralize the pathogen	Cytotoxic T-cells release chemicals that trigger apoptosis of the infected cell

When helper T-cells are presented an antigen, they release chemicals (cytokines) that stimulate the APC to proliferate and continue destroying the pathogen. At the same time, other helper T-cells release cytokines that will selectively activate only those B-cells and T-cells that are specific to the invading pathogen. Many different types of cytokines (including various interleukins) function in immune responses. For our purposes, we will refer to them generically as cytokines. The process of selecting specific immune cells for cloning is known as clonal selection, as shown in Figure 11.1.

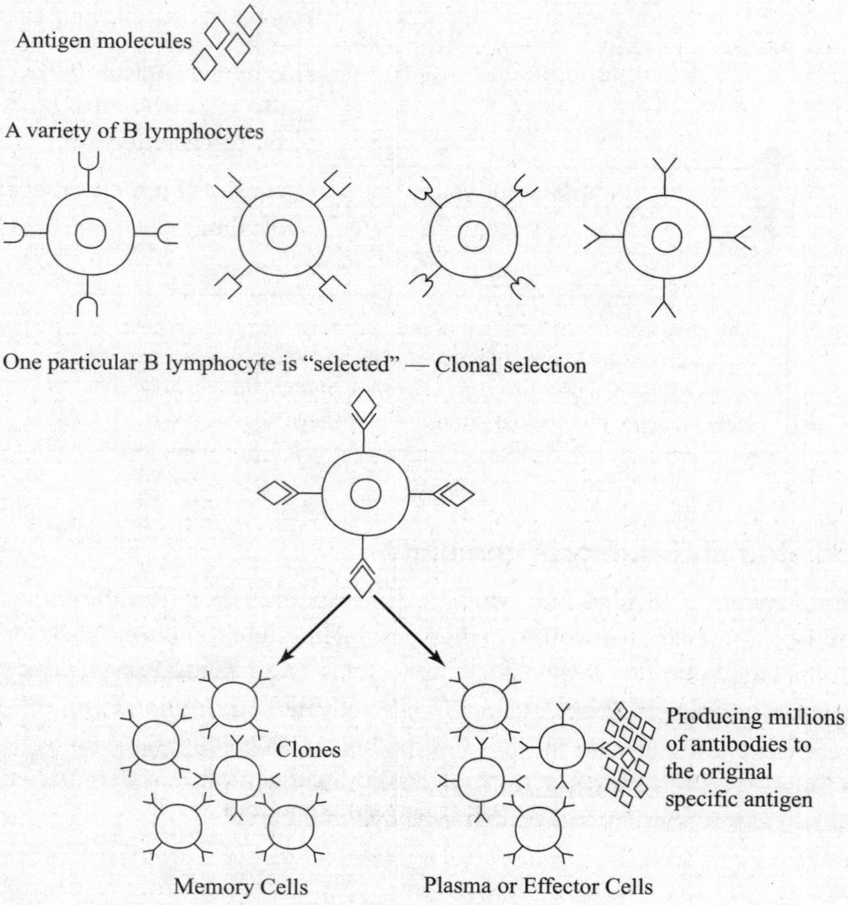

Figure 11.1. Clonal selection

The activation of the humoral response triggers existing B-cells to give rise to effector cells and memory cells. The effector cells of the humoral response are called plasma cells. Plasma cells develop from activated B-cells and produce antibodies that carry out the humoral response. Antibodies are immunoglobins (globular proteins of the immune system) that destroy free pathogens.

The cell-mediated response is initiated at the same time as the humoral response. The cell-mediated response activates its effector cells (cytotoxic T-cells) as well as memory cells. The cytotoxic T-cells begin destroying infected cells by releasing chemicals that lyse the cell membranes of the infected cells. This induces cell death (apoptosis) of the infected cells.

The memory cells that are produced during the first exposure to the pathogen function only if there is a subsequent invasion by the same pathogen. Since the memory cells can quickly recognize the previously encountered pathogen, the immune response is much quicker during the secondary exposure. (Refer to Figure 11.2.)

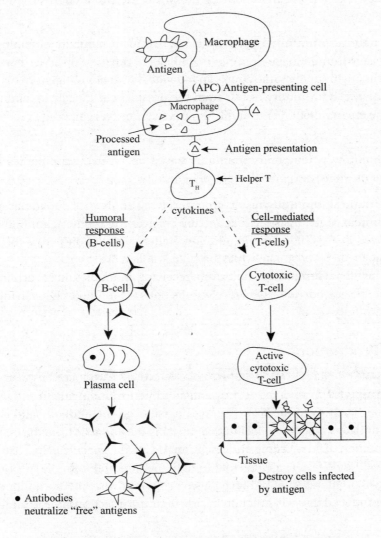

Figure 11.2. Humoral and cell-mediated immunity

Review of Immune Response

- **Challenge:** The body is invaded by a pathogen.
- **Response:** Helper T-cells are alerted of the presence of the pathogen by an APC. Humoral and cell-mediated immunity are activated by helper T-cells.
- **Clonal selection**: Only those B-cells and T-cells that are specific to the antigen are activated (signalled to clone themselves) from the large pool of existing B-lymphocytes and T-lymphocytes.
- **Memory cells**: Memory cells are produced to be ready to destroy any subsequent invasion by the same pathogen quickly. (Note that when the same pathogen invades, memory cells for helper T-cells, B-cells, and T-cells are all activated.)

Active and Passive Immunity

Active immunity is permanent immunity based on the production of antibodies by the infected organism.

- **Active natural immunity** results from the organism naturally obtaining the antigen from the environment. For example, an individual can develop active natural immunity after drinking from the same cup as an individual who has antigens in his or her saliva.
- **Active artificial immunity** results from the organism artificially obtaining the antigen from the environment. For example, immunizations (vaccinations) provide active artificial immunity.

Passive immunity is temporary immunity based on receiving antibodies from another organism that previously experienced active immunity.

- **Passive natural immunity** results from receiving antibodies due to natural processes. For example, while breastfeeding, a baby receives the mother's antibodies, which are transferred in the colostrum. In addition, maternal antibodies cross the placenta and enter the foetal bloodstream while the child is still in the uterus.
- **Passive artificial immunity** results from receiving antibodies due to artificial processes. For example, an individual who receive injections of antibodies is receiving passive artificial immunity.

Antibody Production

Antibodies (immunoglobins) are Y-shaped globular proteins produced by plasma cells during the humoral response. Antibodies destroy antigens by surrounding them and targeting them for destruction. The tips of the Y are the binding sites on the antibody and can effectively clump many antigens together to facilitate removal of the antigens from the body.

The presentation of an invading antigen by an APC to the specific helper T-cell stimulates the helper T-cell to release cytokines that initiate the humoral immune response. The activated B-cell will proliferate and produce plasma cells that will produce antibodies. Memory cells are also produced that will function in case of future infections by the same antigen.

Monoclonal Antibody Production

Monoclonal antibody production is different from natural antibody production in that it entails the use of biotechnology. Monoclonal antibody production allows for the mass production of an antibody specific to any desired antigen. The first step in monoclonal antibody production is the injection of a rodent with a selected antigen. After enough time has elapsed to allow for a full immune response by the rodent, the spleen is removed. The B-cells that were produced in the immune response against the desired antigen are removed from the spleen. These B-cells are fused with cancer (myeloma or tumour) cells in order to produce hybridomas. The hybridomas obtain the ability to keep dividing indefinitely from the cancer cell and obtain the ability to produce antibodies from the B-cell. The antibodies produced by the hybridomas are collected and used for diagnosis and treatment. (Refer to Figure 11.3.) Diagnostic uses include the detection of antibodies to HIV, the detection of cardiac enzymes that signal a heart attack and the detection of HCG (human chorionic gonadotropin) in pregnancy tests. Treatments include targeting cancer cells with chemotherapeutic drugs attached to monoclonal antibodies, emergency treatment of rabies and tissue typing for transplant compatibility.

Monoclonal Antibody Production

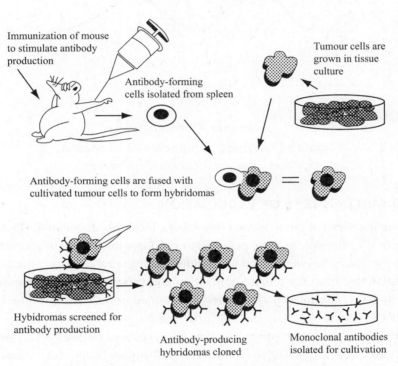

Immunization of mouse to stimulate antibody production

Antibody-forming cells isolated from spleen

Tumour cells are grown in tissue culture

Antibody-forming cells are fused with cultivated tumour cells to form hybridomas

Hybidromas screened for antibody production

Antibody-producing hybridomas cloned

Monoclonal antibodies isolated for cultivation

Figure 11.3. Monoclonal antibody production

The Principle of Vaccination

The goal of vaccination is to trigger a primary immune response that will result in the production of memory cells. A weakened form of the pathogen is injected or inhaled by the individual (active artificial immunity). The weakened form allows the immune system to recognize the antigen and mount an immune response without causing symptoms that result from encountering the active antigen. After the first exposure to an antigen, the immune system takes 5 to 10 days to develop a full immune response and to rid the body of the pathogen. After the second exposure (and all future exposures), the immune system takes only 3 to 5 days to develop a full immune response. In addition, the number of antibodies produced in the second (and later) exposure is more than triple that of those produced during the primary exposure. The much faster secondary response is due to the presence of memory cells that can quickly identify and eliminate the pathogen. This rapid response eliminates the antigen from the body before any symptoms can develop. (Refer to Figure 11.4.)

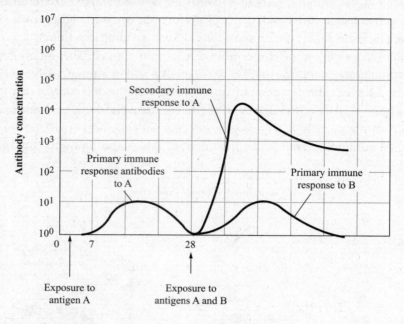

Figure 11.4. Graph of primary and secondary exposure times and of antibody production

BENEFITS AND DANGERS OF VACCINATION

Pathogens are the cause of many serious side effects, including death, in the human population each year. Vaccinations have the potential to eliminate diseases in the population and to prevent their serious side effects. Due to vaccinations, the viral disease caused by smallpox has been eliminated from the human population and the crippling viral disease caused by polio is rarely seen. Vaccinations can protect the human population from the spreading of diseases that can lead to epidemic and pandemic events.

Vaccinations reduce the symptoms and side effects caused by disease and greatly reduce health care costs. In addition, the work force is not disrupted by illnesses caused by disease that could impact local and global economies.

Risks due to vaccinations are minimal but do exist. Some individuals may be allergic to components found in vaccinations and may experience allergic symptoms after exposure to the vaccine. The injection site could become infected or show irritation due to the penetration of the skin that could introduce foreign substances to the body. Rarely, vaccinations can cause more severe side effects mainly due to the immune response of the recipient. Excessive immunological responses to vaccines in rare cases can cause neurological issues. Vaccinations are subjected to great study. Trials are carried out prior to mass delivery in order to ensure the safety of the population receiving the vaccination. (Refer to Table 11.3.)

Table 11.3: Benefits and Dangers of Vaccinations

Benefits of Vaccinations	Possible Dangers of Vaccinations
The total elimination of diseases (for example, smallpox)	Infections at the site of injection
Prevention of epidemics and pandemics	Allergic reactions to vaccine components
Decreased health care costs	Irritation at the vaccination site
Prevention of harmful side effects of the disease	In rare cases, more severe complications have been reported (triggering autoimmune problems, dangerously high fevers, immobilization of injected limb)

Allergies

Allergies result when the immune system responds with an exaggerated immune response to harmless antigens such as food and pollen. The allergic reaction leads to white blood cells releasing histamine in response to the presence of the allergen (allergic substance). Histamine triggers physiological responses, including:

- Constriction of smooth muscles (this can block airways)
- Increasing the permeability of capillaries (causes inflammation)

The constriction of smooth muscles can lead to asthma—the constriction of the smooth muscles in the respiratory system that prevents proper air flow.

11.2 MOVEMENT

Muscles work in conjunction with bones in order to move the body. Bones and exoskeletons provide the framework for muscle attachment and act as levels to produce movement. Bones are held together by ligaments and are attached to muscles by tendons. Nerves trigger the contraction or relaxation of muscles in order to move the bones. Skeletal muscles work in antagonistic pairs in order to move the body.

- **Bones:** Support the body, allow attachment for muscles for body movement, protect vital organs, serve as a warehouse for minerals and function for blood cell formation.
- **Ligaments:** Provide stability at joints by attaching bone to bone.
- **Tendons:** Attach muscle to bone to allow for movement.
- **Nerves:** Send impulses to muscles to signal relaxation or contraction of the muscle.

> **REMEMBER**
>
> Ligaments—attach "like to like" (bone to bone). This phrase may help you remember their function!

Most skeletal muscles, such as the triceps and biceps, are antagonistic to each other. This means that when one is contracted, the other is relaxed. The triceps contract to extend the arm (extensor), and the biceps contract to flex the arm (flexor). When one muscle is contracted, the other must be relaxed. The relaxation and contraction of muscles is under control of the nervous system. The elbow is a synovial joint. Synovial joints allow movement in some but not all ranges.

> **Triceps:** This extensor muscle contains three (tri) bundles of muscle cells (fibres).
>
> **Biceps:** This flexor muscle contains two (bi) bundles of muscle cells (fibres).

Human Elbow Joint

Joints are locations in the body where bones come together. Joints allow movement in different planes based on the type of joint found at the site of connection between the bones. Synovial joints contain synovial fluid. This fluid prevents friction being produced by bones rubbing together. The fluid is, in effect, a cushion. Certain types of joints prevent movement in some directions. Other types of joints allow great freedom of movement in many directions. The elbow joint, as shown in Figure 11.5 and Table 11.4, is an example of a hinged synovial joint. Hinged joints allow extension and flexion of the bones located at the joint (movement along one plane). Ball and socket synovial joints, in contrast, allow much greater rotation and allow movement in all directions. The hip joint is an example of a ball and socket joint.

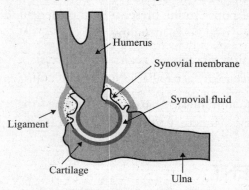

Figure 11.5. Human elbow joint

Table 11.4: Structures of the Elbow Joint

Structure	Function
Cartilage	Reduces friction and cushions impact between the ends of bones
Synovial fluid	Lubricates and prevents friction between the ends of bones
Joint capsule	Seals the joint space and provides stability to the joint
Biceps	Flexor muscle that, when contracted, pulls the arm upwards
Triceps	Extensor muscle that, when contracted, pulls the arm downwards
Humerus	Long, upper-arm bone
Radius	Shorter of the two bones of the forearm
Ulna	Longer of the two bones of the forearm

Skeletal Muscle

Three types of muscle exist in the human body—skeletal, cardiac, and smooth. Each muscle type is composed of myofibrils (actin and myosin) arranged in unique ways in order to carry out distinct functions. Muscle contraction occurs based on the movement of muscle fibres sliding over each other to allow the muscle to shorten and to elongate. This is analogous to the workings of an extension ladder. The ladder is short when the parts slide over each other but can be elongated by extending the parts out past each other.

REMEMBER

The prefixes *sarco-* and *myo-* always refer to muscle!

The basic unit of muscle contraction is the sarcomere. Each muscle cell (fibre) can contain thousands of sarcomeres. Each sarcomere is composed of two myofibrils known as actin and myosin. Myosin myofibrils are thicker than actin myofibrils. The stacking of the thick and thin myofibrils in skeletal muscle produces a banding pattern that can be seen under the microscope, as shown in Figure 11.6. Areas of the sarcomere containing only myosin and some parts of actin myofibrils produce a dark banding pattern since myosin is much thicker than the actin myofibrils. Areas with only actin myofibrils will appear as light bands.

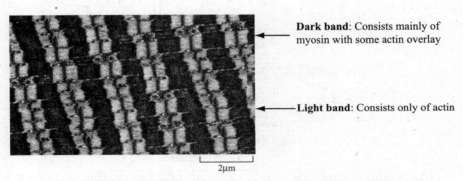

Dark band: Consists mainly of myosin with some actin overlay

Light band: Consists only of actin

2µm

Figure 11.6. Dark and light bands as seen in an electron micrograph

The stacked myofibrils are found within each cell surrounded by the muscle cell membrane—the sarcolemma. The sarcolemma contains infoldings (T-tubules) that reach deep into the cell. These infoldings are called T-tubules because the "T" means "transverse tubule". The T-tubules greatly increase the surface area of the sarcolemma and allow action potentials to travel rapidly throughout the cell.

Since skeletal muscle cells can be very long, each cell contains multiple nuclei (multinucleated) within the sarcolemma in order to regulate the many processes occurring in the elongated cell efficiently. Muscle cells also contain a special type of endoplasmic reticulum, the sarcoplasmic reticulum. The sarcoplasmic reticulum stores and releases calcium needed to initiate muscle contraction. (Refer to Figure 11.7.)

Muscle cells use a great deal of energy to fuel muscle contraction. For this reason, they are packed with mitochondria. When muscle mass increases, the increase is due to the increasing number of mitochondria being produced by each cell. The increase in muscle mass will be sustained only as long as the additional mitochondria are needed. If the additional mitochondria are not being used, they will be broken down and removed from the cell. This will cause the muscle mass to decrease and is referred to in the statement "use it or lose it".

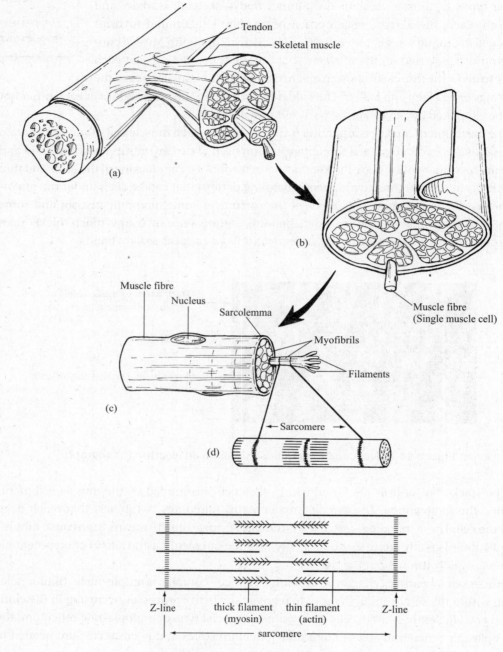

Figure 11.7. Muscle fibre structure

SARCOMERE

The sarcomere is the functional unit of muscle cells. Each muscle cell is composed of many repeating sarcomeres. Each sarcomere is composed of long proteins that facilitate muscle contraction by sliding past each other. The borders of each sarcomere are called the Z lines. The region where only myosin is located is referred to as the H zone. The A band contains myosin with overlapping actin and appears dark in electron micrographs. The I band is the region of the sarcomere with only actin filaments. It appears lighter in electron micrographs.

In Greek, *sarx* means "flesh" and *meros* means "part"; hence, the sarcomere is part of the flesh (muscle).

- **Z lines**: Borders of the sarcomere
- **H zone**: Region where only myosin is located
- **A band**: Region where myosin with overlapping of actin is located (dark band)
- **I band**: Region where only actin is located (light band)

Note that IB will require identification only of the Z lines, actin filaments, myosin filaments with heads and the resulting light and dark bands. You will not be required to know H zone or the A and I bands. However, points may be awarded for knowing the extended components of the sarcomere.

Skeletal Muscle Contraction

The contraction of skeletal muscle is initiated by an action potential sent from the nervous system that arrives at a neuromuscular junction. The steps involved are outlined below and are also shown in Figure 11.8.

1. The action potential arrives at the terminal button of a neuron, triggering the release of a neurotransmitter (most commonly acetylcholine).
2. Neurotransmitter is released into the synapse of the neuromuscular junction (site where a neuron meets with a muscle cell).
3. Neurotransmitter is attached to receptors on sodium-gated ion channels in the sarcolemma. Sodium channels open, depolarizing the muscle fibre (cell).
4. Action potential spreads through the T-tubules of the sarcolemma, initiating the release of calcium from the sarcoplasmic reticulum.
5. Calcium attached to the protein troponin causes tropomyosin to shift, exposing the myosin-binding sites.
6. Each myosin head binds an ATP and splits the ATP into ADP + P_i (inorganic phosphate), causing the myosin head to shift to a "cocked" (bent outwards) position.
7. Each myosin head forms a cross-bridge with actin and releases the P_i. Release of the P_i causes the myosin heads to shift position (bend inwards from the "cocked" position) and slide actin over myosin, shortening the sarcomere.
8. Each myosin head binds a new ATP molecule in order to break the cross-bridge.
9. The cycle repeats until no more action potentials are sent and the calcium is pumped back into the sarcoplasmic reticulum. The removal of calcium from the sarcoplasm (muscle cell cytoplasm) allows tropomyosin to shift its position back to covering the myosin heads. The muscle cell relaxes.

REMEMBER

Actin consists of 2 helical actin molecules with tropomyosin wrapped around them, regulating the exposure of the myosin binding sites.

ANIMAL PHYSIOLOGY

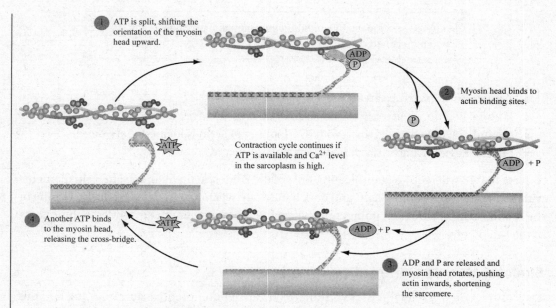

1. ATP is split, shifting the orientation of the myosin head upward.

2. Myosin head binds to actin binding sites.

Contraction cycle continues if ATP is available and Ca²⁺ level in the sarcoplasm is high.

3. ADP and P are released and myosin head rotates, pushing actin inwards, shortening the sarcomere.

4. Another ATP binds to the myosin head, releasing the cross-bridge.

Figure 11.8. Cross-bridge formation and release

Electron Micrographs of Skeletal Muscle

In the electron micrograph shown in Figure 11.9, you can clearly see that the light bands change in width but the dark bands do not. Remember that actin slides over myosin when a muscle cell contracts. The light bands shorten (b) since more of the actin is interacting (forming cross-bridges) with the myosin. When the muscle relaxes (a), the actin slides back away from the myosin with fewer interactions between them, resulting in a wider light band.

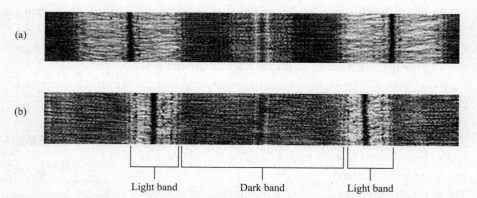

(a)

(b)

Light band   Dark band   Light band

Figure 11.9 Electron micrograph of (a) relaxed and (b) contracted muscle fibres

11.3 THE KIDNEY AND OSMOREGULATION

TOPIC CONNECTIONS

- 1.3 Membrane structure
- 1.4 Membrane transport

The excretory system functions to rid the body of toxic waste through the process of excretion.

> **Excretion:** The removal of the waste products of metabolic processes from the body.

All organisms must regulate water balance within their tissues. Animals are classified as being either osmoconformers or osmoregulators. Osmoconformers maintain solute concentrations in their body tissues that are isotonic to (the same as) the surrounding environment. Saltwater fish, such as sharks, are osmoconformers and can store molecules such as urea in their tissues. By storing solute in their tissues, they can maintain isotonic concentrations of solute between their tissues and the saltwater environment they inhabit. Osmoregulators have adaptations to live in aquatic environments that are not isotonic to their body tissues. Freshwater fish are osmoconformers. They rid their tissues of excess water by constantly excreting very dilute urine. Some saltwater fish are osmoregulators that possess salt-excreting glands to pump out the excess salt they take in from their high solute environment.

Insects possess Malpighian tubules to carry out excretion and osmoregulation. Mammals possess kidneys to carry out excretion and osmoregulation. Malpighian tubules absorb salts, water and waste in order to regulate the concentration of these molecules within the insect's tissues. The kidneys of humans possess nephrons that are tubular in structure and carry out similar functions as the Malpighian tubules.

The Human Kidney

The human excretory system includes the kidneys, ureters, bladder and urethra (see Figure 11.10). The kidneys carry out both excretion and osmoregulation as they filter blood and regulate solute balance. The filtrate produced by the kidney eventually becomes urine that drains into the ureters and collects in the bladder. From the bladder, the urine is released from the body through the urethra.

> **REMEMBER**
>
> Anything involving or related to the kidney is referred to as *renal*.

The outside of the kidney is the renal cortex. The middle of the kidney is the renal medulla. The inner, funnel-shaped section that collects the urine is the renal pelvis. The renal artery brings blood into the kidney, and the renal vein takes blood away from the kidney.

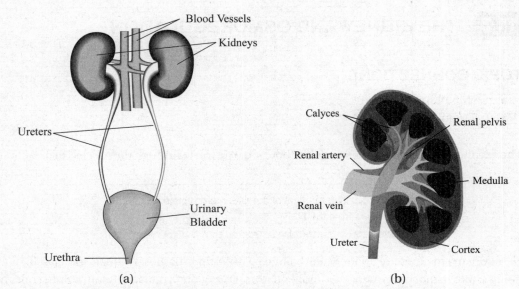

Figure 11.10. (a) The excretory system and (b) the kidney

THE NEPHRON

The functional unit of the kidney is the nephron. The nephron carries out the processes of ultrafiltration, selective reabsorption, secretion and excretion. Each process occurs in a distinct location in the nephron. (Refer to Figure 11.11 and Table 11.5.)

Table 11.5: Functions of the Nephron

Function	Location	Processes
Ultrafiltration	Glomerulus	Filtering of molecules small enough to fit through the fenestrations (openings) in the capillaries from the blood into Bowman's capsule (urea, salts, glucose, amino acids). Large proteins and blood cells are too large to fit through the fenestrations and stay in the blood capillaries.
Selective reabsorption	Proximal convoluted (folded) tubule	Glucose, amino acids and salts are reabsorbed back into the blood capillaries. These are actively pumped out of the proximal tubule, and water follows passively by osmosis.
Secretion	Distal convoluted tubule	Toxins and metabolites of drugs are pumped into the tubule to be excreted from the body.
Excretion	Collecting duct	Urea, some salts, water, ammonia and toxins are excreted from nephron and, ultimately, from the body as components of urine.

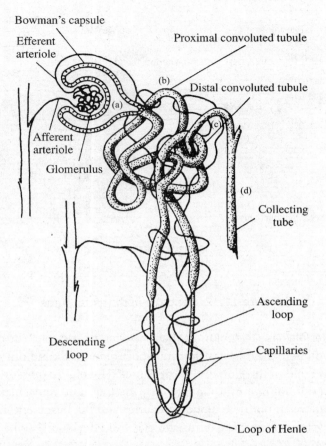

Figure 11.11. The nephron and associated functions: (a) ultrafiltration; (b) selective reabsorption; (c) secretion; and (d) excretion

Ultrafiltration refers to filtration at the molecular level. Ultrafiltration allows material to leave the glomerulus and enter into Bowman's capsule of the nephron. The capillaries found in the glomerulus of the nephron are more porous (fenestrated) than most capillaries. This allows the nonselective movement of molecules based solely on size. Podocyte cells surround the capillaries to aid in stabilizing the cells and to assist in ultrafiltration. Podocytes possess slits (folds in their membrane) that allow the passage of material. A basement membrane is located between the capillaries and podocytes and holds the capillaries to the podocyte cells to facilitate ultrafiltration. Small molecules can pass through the fenestrations in the capillaries. However, large proteins and blood cells cannot. (Refer to Figure 11.12.)

The vessel bringing blood into the glomerulus is the afferent arteriole. The vessel taking blood away from the glomerulus is the efferent arteriole.

> **REMEMBER**
>
> In Latin, *fenestra* means "windows". Hence fenestrations are openings like windows!

> **Afferent:** Going towards; moving into
> **Efferent:** Going away from; moving out of

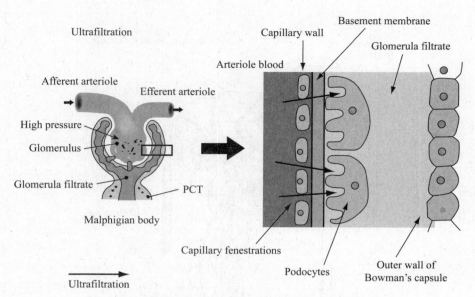

Figure 11.12. Structure of the glomerulus

The afferent arteriole has a wider lumen than the efferent arteriole. The difference in width of the afferent and efferent arterioles that transport blood into and out of the glomerulus causes pressure to build in the glomerular capillaries. The higher pressure in the capillaries allows for effective ultrafiltration of the blood by forcing more molecules to enter into the nephron. Fenestrations in both the blood capillaries and the basement membrane of Bowman's capsule allow the molecules to diffuse quickly from the blood capillaries into the nephron. (Refer to Figure 11.13.)

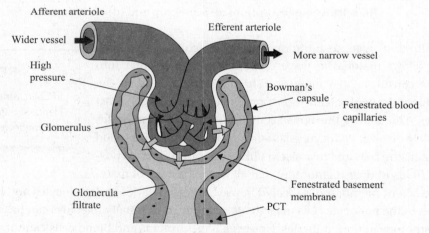

Figure 11.13. Ultrafiltration

Osmoregulation

Organisms must be able to regulate water balance within cells and tissues. In addition to excretion, the kidney also functions for osmoregulation.

> **Osmoregulation:** Control of water balance in the blood, tissues or cytoplasm of cells

The process of selective reabsorption in the proximal convoluted tubule allows selected molecules to be removed from the filtrate and put back into the blood capillaries. The cells that make up the proximal tubule contain microvilli to increase surface area for the movement of molecules through protein channels out of the filtrate. Glucose, amino acids and salts are all actively pumped out of the filtrate and back into the capillaries to prevent these nutrients from being lost from the body. Water passively follows the solutes due to osmosis.

In order to conserve water, the kidney must be able to move water out of the nephrons. The amount of water left in the nephrons for excretion depends on the concentration of water in the body. The loop of Henle is an important structure in maintaining water balance in the body. The processes occurring in the loop of Henle maintain high solute concentrations in the surrounding fluid (hypertonic conditions).

The loop of Henle reaches into the medulla of the kidney where solute concentration is high. This high solute concentration in the surrounding environment causes the osmotic movement of water out of the loop of Henle and back into body tissues. More water can be reabsorbed from the collecting duct when the hormone ADH (antidiuretic hormone) is present. When the hypothalamus detects the solute concentration of the blood as being too high, it triggers the release of ADH from the pituitary. Antidiuretic hormone travels to the collecting duct of the nephron, where it causes more channels for water to open and facilitates increased water movement out of the filtrate into the surrounding capillaries. More water entering the capillaries dilutes the solute in blood and brings the osmotic level back to homeostatic levels. The return of the homeostatic osmotic levels of blood inhibits the release of ADH and the transport of additional water out of the nephron. (Refer to Figure 11.14.)

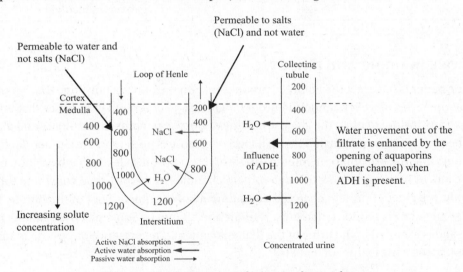

Figure 11.14. Osmoregulation in the nephron

The descending loop of Henle is permeable to water but not salts. The ascending loop of Henle is permeable to salts but not water. This allows for the salt concentration to increase in the descending loop so there is a high concentration that facilitates salt leaving the ascending loop and entering into the surrounding tissue. The differences in permeability to salt and water allow for solute (salts) to build up in the area surrounding the collecting duct so water will naturally diffuse into the medulla from the collecting duct. The long loop of Henle, reaching into the high solute concentration of the medulla, allows for efficient conservation of water in the body. Longer loops of Henle are found in mammals that need to conserve more water. Hence, the need for water conservation and the length of the loop of Henle are positively correlated.

Protein, Glucose and Urea Filtration

Proteins, glucose and urea each exhibit different levels of filtration from the blood. Large proteins are too big to undergo ultrafiltration and are not found in the filtrate. Glucose is found in the filtrate but is actively reabsorbed in the proximal convoluted tubule and is not found in urine. Urea is found in the filtrate. Since it is not selectively reabsorbed, it is present in urine. (Refer to Table 11.6.)

Table 11.6: Proteins, Glucose and Urea in Blood Plasma, Glomerular Filtrate and Urine

Molecule	Present in Blood Plasma	Present in Glomerular Filtrate	Present in Urine
Proteins	Yes	No (too large to undergo ultrafiltration)	No
Glucose	Yes	Yes	No (selectively reabsorbed since it is vital for cellular respiration)
Urea	Yes	Yes	Yes (excreted since it is a toxic molecule)

GLUCOSE IN URINE AND DIABETES

Diabetic patients have lost the ability to produce insulin. Insulin is a hormone that acts as the key that opens up protein channels in order for glucose to enter into cells. Since diabetics do not have any insulin to open the glucose channels, glucose remains in the blood, and glucose levels become dangerously high. This high level of glucose is present in the glomerular filtrate and enters into the proximal convoluted tubule of the nephron. Selective reabsorption of glucose occurs by active transport through protein channels in the proximal convolute tubules. Normally, the glucose protein channels can pump all of the glucose out of the filtrate. However, the high levels found in diabetics saturate the channels. Some glucose remains in the filtrate and is excreted in the urine of the diabetic patient. The presence of glucose in urine is the main identifying feature of diabetes.

NITROGENOUS WASTE

Organisms have evolved to excrete ammonia, urea or uric acid as their main nitrogenous waste product. The nitrogenous waste produced by an organism is directly related to the environment in which the organism has evolved to inhabit.

All organisms produce ammonia as a by-product from the digestion of proteins and nucleic acids. Freshwater organisms can simply excrete the ammonia into the environment and do not need to use energy to convert ammonia to the less toxic molecule urea. Terrestrial animals need to convert the toxic ammonia into the less toxic nitrogenous waste urea since they cannot immediately expel the nitrogenous waste product. Some animals use even more energy and convert urea to uric acid. Uric acid is a precipitate and the least toxic form of nitrogenous waste. Birds as well as other egg-laying organisms produce uric acid in order to allow the waste to settle away from the developing chick inside of eggs. Urea and uric acid both require less water for excretion, making their production an important component of osmoregulation and allowing less water to be lost to the environment.

Renal Arteries Versus Renal Veins

The composition of blood in the renal artery and the renal vein is not the same, as shown in Table 11.7. Blood entering the kidney through the renal artery is high in urea, oxygen and other toxins. The kidney filters out urea and toxins from the blood entering the kidney. The cells in the kidney use the oxygen transported into the tissues for cellular respiration. The carbon dioxide produced from cellular respiration is carried away from the kidneys by the blood in the renal vein.

Table 11.7: Composition of Blood in the Renal Artery and Renal Vein

	Renal Artery	Renal Vein
Oxygen levels	High	Low (used for cellular respiration by cells in the kidney)
Carbon dioxide levels	Low	High (given off by kidney cells during cellular respiration)
Urea (nitrogenous waste)	High	Low (removed during ultrafiltration and remains in the filtrate for excretion from the body)
Glucose	Slightly high	Slightly low (most is actively reabsorbed in the proximal convoluted tubule—some is used by the kidney for cellular respiration)

11.4 SEXUAL REPRODUCTION

TOPIC CONNECTIONS

- 3.3 Meiosis
- 6.6 Hormones, homeostasis and reproduction

Spermatogenesis begins with the mitotic division of the germinal epithelial cells to produce many diploid ($2n$) primary spermatocytes. The primary spermatocytes undergo meiosis I to produce haploid (n) secondary spermatocytes. Meiosis II of the secondary spermatocytes produces 2 haploid (n) spermatids. Sertoli cells nourish and help the development of the spermatids into mature spermatozoa (n). The mature spermatozoa are carried in fluid flowing from the seminiferous tubules into the epididymis, where they gain motility. (Refer to Figures 11.15, 11.16 and 11.18 as well as Tables 11.8 and 11.9.)

The Testes and Spermatogenesis

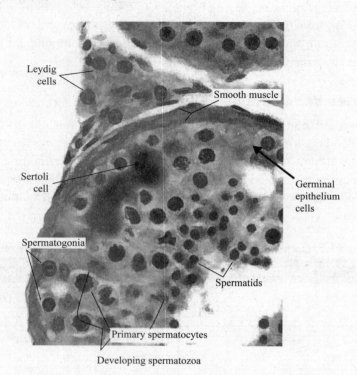

Figure 11.15. Light microscope of testicular tissue

Table 11.8: Cells of the Testes

Structure	Function	Location
Interstitial cells—Leydig cells	Produce testosterone when stimulated by LH (luteinizing hormone)	Outside the testes in the interstitial fluid
Germinal epithelium cells	Cells that make up the wall of the seminiferous tubules and function to initiate sperm production	Innermost layer of the testis tissue
Developing spermatozoa	Undergo meiosis and differentiate into mature, functional sperm	Towards centre lumen of the testicular tissue
Sertoli cells	Support and nourish the developing sperm when stimulated by FSH (follicle-stimulating hormone)	Adjacent to developing spermatozoa (much larger than surrounded developing spermatozoa cells)

ANIMAL PHYSIOLOGY

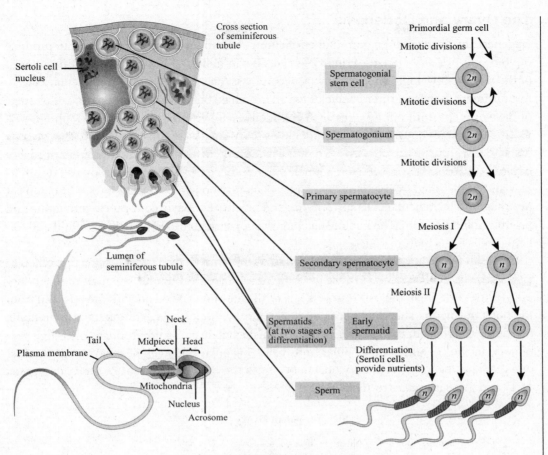

Figure 11.16. Spermatogenesis

Table 11.9: Role of Hormones in Spermatogenesis

Hormone	Source	Target Tissue	Response
LH (luteinizing hormone)	Pituitary gland	Leydig cells (interstitial cells) of the testes	Release of testosterone from Leydig cells
FSH (follicle-stimulating hormone)	Pituitary gland	Primary spermatocytes	Production of secondary spermatocytes from meiotic division of primary spermatocytes
Testosterone	Leydig cells (interstitial cells surrounding seminiferous tubules)	Secondary spermatocytes	Production of mature sperm from secondary spermatocytes

ANIMAL PHYSIOLOGY

The Ovary and Oogenesis

Oogenesis begins when the germinal epithelium cells undergo mitotic divisions to produce diploid ($2n$) primary oocytes. Primary oocytes begin meiosis I but are halted at prophase I of the meiotic division. Primary oocytes become associated with a layer of follicular cells to form primary follicles. Primary follicles are all produced before birth and remain at this stage of development until puberty begins. Each menstrual cycle initiates the development of a mature secondary oocyte by stimulating the primary oocytes to finish meiosis I. This meiotic division involves the unequal division of the cytoplasm, resulting in the formation of a large haploid (n) secondary oocyte and a small polar body cell (n). The polar body cell degenerates and is absorbed by the body. The large secondary oocyte begins meiosis II but stops at prophase II. The follicular cells surrounding the secondary oocyte continue to proliferate and produce follicular fluid. The haploid secondary oocyte surrounded by follicular fluid and follicular cells is the mature follicle.

The mature secondary oocyte is released from the ovary at ovulation, leaving the follicular cells in the ovary. The follicular cells develop into the corpus luteum. The mature secondary oocyte does not finish meiosis II unless it is fertilized. Unfertilized oocytes are released from the body during the shedding of the endometrium without having completed their meiotic divisions. If fertilized, the secondary oocyte finishes meiosis II and produces one large secondary oocyte (n) and another polar body cell. This polar body cell degenerates as the first one did, and the large secondary oocyte becomes the mature ovum (egg). (Refer to Figures 11.17 and 11.18 as well as to Table 11.11 .)

Structure of an Ovary

Primordial follicle   Primary follicle   Secondary follicle   Mature follicle

Secondary oocyte

Rupturing follicle

Germinal epithelium

Corpus albicans

Mature corpus luteum   Corpus luteum

Figure 11.17. Ovarian tissue

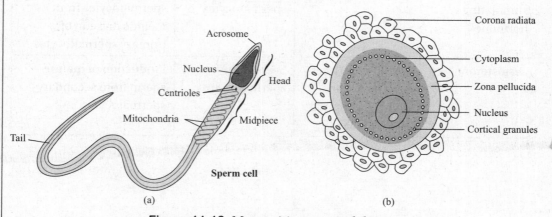

Acrosome

Nucleus

Head

Centrioles

Mitochondria   Midpiece

Tail

Sperm cell

(a)

Corona radiata

Cytoplasm

Zona pellucida

Nucleus

Cortical granules

(b)

Figure 11.18. Mature (a) sperm and (b) egg

ANIMAL PHYSIOLOGY

Production of Semen

The epididymis, seminal vesicle and prostate gland are all involved in the production of semen. Semen is the fluid that contains mature sperm and that is released at ejaculation. (Refer to Tables 11.10 and 11.11.)

Table 11.10: Production of Semen

Gland	Location	Role
Epididymis	Above testes	Site of maturation and motility (ability to swim) development of sperm. Sperm are stored here until released at ejaculation.
Seminal vesicle	Slightly above the prostate gland, posterior to the bladder	Produces and stores a fluid rich in nutrients that are used by the sperm (including fructose for energy for motility). Produces and stores mucus to help protect the sperm in the vagina.
Prostate	Sits under the bladder	Produces and stores a fluid rich in mineral ions and an alkaline solution to neutralize the acidity of the vagina.

Comparing Sperm and Egg Production

There are several distinctions between spermatogenesis and oogenesis. The numbers of gametes produced, the timing of production of the gametes, the release of the gametes and the rate of production of gametes are outlined in Table 11.11.

Table 11.11 Spermatogenesis and Oogenesis

	Spermatogenesis	Oogenesis
Number of gametes produced	Millions each day	One every 28 days (rarely a few more)
Time of formation of gametes	After puberty	Begins during foetal development and continues at puberty
Time of release of gametes	Released at ejaculation	Released at ovulation (day 14 of the menstrual cycle)
Timeline for production	Begins at puberty and continues throughout life	Begins before birth and ends at menopause (average age 51)
Number produced per meiotic division	4 functional sperm produced for each meiotic division	1 functional egg produced for each meiotic division (3 polar bodies are produced and reabsorbed)

REMEMBER

The testes produce over 12 billion sperm each month while normally only 1 egg is produced in the same time period!

Preventing Polyspermy

Fertilization is possible only after many sperm have reached the egg and have pushed away the follicular cells that make up the corona radiata. When the sperm span the corona radiata, they release digestive enzymes from their acrosomes (acrosomal reaction), digesting the zona pellucida. This allows the sperm to reach the egg cell membrane. Protein receptors located on the egg cell membrane recognize and attach to receptors on a single sperm cell membrane, allowing the sperm nucleus to enter the egg cytoplasm. The entry of the sperm nucleus into the egg triggers the cortical reaction. During the cortical reaction, the cortical granules fuse with the egg cell membrane and release their contents by exocytosis into the perivitelline space (space between the vitelline layer and the cell membrane). The contents released include enzymes that catalyse the cross-linking of glycoproteins found in the zona pellucida. This process causes the zona pellucida to become hard and resist the entry of any additional sperm (polyspermy).

HCG and Pregnancy

After fertilisation, the zygote ($2n$) undergoes mitosis to form a hollow ball of cells known as a blastocyst. The blastocyst travels to the uterus, where it implants into the endometrial lining. The corpus luteum (formed from the tissue left in the ovary after ovulation) secretes progesterone to maintain the lining. The corpus luteum naturally begins to disintegrate, and the production of progesterone stops. If fertilization did not occur, this would trigger the onset of the menstrual cycle with the shedding of the endometrium. In order to prevent the endometrium (along with the developing foetus) from being shed, the early embryo secretes the hormone human chorionic gonadotropin (HCG). Human chorionic gonadotropin stimulates the corpus luteum to remain intact and continue to secrete progesterone. At around 12 weeks of gestation, the embryo takes over the role of HCG production. The hormone levels of HCG continue to rise throughout early pregnancy. (Refer to Figure 11.19.)

If a foetus is present, it will secrete HCG that will travel in the blood to reach its target tissue—the corpus luteum. Some of the hormone HCG is released in the urine of pregnant women and can be detected early in pregnancy with the use of over-the-counter pregnancy test kits.

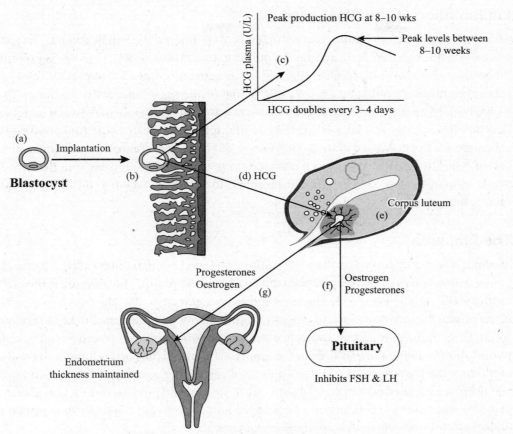

Figure 11.19. HCG and early pregnancy

FEATURED QUESTION

May 2004, Question #33

Where is human chorionic gonadotropin (HCG) produced?

A. Ovary
B. Anterior pituitary
C. Embryo
D. Posterior pituitary

(Answer is on page 554.)

Implantation of the Blastocyst

Sexual intercourse (copulation) releases millions of sperm into the female vagina. The sperm must swim through the cervix and up the uterus to enter the oviducts. If an egg is present in the oviduct, the sperm (n) fertilizes the egg (n), producing a diploid ($2n$) zygote. The zygote carries out mitosis, producing a 2-cell embryo that divides again into a 4-cell embryo. These early divisions are a special type of cellular division in which the embryo does not increase in size, but the existing cytoplasm is divided into multiple cells. The first structure produced is a solid ball of cells known as a morula. The process of mitotic divisions continues until a hollow ball of cells (16–32 cells) known as a blastocyst is produced. During these mitotic divisions, the developing blastocyst travels from the oviducts to the uterus, where it will implant into the uterine lining (endometrium).

The Placenta

At about 8 weeks of gestation, the embryo has developed enough to be called a foetus. The foetus is attached to a disk-shaped placenta by an umbilical cord. The placenta is the lifeline for the foetus, supplying it with nutrients and removing wastes. The placenta also functions in the production of oestrogen and progesterone in order to maintain the thickened endometrium. Early in pregnancy, the corpus luteum secretes the hormones oestrogen and progesterone. After implantation, HCG from the embryo stimulates the corpus lutuem to continue to produce the hormones. At about the middle of gestation, the corpus luteum disintegrates and the placenta takes over the role of secreting oestrogen and progesterone. It is vital that the placenta takes over the production of oestrogen and progesterone in order to maintain the thickened endometrium and sustain the pregnancy.

The placenta is attached to maternal tissues by many villi (folded projections) that embed the placenta into the uterine wall. The villi allow for a very large surface area to facilitate the exchange of nutrients and wastes between the mother and the foetus. Maternal and foetal blood vessels are very close to each other in the placenta, but they do not directly connect. This allows material to be exchanged between the mother and foetus without allowing direct transfer of blood cells. For example, carbon dioxide diffuses from the foetal blood into the maternal blood, and oxygen diffuses from the maternal blood into the foetal blood. Nutrients also pass into the foetal blood, and wastes pass out of the foetal blood. Materials are exchanged between the maternal and foetal blood in order to provide the foetus with nutrients and to remove toxic wastes.

Amnion and Amniotic Fluid

The sac that holds the foetus consists of two layers—the amnion and the chorion. The amnion contains the foetus and the amniotic fluid, while the chorion contains part of the placenta. The amniotic fluid protects the foetus by cushioning it from any physical trauma, gives the foetus space to move for muscle development, supplies a fluid for lung development and helps to maintain a stable temperature for the foetus.

ANIMAL PHYSIOLOGY

Birth and Positive Feedback

Once the foetus has reached full term (about 38 weeks of pregnancy) the process of birth begins. The foetus drops low near the birth canal and puts pressure onto the cervix. The increased pressure and stretching of the cervix triggers the release of oxytocin from the pituitary gland. Oxytocin causes contractions of the uterus. These contractions stimulate the release of more oxytocin. The release of more oxytocin results in harder and longer uterine contractions. This is an excellent example of positive feedback. The uterine contractions cause the cervix to dilate and become thinner as labor progresses. When the cervix is dilated to about 10 centimetres, the foetus is ready to be delivered through the vagina.

Oestrogen levels rise steadily throughout pregnancy and reach their highest level just before birth. High levels of oestrogen increase the number of oxytocin receptors present on uterine cells. The increase in oxytocin receptors leads to more muscle cell contraction within the uterus, speeding up the process of birth.

> **Positive feedback:** A change in a homeostatic level is amplified; disrupts homeostasis; a very rare type of feedback in humans (for example, childbirth)
>
> **Negative feedback:** A change in a homeostatic level is counteracted; maintains homeostasis; the most common type of feedback (for example, thermoregulation)

1. Which of the following are examples of active immunity?

   I. Receiving a vaccination for chicken pox
   II. Eating food contaminated with bacteria
   III. Transfer of antibodies to the foetus from the placenta

   A. I only
   B. I and II only
   C. II and III only
   D. I, II and III

2. A hybridoma is formed from the fusion of:

   A. A B-cell and a T-cell
   B. A B-cell and an antibody
   C. A T-cell and an antibody
   D. A B-cell and a tumour (cancer) cell

3. Bones are connected to bones by:

   A. Ligaments
   B. Tendons
   C. Joints
   D. Synovial fluid

4. Which of the following ions is directly responsible for initiating muscle contraction?

   A. Sodium
   B. Potassium
   C. Calcium
   D. Magnesium

5. Which of the following are involved in the process of ultrafiltration in the kidney?

   I. Fenestrated capillaries
   II. Increased pressure
   III. Basement membrane

   A. I only
   B. I and II only
   C. II and III only
   D. I, II and III

6. Antidiuretic hormone (ADH) acts on the:

   A. Collecting duct, making it more permeable to water
   B. Collecting duct, making it less permeable to water
   C. The loop of Henle, making it more permeable to water
   D. The loop of Henle, making it less permeable to water

7. Which hormone is responsible for the production of testosterone?

  A. FSH (Follicle-stimulating hormone)
  B. LH (luteinizing hormone)
  C. ADH (antidiuretic hormone)
  D. Progesterone

8. HCG (human chorionic gonadotropin) is released from the:

  A. Pituitary
  B. Ovary
  C. Foetus
  D. Corpus luteum

9. Sperm gain motility in the:

  A. Seminal vesicles
  B. Epididymis
  C. Prostate
  D. Vas deferens

10. What are the relative levels of progesterone and oxytocin during childbirth?

	Progesterone	Oxytocin
A.	Increasing	Increasing
B.	Decreasing	Decreasing
C.	Increasing	Decreasing
D.	Decreasing	Increasing

11. Which of the following is a form of active artificial immunity?

  A. Antibodies crossing the placenta
  B. Antibodies injected as an antivenom
  C. Vaccinations
  D. Becoming ill after drinking from the same cup as an infected individual

12. What is stored and released from the sarcoplasmic reticulum in order to initiate muscle contraction?

  A. Calcium
  B. Magnesium
  C. Sodium
  D. Potassium

13. Which of the following are functions of the kidney?

    I.   Osmoregulation
    II. Excretion
    III. Production of urea

    A.   I only
    B.   I and II only
    C.   II and III only
    D.   I, II and III

14. Which part of a sarcomere contains myosin fibres?

    A.   Light bands
    B.   Dark bands
    C.   Light and dark bands
    D.   Z lines

15. Tendons hold:

    A.   Bone to bone
    B.   Muscle to bone
    C.   Connective tissue to bone
    D.   Muscle to muscle

16. The graph below shows changes in HGH (human growth hormone) levels over an average male's lifetime.

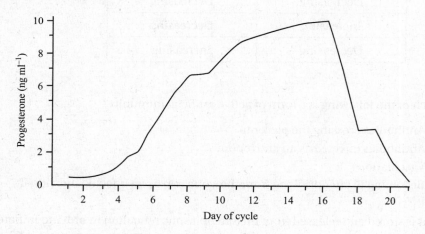

(a) State the day of the cycle when progesterone levels are the highest. (1)
(b) Calculate the percent change in progesterone levels between day 5 and day 16. (2)
(c) Describe the relationship between day of cycle and progesterone levels. (2)
(d) State the hormone that induces ovulation and the day of the cycle this event usually occurs. (2)

17. Annotate the diagram of the nephron to show the name of the structure and the process occurring at each location. (4)

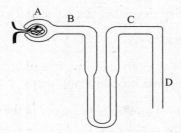

18. Outline the process of blood clotting. (4)

19. Define excretion. (2)

20. Compare spermatogenesis and oogenesis (4).

21. (a) Draw the structure of a sarcomere. (4)
    (b) Outline the role of bone, tendons, ligaments and nerves in human movement. (6)
    (c) Explain the principle of skeletal muscle contraction. (8)

ANSWERS EXPLAINED

1. **(B)** Antibody transfer through the placenta is passive immunity.

2. **(D)** A hybridoma results from the fusion of a B-cell and a tumour (cancer) cell.

3. **(A)** Ligaments attach bone to bone.

4. **(C)** Calcium released from the sarcoplasmic reticulum attaches to troponin, causing tropomyosin to move and expose myosin binding sites.

5. **(D)** Fenestration allows more material to move out of the capillaries. Increased pressure forces more filtrate out of the capillaries. The basement membrane holds the capillaries close to Bowman's capsule to maintain a close relationship for diffusion of the filtrate to occur.

6. **(A)** ADH acts on the collecting duct of the nephron, increasing its permeability for water.

7. **(B)** LH is responsible for triggering the Leydig cells of the testes to release testosterone.

8. **(C)** HCG is released from the early foetus in order to stimulate the corpus luteum to continue to release progesterone.

9. **(B)** Sperm gain motility in the epididymis.

10. **(D)** Progesterone levels decrease (stimulating shedding of the endometrium), and oxytocin levels increase (stimulating uterine contractions).

11. **(C)** Vaccinations result in artificial active immunity.

12. **(A)** The sarcoplasmic reticulum stores and releases calcium. When released, the calcium initiates muscle contraction.

13. **(B)** The kidney functions for osmoregulation and excretion. Urea is produced by the liver, not the kidneys. The kidneys excrete urea.

14. **(B)** The dark bands contain myosin fibres. Myosin fibres are thicker than actin, and their presence in a sarcomere shows up as a dark banding pattern.

15. **(B)** Tendons hold muscle to bone.

16. (a) Day 15 (+/− 1)

    (b) 10 ng/mL^{-1} (day 16) − 2 ng/mL^{-1} (day 2) ÷ 2 ng/mL^{-1} (day 2) × 100 = 400% percent change (increase). Remember that percent change is equal to the new value minus the old value divided by the old value times 100.

    (c)
    - Change exhibits a bell curve pattern/gradually increases followed by gradual decrease
    - Highest level is seen at day 15 (+/−1)/peaks between days 14–16
    - Lowest level is day 0–1 or day 21
    - The decrease is more rapid than the increase
    - Steady rise between day 0 and 21
    - Steady decrease between day 16 and 21

    (d) The hormone is luteinizing hormone (LH), and it triggers ovulation around day 14.

17. One point is awarded for each correctly labelled part with the process identified.
    - *A:* Glomerulus/Bowman's capsule—ultrafiltration
    - *B*: Proximal convoluted tubule—selective reabsorption
    - *C*: Distal convoluted tubule—secretion
    - *D*: collecting duct—excretion

18.
    - Begins with damaged tissue releasing clotting factors
    - Platelets attracted to site
    - Platelets release clotting factors
    - Prothrombin is converted into thrombin
    - Thrombin catalyses the conversion of fibrinogen into fibrin
    - Clot is sealed/thrombus formed
    - Calcium/vitamin K are needed/cofactors for conversions
    - Clots prevent blood loss/infection

19. Excretion is the removal of waste products produced by metabolic processes/the body. One point is awarded for including an example: urea/toxins; water regulation; excess salt removal

20. One point is awarded for each correctly matched comparison.

Spermatogenesis	Oogenesis
Begins at puberty	Begins before birth
Continues throughout life	Ends with menopause
Produces millions of sperm daily	Produces one egg per month
Each meiotic division results in 4 functional sperm	Each meiotic division results in 1 functional egg and 3 polar body cells

(a) One point is awarded for each correctly drawn and labelled part.
   - Z lines
   - Actin
   - Myosin
   - Myosin heads
   - Myosin binding sites
   - Dark bands
   - Light bands

(b)
   - Bones function for the attachment of muscles
   - Bones function for structure of body/support
   - Tendons hold muscles to bone to allow for movement/contraction
   - Ligaments hold bone to bone to keep structure stable/allow movement of bones
   - Joints function for range of motion
   - Nerves send impulses/action potentials to muscles to signal contraction/relaxation
   - Skeletal muscles are antagonistic/one contracts and the other relaxes
   - Neurotransmitters convey message

- Acetylcholine is the main neurotransmitter at the neuromuscular synapse/synapse
- One point is awarded for a named example of a bone/muscle/tendon

(c) One point is awarded for each process listed in the correct order of occurrence.

- Action potential/impulse arrives at the synapse/neuromuscular junction
- Acetylcholine is the main neurotransmitter involved in muscle contraction
- Depolarization of sarcolemma/muscle cell membrane
- Action potential travels through the T-tubules/infoldings in the cell membrane
- Calcium is released
- Calcium is released from the sarcoplasmic reticulum
- Calcium attaches to binding sites (on troponin)
- Tropomyosin moves/shifts position, exposing myosin binding sites
- Cross-bridges from/myosin heads bind to binding sites on actin
- Myosin head releases P_i (inorganic phosphate)/phosphate
- Myosin head shifts and moves actin
- Myosin head must bind ATP to break cross-bridge/detach from actin
- Myosin splits the ATP into ADP + P_i, returning to "cocked"/binding position
- Cross-bridges continue to form until action potential is no longer sent/stops
- Calcium is pumped back into sarcoplasmic reticulum
- Calcium no longer binds to troponin; tropomyosin shifts back to original position
- Binding sites covered by troponin
- Contraction stops/actin slides back over myosin/muscle relaxes

OPTIONAL TOPICS

Neurobiology and Behaviour

CENTRAL IDEAS
- Neural modification and development
- Brain functions
- Perception of stimuli
- Behavioural patterns
- Neural transmission and psychoactive drugs
- Natural selection and behaviour

TERMS
- Addiction
- Ethology
- Innate behaviour
- Learned behaviour
- Neural tube
- Photoreceptors
- Stimulants

INTRODUCTION

Neural development begins during embryogenesis. Modifications within neural tissues continue throughout life. The human brain develops from embryonic neural tissue to consist of specific regions that carry out specific functions. The perception of stimuli relies on the proper workings of the neural connections within the brain. Responses to environmental stimuli result from neural activity and can be innate or learned from experience. Psychoactive drugs can change the way in which neurons communicate and can cause abnormal behaviour in organisms.

NOS Connections
- Scientists use models to represent scientific structures and processes (anatomy and physiology of the brain and neurons, behavioural models).
- Ethical issues exist in scientific experimentation (animal studies, such as drug addiction and brain functions).
- Collaboration and cooperation are necessary in scientific investigations and discoveries (brain functions, pharmaceutical drug development, behavioural studies).
- Technology enhances scientific observation, understanding and discoveries that can enhance human life (hearing aids, cochlear implants).

A.1 NEURAL DEVELOPMENT

TOPIC CONNECTIONS

- 6.5 Neurons and synapses

After fertilization, the zygote undergoes mitotic cellular division and a blastula develops. The blastula is a hollow ball of cells that further develops into the gastrula. The gastrula consists of three distinct germ layers—endoderm, mesoderm and ectoderm. A neural crest is derived from the ectoderm and is sometimes considered a fourth embryonic germ layer. The development of the neural tube is shown in Figure A.1.

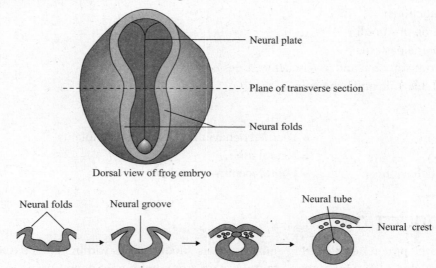

Figure A.1. Development of the neural tube in *Xenopus* (tadpole)

The ectoderm tissue folds inwards, creating a neural fold to begin the formation of the neural tube from the neural plate. The neural crest develops from cells that are released from the neural fold during formation of the neural tube. The migratory cells derived from the neural crest develop into structures such as bone, smooth muscle and cartilage.

The neural tube continues to elongate in early embryonic development and develops into the brain and spinal cord. As neurons develop from differentiation of the neural tube tissue, they migrate to their final location and complete their maturation.

Chemical stimuli initiate the growth and elongation of an axon from each developing neuron. The axons of developing neurons can grow to reach great distances outside of the central nervous system in the developing human foetus. Figure A.2 shows the development of the adult axon.

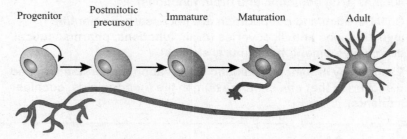

Figure A.2. Formation of a mature adult neuron

Neurons continue to develop and form multiple synaptic connections with other neurons. As a child develops, synapses continue to form between the developing neurons. Synapses that are not active are not maintained. Neurons that are not being used are lost from the body in a process known as "neural pruning". The removal of nonactive synapses, the pruning of inactive neurons and the formation of new synapses allows the brain to respond to new experiences.

REMEMBER

The ability of the nervous system to respond to stimuli and change the number of synapses present is often referred to as "plasticity" of the nervous system.

Spina Bifida

Failure of the neural tube to close completely leads to spina bifida. The severity of spina bifida varies based on the size of the opening in the spinal cord. If the opening is wide enough, the spinal cord (nerves) can protrude through the spinal bones (see Figure A.3). Paralysis and loss of feeling in the lower extremities can result from damage to the protruding nerves. Research has shown that a diet lacking folic acid can contribute to higher incidences of spina bifida.

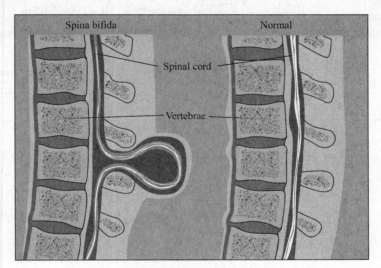

Figure A.3. Spina bifida

Strokes and Neural Activity

The nervous system has the capacity to reorganize neural pathways and repair damaged tissue. Reorganization and repair to damaged tissue can allow the return of lost neural function resulting from a stroke. In some cases, areas of the brain previously not responsible for a specific neurological function take over the control of the function previously controlled by the damaged part of the brain. The recovery of neurological functions lost due to a stroke is further evidence of the plasticity of the nervous system.

A.2 THE HUMAN BRAIN

The brain forms from growth and development that occur in the anterior portion of the neural tube. The brain and spinal cord together make up the central nervous system (CNS). Nerves that branch from the central nervous system make up the peripheral nervous system (PNS). Different regions of the brain control specific neurological functions. The central nervous system (brain and spinal cord) is divided into autonomic and somatic divisions. The autonomic division is not under conscious control (involuntary). Autonomic responses in the body are mainly controlled by the brain stem.

- **Voluntary**: Under conscious control. (Somatic—you control the movements.)
- **Involuntary**: Not under conscious control. (Autonomic—you do not control the movements.)

The somatic nervous system controls voluntary functions such as skeletal muscle movement. The autonomic nervous system controls involuntary functions such as smooth muscle contraction (intestines) and cardiac muscle contraction (heart). It also regulates the release of products from glands. The autonomic nervous system is divided into the sympathetic and parasympathetic systems. The two branches of the autonomic nervous system are antagonistic to each other. They have opposite effects on the body (see Table A.1).

Table A.1: Sympathetic and Parasympathetic Control

Controlled Response	Sympathetic	Parasympathetic
Heart rate	Speeds up— Supplies more blood flow to body muscles to prepare for increased muscle contraction	Slows down— Returns heart rate to normal after stressful situation to allow body to rest and recover
Breathing	Speeds up— Increases oxygen supply to meet increased demands of contracting muscles	Slows down— Returns breathing to normal after stressful situation to allow the body to rest and recover
Iris	Dilates pupil (contracts circular iris muscles)— More light available to ensure optimal visual processing in order to analyse the environment and plan actions needed for escape	Constricts pupil (constricts radial iris muscles)— Constricts pupils back to normal size based on light availability in the environment. This prevents too much light from entering the eyes and damaging the retinal cells
Digestion (blood flow to gut, saliva production, production of gastric juices)	Decreases— Less energy is used for digestion so more is available for the muscles to meet the increased need for muscle contraction	Increases— More energy used for digestion to replace the stored energy lost from the body in order to be prepared for future needs

- **Sympathetic:** "Fight or flight" response. (Use this phrase to help you remember what the sympathetic system does.) The body prepares to run away from the stressful event or to fight with the stressor (animal, etc.).
- **Parasympathetic:** "Rest and digest" response. This slows down the increased physiological responses, letting the body rest and recover. It increases digestion to replace the lost energy from the increased needs of the stressful situation.

Human Brain Structures

The human brain consists of different structures that control specific functions. Most of these functions are autonomic functions that occur without conscious thought. Since it takes a lot of energy to fuel metabolic neurological processes, the brain uses more energy (ATP) than any other organ in the human body. The main structures of the brain are shown in Figure A.4, and the functions of brain structures are outlined in Table A.2.

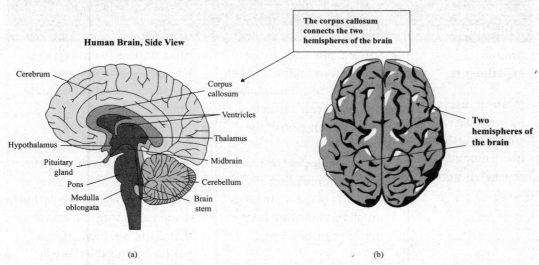

Figure A.4. (a) Human brain structures and (b) two hemispheres of the brain

Table A.2: Function of Brain Structures

Brain Structure	Function
Medulla oblongata	▪ Controls breathing and heart rate ▪ Regulates digestion and swallowing ▪ Controls vomiting reflex
Cerebellum	▪ Controls movement and balance

Hypothalamus	▪ Coordinates actions of the endocrine (hormone) and nervous system (neurons)
	▪ Produces some of the hormones stored in the pituitary
	▪ Controls the release of the pituitary hormones
	▪ Maintaining homeostasis
	   □ Temperature regulation (thermoregulation)
	   □ Blood glucose levels
	   □ Blood solute concentration (osmoregulation)

Pituitary gland (anterior and posterior pituitary)	▪ Stores and releases hormones when stimulated by the hypothalamus (some pituitary hormones are produced by the hypothalamus and stored in the pituitary, and some pituitary hormones are produced by the pituitary)
Cerebral hemispheres	▪ Control higher-level functions such as learning, emotions, problem solving and personality

Identifying the Role of Brain Parts

Scientists have discovered how the brain is involved in controlling specific body functions through experiments carried out on animals, the effect of lesions on brain functions, autopsy and using functional magnetic resonance imaging (fMRI). (See Table A.3.)

Table A.3: Identification of Brain Functions

Method of Identification	Information Gained	Example
Animal experiments *Many individuals have ethical issues concerning the use of animals for research purposes.	■ Surgical procedures that stimulate or inhibit parts of the brains of animals have been performed to determine the function of the brain part. ■ The physiological response to the manipulation of various parts of the brain gives insight into the function of that specific brain part.	■ Animals have been used to investigate how repeated drug use affects the workings of the brain. ■ Repeated exposure to psychoactive drugs has been shown to lead to addiction in laboratory rats. The rats are given the choice of a lever that dispenses a drug and a lever that dispenses food. After a period of time, the rats will chose the lever giving the drug over the lever giving food.
Lesions—Destroyed regions of the brain due to stroke, disease, accidents or tumours	■ The loss of physiological function when parts of the brain develop lesions allows scientists to determine the function of the brain region that has been damaged.	■ Damage to the left hemisphere of the brain causes language difficulties—evidence that the left hemisphere plays a role in language.
Functional magnetic resonance imaging (fMRI)—Neuroimaging that detects blood flow in the brain	■ fMRI can be used to gain images of brain activity during different physiological functions. Active regions receive more blood flow than inactive regions. The fMRI detects the increased blood flow to the area of the brain involved in processing the information.	■ By exposing subjects to different sights and sounds or by having them touch various objects while in the fMRI machine, scientists can identify regions of the brain processing the information based on increased blood flow to regions involved in each process.
Autopsy	■ Autopsy of the brain can reveal damaged areas and changes in brain structure (often used to diagnose Alzheimer's and FTD—frontotemporal degeneration after death).	■ Information gained from brain changes seen in an autopsy can be compared to neurological changes seen in the patient before death.

The Pupil Reflex

The dilation and constriction of the pupil is achieved by the action of muscle cells in the iris. The circular muscles of the iris contract to dilate the iris of the eye (make the pupil larger). The radial muscles of the iris contract to constrict the iris of the eye (make the pupil smaller). The size of the pupil allows the perfect amount of light into the eye in order to activate the photoreceptors and form images of the environment. The photoreceptors send messages to the brain stem (medulla oblongata) that relate information about the intensity of the light in the environment. If the light intensity is strong, the medulla sends messages to the iris of the eye to contract the radial muscles and make the pupil smaller. This limits the amount of light entering the eye and protects it from damage by the intensified rays. If the light intensity is low, the medulla sends messages to the iris to contract the circular muscles of the iris and dilate the pupil to allow maximum light into the eye from the low levels available.

Brain Stem Death and the Pupil Reflex

The medulla oblongata controls breathing and heart rate as well as pupil dilation. The fact that the medulla oblongata controls the state of pupil dilation is useful when determining brain stem death. When evaluating damage from injuries that may affect the brain stem, doctors first check the dilation response of the pupils. This is a quick and easy method since it requires simply shining light into the patient's eyes. If the pupils do not respond to the light, the doctor is alerted to the possibility of brain stem death and can act accordingly.

The Cerebral Cortex

The largest structure in the brain is the cerebral cortex. When compared to the cerebral cortex in other animals, the human cerebral cortex is the most developed. The cerebral cortex is divided into two hemispheres that contain extensive folding in order to fit inside the cranium. The two hemispheres of the cerebral cortex carry out higher-order functions such as memory, attention, language and consciousness. The left and right hemispheres each control specific physiological functions. Sensory information received from the right side of the body is interpreted in the left hemisphere. Sensory information received from the left side of the body is interpreted in the right cerebral hemisphere. The main functions of the cerebral hemispheres are outlined in Table A.4.

Table A.4: Cerebral Hemispheres

Left Cerebral Hemisphere (Right-Hand Control)	Right Cerebral Hemisphere (Left-Hand Control)
■ Involved in scientific and mathematical skill (reasoning and analytical) ■ Receives visual information from sensory receptors on the right side of the body ■ Controls muscle contractions on the right side of the body	■ Involved in creativity and imagination (artistic and musical) ■ Receives visual information from sensory receptors on the left side of the body ■ Controls muscle contractions on the left side of the body

The brain is divided into regions that are involved in the control of specific neurological functions. Although specific brain activities are associated with specific areas of the brain, the control of some brain activities actually involves many regions of the brain. Examples of distinct regions of the brain associated with the control of specific functions include, but are not limited to, the visual cortex, Broca's area, Wernicke's area and the nucleus accumbens. The brain regions and their specific functions are outlined in Table A.5.

Table A.5: Specific Areas of Brain Functions

Brain Region	Specific Functions
Visual cortex	■ Interpret visual stimuli ■ Flips image input to upright visuals ■ Plays a role in producing images of memories and imagination
Broca's area	■ Language ■ Language comprehension ■ Possible role in body gesture–word relationship
Wernicke's area	■ Speech ■ Understanding the spoken language
Nucleus accumbens	■ Reward pathway ■ Dopamine release

A.3 PERCEPTION OF STIMULI

TOPIC CONNECTIONS

■ 3.4 Inheritance

Sensory Receptors

The ability of an organism to respond to stimuli depends on the ability to sense the presence of the stimuli in the environment (internal or external). The human body possesses receptors that are designed to detect various forms of stimuli. The main types of sensory receptors are outlined in Table A.6.

Table A.6: Human Sensory Receptors

Type of Receptor	Stimuli Detected	Examples
Mechanoreceptor	■ Pressure ■ Gravity	■ Detection of sound waves in ear ■ Detection of pressure applied to skin
Chemoreceptor	■ Dissolved chemicals ■ Vapour chemicals	■ Detection of chemicals dissolved in saliva (taste) ■ Detection of chemicals in vapour (smell-olfactory receptors)
Thermoreceptor	■ Temperature (warm and cold)	■ Detection of warm and cold by the skin ■ Detection of the temperature of blood as it travels through the hypothalamus
Photoreceptor	■ Electromagnetic radiation (light)	■ Detection of light (rod cells of the eye) ■ Detection of colour (cone cells of the eye)

Glaucoma: Increased pressure in the eye that can damage the optic nerve

Cataracts: Clouding of the lens that can impair light from entering the eye

The Human Eye

The human eye contains photoreceptors that detect electromagnetic radiation in light. The structure of the eye allows light to be focused onto the photoreceptors in order to obtain optimal visual processing (see Figure A.5). The functions of structures of the eye are outlined in Table A.7.

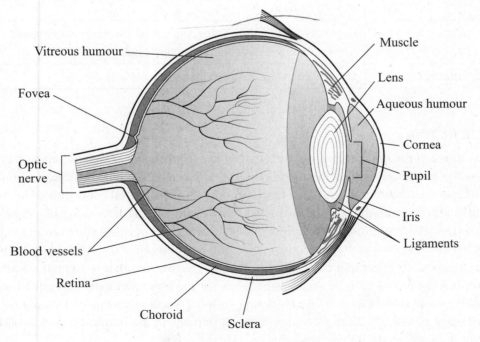

Figure A.5. The human eye

Table A.7: Structures of the Eye

Structure	Function
Sclera	Tough outer layer that protects the eye ("whites of the eye")
Cornea	Clear continuation of the sclera at the front of the eye that allows light to pass through
Conjunctiva	Moisture-producing outer lining of the eye and inner lining of the eyelids
Eyelid	Covers the eyes for protection and spreads moisture over the eye
Choroid	Dark layer of tissue that is full of capillaries to nourish the retina of the eye
Aqueous humour	Thin fluid in the front of the eye that helps inflate the eye by providing pressure and contains nutrients for the cells of the eye
Pupil	Opening in the centre of the iris that controls the amount of light that enters the eye
Lens	Clear structure that focuses light on the fovea of the retina
Iris	Coloured part of the eye that contracts and relaxes to change the size of the pupil
Vitreous humour	Gelatinous thick fluid in the back of the eye that supports the eye by providing pressure
Retina	Contains rods and cones that detect colour—photoreceptors collect light and create vision
Fovea	Location of most of the cones—most accurate visual region (light should focus here)
Optic nerve	Contains all the neurons leaving the retina, and carries visual messages to the visual cortex of the brain
Blind spot	Site where the optic nerve leaves the back of the eye

THE RETINA

The retina of the eye contains photoreceptors (rods and cones) as well as neurons. Rods are sensitive to changes in light intensity (light and dark changes) but do not detect colour. Rods are far more numerous in the retina than are cones. Cones are most concentrated in the fovea and they detect colour. Three types of cones absorb wavelengths in the range of red, green and blue. The selective stimulation of these three different types of cones allows colour to be perceived and is known as the trichromatic theory. If one or more of the three types of cones are not functioning correctly, colour blindness results. Remember that red–green colour blindness is a sex-linked trait. (The genes that code for the photopigments are located on the X chromosome.) Hence, colour blindness appears more frequently in males. Damage to the eye can result in loss of colour perception as well. Damage to the retina can lead to blindness. Table A.8 outlines the differences between rods and cones.

Light enters the retina. First it passes through the ganglion cells and then bipolar cells before passing through the rods and cones. When light reflects off the back of the eye (behind the rods and cones), it is detected by the rods and cones and transformed into an electrical message sent out through the bipolar cells to the ganglion cells and finally to the visual cortex of the brain.

Table A.8: Comparison of Rods and Cones

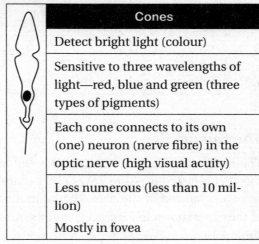

Rods	Cones
Detect dim light (black and white)	Detect bright light (colour)
Sensitive to all wavelengths of light (one type of pigment)	Sensitive to three wavelengths of light—red, blue and green (three types of pigments)
Groups of rods connect to one neuron (nerve fibre) in the optic nerve (low visual acuity)	Each cone connects to its own (one) neuron (nerve fibre) in the optic nerve (high visual acuity)
More numerous (over 100 million) Spread out in retina	Less numerous (less than 10 million) Mostly in fovea

EDGE ENHANCEMENT AND CONTRALATERAL PROCESSING

The processing of visual stimuli begins in the retina. The rods and cones convert light energy into action potentials that are ultimately sent to the occipital lobe of the brain where visual information is processed. Edge enhancement increases the sharpness of the image. Contralateral processing allows the brain to interpret distances and relative sizes of images.

Bipolar cells are responsible for edge enhancement by sending messages to the ganglion cells when activated by a photoreceptor. An activated bipolar cell will send an activating message to a ganglion cell. At the same time, the bipolar cell may send inhibitory messages to bipolar cells that are located in the same receptive field. The selective activation of ganglion cells in receptive fields creates a sharper image.

Contralateral processing refers to the fact that the ganglion cells leaving the back of the eye that are near the nose cross over to opposite sides of the brain. The ganglion cells leaving the right eye cross to the left half of the brain, and the ganglion cells leaving the left eye cross to the right half of the brain. The point where the ganglion cells cross is referred to as the optic chiasma. The crossing-over of information at the optic chiasma allows for the perception of distances of objects as well as the interpretation of the sizes of objects seen. Table A.9 outlines the processes of edge enhancement and contralateral processing.

REMEMBER

Chiasma means "crossing". Remember the crossing-over site in meiosis between two homologous chromosomes is also called the chiasma.

Table A.9: Edge Enhancement and Contralateral Processing

Visual Processing	Main Site of Processing	Function	Result of Processing
Edge enhancement	Retina of the eye	Selectively activates only some of the ganglion cells	Enhances the sharpness of images (where edges are located)
Contralateral processing	Optic chiasma in brain	Sends ganglion cells entering brain close to the nose to the opposite side of the brain	Allows the interpretation of distances and sizes of objects

The Human Ear

The body contains different structures to detect various stimuli. The ears contain receptors that transmit pressure waves into auditory sounds. The brain receives information from receptors located in the ears. The brain interprets the sounds being heard. The structure of the ear is shown in Figure A.6, and the function of each structure in the ear is outlined in Table A.10.

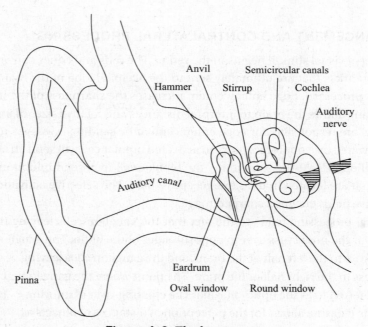

Figure A.6. The human ear

Table A.10: Structures of the Ear

Structure	Function
Pinna	Collects sound waves
Eardrum (tympanic membrane)	Transmits sound waves to the inner ear bones
Bones of the middle ear (hammer, anvil, stirrup)	Transmit and amplify (by up to 20 times) the sound waves received from the eardrum
Oval window	Receives pressure from the middle ear bones (directly from the stirrup) and transmits pressure waves into the inner ear
Round window	Absorbs pressure waves after they travel through the cochlea, preventing the waves from moving backwards
Semicircular canals	Help to maintain balance by sensing the position of the body (hair cells detect movement)
Auditory nerve	Made up of neurons that are activated in the cochlea by pressure waves bending hairs in the inner ear; sends auditory messages to the brain
Cochlea	Part of the ear that contains hairs that are attached to neurons and activated by pressure waves; transforms the pressure waves into neural impulses

PERCEPTION OF SOUND

The pinna collects sound waves and directs them to the eardrum (tympanic membrane). The sound waves cause the eardrum to vibrate and put pressure onto the inner ear bones. The inner ear bones absorb the vibrations and amplify the intensity of the vibrations. The last inner ear bone (stirrup) vibrates on the oval window, causing the oval window to move and produce pressure waves that travel through the cochlea. As the pressure waves travel through the cochlea, they cause hair cells to bend. The hair cells are of varying heights and are in a jelly-like substance. Larger sound waves vibrate more hair cells that, when activated by the sound wave, produce louder sounds. The more hair cells that are activated (bent) the louder the sound perceived by the brain. The more frequent the sound waves (higher frequency) the higher the pitch of the sound perceived.

NOTE: *A.4, A.5 and A.6 are for HL students only, and material from these sections will not appear on the SL exam.*

A.4 INNATE AND LEARNED BEHAVIOUR

The development of behaviours exhibited by organisms can be genetically programmed or based on experiences in the environment.

- **Innate behaviour:** Genetically programmed (inherited) behaviour that develops independently of previous experience in the environment.
- **Learned behaviour:** Behaviour that develops based on environmental experiences and is not genetically programmed.

Animals must be able to sense and respond to stimuli in order to maintain homeostasis in constantly changing environments. Reflexes allow involuntary (autonomic) and instant reactions to stimuli. Reflexes involve specific pathways of neural communication known as reflex arcs. A typical reflex response involves the following:

1. Receptors detect the presence of the stimuli.
2. Sensory neurons send impulses (action potentials) to relay neurons in the central nervous system (brain or spinal cord).
3. Relay neurons send messages from the sensory neuron to the motor neuron
4. Motor neurons send impulses (action potential) to the muscle cells to contract.

Neurotransmitters are released from the sensory neuron, relay neuron and motor neuron in order to transmit the impulse across the synapse. The pain reflex arc is shown in Figure A.7.

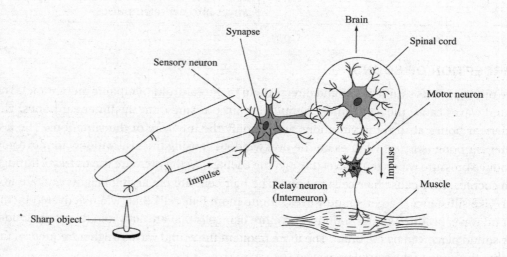

Figure A.7. Pain reflex arc

Pain receptor cells in the skin detect the pain stimulus, and the message is sent to the sensory neuron. The sensory neuron sends the message to the relay neuron in the spinal cord. The relay neuron is located in the grey matter of the spinal cord. Although parts of the sensory and motor neuron are found in the spinal cord, they reach into the outer white matter of the spinal cord.

- **Grey matter**: Consists mainly of dendrites of neurons.
- **White matter**: Consists mainly of axons of neurons. The myelin sheath of the axon is waxy and gives the white matter its lighter "white" appearance.
- **Dorsal root**: Composed of sensory neurons entering the spinal cord (afferent).
- **Ventral root**: Composed of motor neurons leaving the spinal cord (efferent).

The relay neuron sends the message to the motor neuron. The message from the motor neuron to the muscle causes the muscle to contract and pull the hand away from the source of pain.

Reflex Conditioning

Classical conditioning describes the process of learning to associate one stimulus with another previously unrelated stimulus (forming new associations between stimuli). The most famous experiment showing classical conditioning is an experiment carried out by Ivan Pavlov. Pavlov noted that dogs would salivate when presented with food. He began ringing a bell at the exact time he presented his dogs with food. After a period of time, Pavlov would ring the bell without presenting food. Interestingly, the dogs began to salivate solely based on the ringing of the bell. The dogs had been conditioned to associate the ringing of bell with the food.

> Unconditioned stimulus (UCS)—Food
>
> Unconditioned response (UCR)—Salivation
>
> Conditioned stimulus (CS)—Ringing bell
>
> Conditioned response (CR)—Salivation due to bell ringing

Learning and Inheritance in Birdsong

Birdsong is a species-specific behaviour that is used by male birds to attract females for mates and alert other male birds of their claim to a territory. Birds are innately born with the ability to know how to sing the song of their species. Young birds sing a very rudimentary version of the species song. The young birds learn to sing the song better by listening to older birds of their species that have learned to sing the song in a more sophisticated, adult manner. The bird that can best sing the song will be the bird chosen for a mate by any listening females. Birdsong is an excellent example of a behaviour that is both innate (genetically programmed) and learned (based on experiences in the environment).

Imprinting Behaviour

Imprinting is a combination of innate and learned behaviours. Imprinting involves learning that occurs at a specific programmed time or stage of development in the life of an animal. The learning associated with imprinting is rapid and is not related to the consequences of exhibiting the behaviour. The best example of imprinting is young birds imprinting on the first moving object they encounter after birth. Konrad Lorenz investigated imprinting with geese and showed that newly hatched offspring would imprint on the first moving object they encountered. Lorenz was able to get the newly hatched geese to imprint on him as well as inanimate objects. Imprinting on a mother figure allows offspring to imitate the behaviours of the mother and learn behaviours that aid in survival.

Operant Conditioning

Learning based on trial and error is known as operant conditioning. The conditioning occurs as organisms learn to associate a specific stimulus with a specific consequence. Behaviours that are rewarded will be exhibited more frequently (become reinforced), and behaviours that are punished will be exhibited less frequently. Research conducted by B. F. Skinner on rats and other animals is most often associated with operant conditioning. Skinner would present the animals with rewards or punishments in order to modify their behaviour. His investigations involved the use of boxes (known as "Skinner boxes") that contained methods for both reward (food) and punishment (electrical shock). By manipulating the frequency of rewards and punishments based on the presence of desired behaviours, Skinner was able to change the behaviours of the rats.

A.5 NEUROPHARMACOLOGY

Neurotransmitters and Synapses

Neurotransmitters are vital in order for information to be transferred across a synapse to another neuron or a muscle cell. Neurotransmitters are released from the presynaptic neuron and are received by the postsynaptic neuron. Neurotransmitters can either excite or inhibit action potentials from being sent by postsynaptic cells.

- **Excitatory neurotransmitters:** Depolarize the postsynaptic cell sending the action potential forward.
- **Inhibitory neurotransmitters:** Hyperpolarize the postsynaptic cell inhibiting the action potential from being sent.

Neural Transmission and Summation

In order for an action potential to be sent, a neuron or muscle cell must depolarize to a level of −50 to −55 millivolts (threshold). The sodium-potassium pump maintains the cell membrane at −70 millivolts by pumping sodium out of the cell and potassium into the cell, creating a membrane potential (net difference in charge between the inside and outside of the cell). Neurons and muscle cells can depolarize by opening sodium channels, allowing sodium to move into the cell due to the higher concentration of the ion outside. If enough sodium moves into the cell to allow the membrane to reach threshold (−50 to −55 mV), the cell will initiate an action potential. If a cell opens potassium channels, potassium moves out due to the higher concentration of potassium inside. The loss of potassium from inside the cell hyperpolarizes the cell. In other words, the cell becomes more negative inside. As a result, no action potential will be sent.

> ### REMEMBER
>
> **Excitatory neurotransmitters** attach to and cause sodium-gated ion channels to open—depolarize.
>
> **Inhibitory neurotransmitters** attach to and cause potassium-gated ion channels to open—hyperpolarize.

One postsynaptic cell may be receiving messages from many presynaptic cells. Some of the neurotransmitters being released from the presynaptic cells might be excitatory, while other presynaptic cells may be releasing inhibitory neurotransmitters. A postsynaptic cell may be receiving excitatory neurotransmitters that would normally cause the cell to reach threshold. At the same time, though, that postsynaptic cell may also be receiving inhibitory neurotransmitters from another presynaptic cell that cancels out the message from the excitatory neurotransmitters. The sum of the actions of all the neurotransmitters being released by the presynaptic cells must reach threshold (–50 to –55 mV) before an action potential can be sent. This phenomenon is known as summation.

Fast-Acting and Slow-Acting Neurotransmitters

Neurotransmitters are classified as being either fast-acting neurotransmitters (opening gated ion channels) or as slow-acting neurotransmitters (using second messengers or proteins and not directly opening gated ion channels). Fast-acting neurotransmitters initiate cellular responses within milliseconds. Slow-acting neurotransmitters can take a few hundred milliseconds up to a few minutes to initiate cellular responses. Slow-acting neurotransmitters modulate fast-acting neurotransmitters by controlling the release of fast-acting neurotransmitters into the synapse or by regulating the effects of the fast-acting neurotransmitter. Slow-acting neurotransmitters are involved in learning and in the formation of memory by changing neural pathways.

Psychoactive Drugs and the Brain

Psychoactive drugs affect the brain by acting on neurotransmission at the synapse. Psychoactive drugs alter the workings of the brain by affecting mood, thinking processes, personality and behaviour as well as by influencing levels of consciousness. Inhibitory psychoactive drugs calm the brain, and excitatory neurotransmitters stimulate the brain. Excitatory neurotransmitters can cause the inhibition of inhibitory neurotransmitters or the activation of excitatory neurotransmitters. Inhibitory neurotransmitters can cause the inhibition of excitatory neurotransmitters or the activation of inhibitory neurotransmitters. Psychoactive drugs that stimulate the brain are known as stimulants, and psychoactive drugs that calm the brain are known as depressants. Psychoactive drugs can act on the synapse by:

- Inhibiting the neurotransmitter from being released.
- Mimicking the neurotransmitter (due to the similar structure of the psychoactive drug to that of the neurotransmitter) and binding to the same channels the neurotransmitter normally acts on.

  □ The psychoactive drug can bind to receptors and prevent the neurotransmitter from being able to bind—inhibiting the action of the neurotransmitter.
  □ The psychoactive drug can bind to receptors and activate the channel in the same manner as the neurotransmitter would—stimulating the same action of the neurotransmitter without the presence of the neurotransmitter.
  □ The psychoactive drug can interfere with the enzymatic breakdown of the neurotransmitter in the synapse, keeping the action of the neurotransmitter going much longer than normally would occur without repeated action potentials being sent.
  □ The psychoactive drug can interfere with the process of reuptake (removing the neurotransmitter from the synapse) and keep the action of the neurotransmitter going much longer than normally would occur without repeated action potentials being sent.

REMEMBER

The inability to regulate dopamine levels has been associated with various disorders:

- Parkinson's disease
- Schizophrenia
- ADHD (attention deficit hyperactivity disorder)
- Depression

Excitatory and Inhibitory Psychoactive Drugs

Psychoactive drugs are classified based on their effect on the nervous system. The effect of the psychoactive drug depends on how the drug affects neurotransmission. Psychoactive drugs affect the action of neurotransmitters, such as dopamine, GABA, serotonin and acetylcholine. Neurotransmitters can have different roles depending on the synapse with which they are involved. Some neurotransmitters can be excitatory or inhibitory depending on the channels they open on the postsynaptic membrane. Each type of neurotransmitter can be involved in many different processes in the body. The neurotransmitters listed below are identified solely by their main function in neurotransmission. Table A.11 outlines the effects of psychoactive drugs on neural transmission.

- **Dopamine**: Excitatory neurotransmitter responsible for activating the reward pathway—gives feeling of euphoria, well-being and overall happiness.
- **GABA:** Inhibitory neurotransmitter that stops neural transmission.
- **Serotonin**: Inhibitory neurotransmitter that has many roles in the body, including balancing mood and regulating sleep.
- **Acetylcholine**: Excitatory neurotransmitter that is involved in regulating muscles and glands.

Table A.11: Psychoactive Drugs

Type of Psychoactive Drug	Examples	Effect on Neural Transmission
Excitatory psychoactive drugs	Nicotine, cocaine, amphetamines	Stimulate action potentials to be sent (excite)
Inhibitory psychoactive drugs	Benzodiazepines, alcohol, tetrahydrocannabinol (THC)	Inhibit action potentials from being sent (calm)

Effects of THC and Cocaine

Tetrahydrocannabinol (THC) is a chemical found in cannabis (marijuana). The presence of THC in the body inhibits the transmission of neural impulses. Cocaine is produced from the leaves of the coca plant. When present in the body, cocaine stimulates neural transmission. Both THC and cocaine are psychoactive drugs that alter the workings of the mind. Cocaine acts as a stimulant since it blocks reuptake of dopamine, allowing the neurotransmitter to stay in the synapse and continue to send excitatory messages. Tetrahydrocannibinol (THC) acts as a depressant since it binds to cannabinoid receptors, inhibiting GABA from being released. Without GABA (an inhibitory neurotransmitter) in the synapse, more dopamine is released. Dopamine is a neurotransmitter that is responsible for activating the reward pathway. The reward pathway creates a feeling of euphoria. Euphoria is a sense of extreme happiness and well-being. Table A.12 outlines the main effects of THC and cocaine on neural transmission. Figure A.8 illustrates these effects.

Table A.12: Cocaine and THC

	THC	Cocaine
Source	Cannabis plant (leaves and buds)	Coca plant (leaves)
Type of psychoactive drug	Inhibitory (inhibits the inhibitory neurotransmitter GABA)	Excitatory (keeps dopamine in the synapse for longer periods of time)
Mode of action	Binds cannabinoid receptors, blocking GABA release (GABA is inhibitory for dopamine); blocking the release of GABA leads to an increased release of dopamine	Blocks the removal of dopamine (excitatory neurotransmitter) from the synapse by inhibiting reuptake of dopamine by blocking the transporter on the presynaptic membrane
Effect on the body	■ Calm and mellow feelings ■ Increased hunger ■ Reactions can vary, with some individuals becoming paranoid or anxious and others experiencing euphoria	■ Talkative ■ Increased energy ■ Decreased hunger ■ Euphoria (sense of pleasure, happiness and well-being)

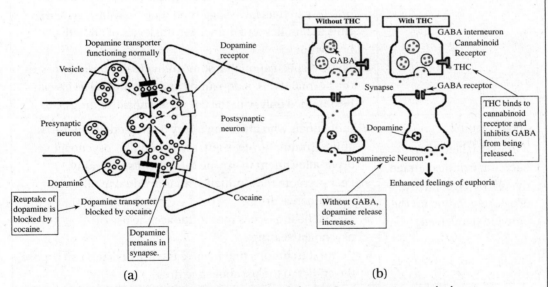

Figure A.8. Effect of (a) cocaine and (b) THC on neurotransmission

> The body produces natural painkillers known as endorphins. Endorphins are released from the pituitary gland during exercise, during sexual activity, and when the body is experiencing excitement or pain. Endorphins act on the same receptors as opiates and have the same effects on the body as opiates such as morphine.

■ **Endorphins:** Endogenous (made by the body) morphine-like proteins that act on the CNS, affecting emotions and decreasing the perception of pain.

Addiction

Individuals using psychoactive drugs can become addicted to the effects of the drug on the physiological processes in the body. Addiction to drugs makes the addict crave and constantly seek out the drug. The level of addiction and risk for addiction vary from individual to individual. Not all drugs are addictive. Among the addictive drugs, the level of addiction varies. Some psychoactive drugs are extremely addictive since they can rewire the brain in a manner that creates the need for the drug in order for the brain to function normally.

Dopamine is the main neurotransmitter involved in addiction since it is involved in the reward pathway. During times of reward, the body releases dopamine naturally. Dopamine is the "feel good" neurotransmitter. Addictive drugs almost always involve processes that increase dopamine levels. Drug addicts become addicted to the euphoria that is associated with increased levels of dopamine. Table A.13 outlines the known causes of addiction.

Table A.13: Causes of Addiction

Cause	Reason
Dopamine levels	■ Dopamine induces euphoria, and the addict becomes addicted to living in this "feel good" state of mind. The body adjusts to the new levels and craves more drugs to achieve the same results, leading to increased drug use.
Genetic predisposition	■ Since some individuals become addicted and others do not, researchers have suggested there may be a gene that runs in families and influences the levels of addictive behaviour. ■ There is speculation that some individuals may not possess the normal levels of dopamine and that normal levels can be obtained only with the use of psychoactive drugs.
Social factors: Some social factors can encourage or discourage the use of psychoactive drugs, depending on the individual involved.	■ Individuals who are living a very depressed life (poverty, abuse, traumatic life events) may find that psychoactive drugs allow them to escape their depressed reality. ■ Peer pressure may influence an individual to take psychoactive drugs. ■ Mental health issues may influence the use of psychoactive drugs. ■ Cultural traditions that involve the use of drugs can expose the addict to the psychoactive drug.

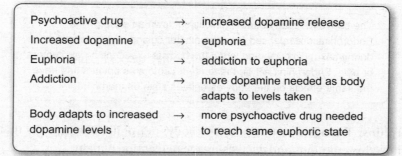

Psychoactive drug	→	increased dopamine release
Increased dopamine	→	euphoria
Euphoria	→	addiction to euphoria
Addiction	→	more dopamine needed as body adapts to levels taken
Body adapts to increased dopamine levels	→	more psychoactive drug needed to reach same euphoric state

Anaesthetics

Anaesthetics cause anaesthesia, the reversible loss of sensation and awareness. Anaesthetics interfere with neural transmission by blocking sensory perception in the central nervous system. Anaesthetics are classified as local or as general based on the level of awareness they impact. Local anaesthetics affect awareness in localized regions. For instance, local anaesthetics are often used in dental procedures to numb specific areas of the mouth. General anaesthetics are used during general surgery to ensure that the patient is unaware of the pain produced by the procedure. The general anaesthetic puts the patient into a temporary and reversible state of unconsciousness.

A.6 ETHOLOGY

TOPIC CONNECTIONS

- 5.2 Natural selection

The study of animal behaviours observed in natural conditions is known as ethology. Natural selection has led to the development of characteristic animal responses to certain stimuli and can change the frequency of animal behaviours. New behaviours evolve when behaviours increase the chances of an organism's survival. Select behaviours help organisms survive. Organisms in the population that exhibit the behaviour are able to reproduce and pass on the behaviour to the offspring. Natural selection for the behaviour leads to the behaviour becoming part of the populations' natural responses. Natural selection can lead to new behaviours developing in populations and to the loss of existing behaviours from populations. Learned behaviours change (increase or decrease) within a population much more rapidly than innate behaviours can change.

Migratory Behaviour

Migration is a genetically programmed behaviour in bird populations. Historically, the migration pattern of the blackcap bird (*Sylvia atricapilla*) was from their home in Germany during the summer to the warmer climate of Spain for the winter. In the past 50 years, a small percentage of the population has begun to migrate to Britain instead of Spain. The advantages of migrating to Britain instead of to Spain are that Britain has a warmer climate than Spain and more food is available in Britain (much of which is from humans feeding the population). In addition, the migration to Britain is a much shorter distance and allows the birds to return more quickly to Germany to take advantage of the summer food availability before other organisms consume it. This new migratory behaviour has become genetically programmed in the populations of blackcaps migrating to Britain. The genetically programmed migratory behaviour is evident by the fact that eggs laid by birds that migrate to Spain produce offspring that naturally migrate in the direction of Spain (southwest). Eggs laid by blackcap birds that migrate to Britain produce offspring that naturally migrate toward Britain (see Table A.14).

Foraging Behaviour

The California garter snake (*Thamnophis elegans*) is found along the coastline and inland areas of California. Inland garter snakes feed mainly on frogs and fish, avoiding the consumption of banana slugs. The coastal garter snakes feed mainly on banana slugs. When newly hatched inland garter snakes are offered banana slugs, they will rarely choose to eat them. Newly hatched coastal garter snakes will quickly take and consume banana slugs. It is believed that the ability to smell the banana slug and recognize it as food is genetically programmed. The garter snakes that inhabit inland areas of California are thought to lack the genetic ability to smell the banana slug and hence do not recognize it as food. The ability to recognize the banana slug as food allowed the few garter snakes with mutations for smelling the banana slug to exploit a new food source. The ability to feed on the banana slugs led to the survival of the coastal garter snakes that could recognize the slugs as food. Garter snakes that survived were able to reproduce. The population that feeds on mainly banana slugs increased in the coastal regions of California. Prior to the evolution of the ability to recognize banana slugs as food, garter snakes were rarely found on the coastline of California. Natural selection for snakes able to feed on the banana slugs led to the current large population of coastal garter snakes in California (see Table A.14).

Table A.14: Animal Responses Due to Natural Selection

Animal	Behaviour Response	Evolutionary Advantage
European blackcap	New migratory pattern	■ Shorter distance to travel ■ Warmer weather ■ More food available
California garter snake	New food source	■ Ability to inhabit new niche ■ High protein provided by eating banana slugs

Altruistic Behaviours

Altruism involves giving of one's own level of fitness (ability to survive and reproduce) in order to ensure the increased fitness of another individual. An altruistic behaviour is an unselfish behaviour that benefits another organism in the population and either has no benefit or decreases the success of the altruistic individual. Altruistic behaviours in social societies are thought to have developed due to the benefit of helping a closely related individual or individuals survive in order to maintain population numbers. This unselfish behaviour is known as kin selection since the behaviour results in the survival of closely related individuals. In social animal societies, altruism often involves giving up one's own reproductive abilities to increase the reproductive abilities of other members of the society. Altruistic behaviour can be seen in the honeybee and naked mole rat societies. The workers work for the good of the reproducing queen while giving up their own reproductive abilities.

A nonreproductive example of altruistic behaviour can be seen in the vampire bats of Costa Rica. These bats feed at night on the blood of mammals. Bats that fail to find food receive food from bats that found food. The bats that found food share food by regurgitating blood into the mouth of the bat that did not find blood food. This behaviour benefits the recipient at the cost of the regurgitating bat who loses some of its nutrition. However, the bat that found food one night might not find food another night and might be fed by the same bat with which it shared food the night before. This reciprocal feeding behaviour helps ensure the survival of both bats and of the society as a whole.

PRACTICE QUESTIONS

1. The graph below shows seal diving behaviour.

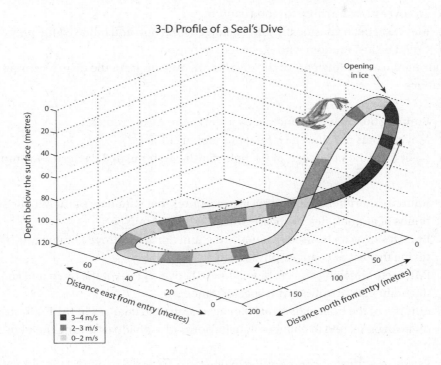

3-D Profile of a Seal's Dive

(a) State the maximum depth achieved by the seal. (1)

(b) Suggest reasons for the seal changing its distance from the ice hole north on the way down and east on the way up. (2)

(c) State the direction of movement at which the seal swam the fastest. (1)

(d) Suggest reasons for the location of the fastest swimming speed. (2)

2. Explain two methods that have been used to identify functions of the brain (3).

3. Outline Pavlov's conditioning in dogs. (4) *HL only*

4. Compare the sympathetic and parasympathetic divisions of the nervous system. (3)

ANSWERS EXPLAINED

1. (a) 120 meters—units are required

   (b)
   - To cover more territory for food sources
   - May have been chasing prey that changed direction and followed the prey
   - Could be just random with no benefit or reason

   (c) During the ascending part of the dive/on the way up from the dive/returning to the surface

   (d)
   - Running out of oxygen and needs to get to the surface quickly
   - Movement is not slowed by foraging/looking for food
   - Light from the hole/sight of the hole may have stimulated the faster swimming

2.
   - Animal experiments with surgical procedures that stimulate or inhibit parts of the brains of animals determine the function of the brain part.
   - Brain lesions can be used to identify the function of destroyed brain parts by identifying the lost or changed physiological function.
   - fMRI neuroimaging detects blood flow to active regions of the brain and can identify brain parts involved with specific functions.
   - Autopsy of the brain can reveal damaged areas and changes in the brain structure that can be related to changes in behaviours of individuals during their life.

3.
   - Pavlov investigated conditioning in dogs.
   - Dogs salivated (UCR—unconditioned response) when presented with food (UCS—unconditioned stimulus).
   - Pavlov rang a bell when presenting dogs with food.
   - After a period of time, the dogs began to salivate (CR—conditioned response) at the sound of the bell (CS—conditioned stimulus) without being presented with food.
   - The dogs learned to associate two unrelated stimuli with one response.

4.
   - The sympathetic and parasympathetic divisions of the nervous system are antagonist.
   - Both are autonomic/not under voluntary control/involuntary.
   - The sympathetic increases heart rate, while the parasympathetic slows down heart rate.
   - The sympathetic increases breathing, while the parasympathetic slows down breathing.
   - The sympathetic dilates the pupils, and the parasympathetic constricts the pupils/ returns them to normal state for the environment.
   - The sympathetic slows digestion, and the parasympathetic speeds up digestion.
   - The sympathetic gets the body ready to face a stressful situation, and the parasympathetic lets the body return to the normal state.

Biotechnology and Bioinformatics

CENTRAL IDEAS

- Microorganisms and industry
- Biotechnology and agriculture
- Biotechnology and environmental protection
- Biotechnology and medicine
- Bioinformatics

TERMS

- Biofarming
- Biofilm
- Biogas
- Expressed sequence tag (EST)
- Fermentation
- Open reading frame (ORF)
- Recombinant DNA
- Transgenic organisms

INTRODUCTION

The use of living organisms to make useful products or modify products is known as biotechnology. The universality of DNA allows the transfer of genes from one organism to another. Gene transfer is used for processes such as the production of medicines, food and plant products. The diverse metabolic activities of microorganisms make them useful in the production of many products and processes within the field of biotechnology.

NOS Connections

- Cooperation and collaboration are vital to scientific progress (genetic data bases (bioinformatics), biotechnology advancements).
- Improvements in scientific tools and technologies enhance scientific discoveries and processes. (PCR, EST, gene probes).
- Risks and benefits of scientific research must be considered (transgenic organisms).
- Accidental discoveries are part of the natural process of science (the discovery of penicillin by Fleming).

BIOTECHNOLOGY/BIOINFORMATICS

B.1 MICROBIOLOGY: ORGANISMS IN INDUSTRY

TOPIC CONNECTIONS

- 2.1 Molecules to metabolism
- 4.3 Carbon cycling
- 6.3 Defence against infectious disease

Microorganisms

Microorganisms are small, single-celled or multicellular organisms that possess diverse metabolic pathways. Their small size and rapid reproductive growth rate make them useful for many industrial processes. Pathway engineering involves modifications and manipulations of microorganisms in order to control their genetic and regulatory activities.

Fermentation

Fermentation takes place in anaerobic (without oxygen) conditions and produces metabolites such as acids (lactate), alcohol (ethanol) and gases (carbon dioxide). Microorganisms carry out anaerobic cellular respiration in order to produce usable energy (ATP) from organic compounds (glucose). Human manipulation of fermentation with pathway engineering is used in the production of many diverse products.

Fermenters are human-made devices used to control the anaerobic processes of fermentation. The processes occurring within fermenters must be monitored in order to optimize the production of desired metabolites. Probes are used to monitor the environment and productivity of the microorganisms in the fermenting chamber. Information from probes is used to make decisions about adjustments that need to be made to the fermenter in order to optimize production of the desired metabolites. Valves on the chamber allow for the release of gases (carbon dioxide) produced during anaerobic respiration (see Figure B.1).

Fermentation can be conducted by either batch or continuous culture methods. In batch methods, the processes of fermentation are maintained in a sealed environment. Temperature is monitored within the fermenter, and a valve allows the escape of fermentation gases. The microorganisms used in batch fermentation go through all stages of growth (exponential, carrying capacity and growth decline). The desired metabolites are collected only when the concentrations of the desired metabolite have reached the maximum level.

Continuous culture methods are open systems. The microorganism populations are maintained at the exponential growth phase. The temperature, pH and oxygen levels are all monitored to ensure optimal growth conditions throughout the long-term production of metabolites. Nutrients are continuously added to ensure rapid growth of the population. The metabolic products produced by continuous culture fermentation are continuously removed.

The waste products of fermentation limit the reproductive capabilities of microorganisms and must be removed to maintain the carrying capacity of the population with continuous culture fermentation. Waste products are not removed in batch fermentation and contribute to the decline in the microorganism population and decreased productivity of the desired metabolites. Table B.1 compares batch and continuous culture fermentation.

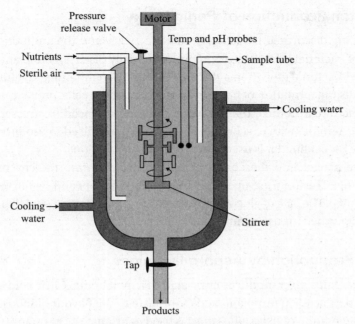

Figure B.1. Fermenting chamber

Labels in figure:
- Pressure release valve
- Motor
- Temp and pH probes
- Nutrients
- Sample tube
- Sterile air
- Cooling water
- Cooling water
- Stirrer
- Tap
- Products

Table B.1: Batch and Continuous Culture Fermentation

Batch Fermentation	Continuous Culture Fermentation
Closed system	Open system
Microorganisms and nutrients are combined and left until desired level of product is produced.	Microorganisms and nutrients are combined and monitored as desired product is continually removed.
Population goes through all growth phases (exponential, carrying capacity, decline).	Population is maintained at exponential growth phase.
One-time process of product production and collection	Ongoing process of continual product production and collection
Limited monitoring of very few variables (temperature)	Continual monitoring of several variables (temperature, pH, oxygen levels)
Examples of product: penicillin	Examples of product: citric acid

BIOTECHNOLOGY/BIOINFORMATICS

Aerobic Batch Production of Penicillin

Alexander Fleming discovered penicillin in the late 1920s, and the antibiotic is widely used in the control of bacterial infections. Fleming accidentally discovered the antibiotic when he noted decreased bacterial growth in a discarded petri dish containing the fungi *Penicillium*. Penicillin inhibits the formation of peptidoglycan in bacterial cells, preventing the formation of cell walls. The fungi produce the antibiotic only when placed in stressed environments lacking glucose. Aerobic (with oxygen) batch fermentation is used to produce large amounts of the antibiotic penicillin that is used to treat bacterial infections.

Nutrients are placed into the chamber with the *Penicillium*, thereby promoting rapid growth of the fungi. The fungi use up the glucose in the chamber and switch to lactose as their main energy source. The lack of glucose stimulates the production of penicillin. The penicillin is collected and purified for use as an antibiotic.

Citric Acid Production by Aspergillus niger

The fungus *Aspergillus niger* produces citric acid as a metabolite. Citric acid is used as a preservative (due to its acidic nature) and adds sour taste to the flavour of foods. The process of continuous fermentation of *Aspergillus niger* is used to produce large quantities of citric acid. The continuous culture is constantly fed with sucrose to promote growth of the fungi and the resulting metabolite citric acid.

Production of Biogas

Methane is produced naturally in marshes by anaerobic methanogens. The production of methane from manure and cellulose-containing plant material is accomplished by simulating the anaerobic conditions present in marshlands. Bioreactors are built to simulate the anaerobic conditions found in marshes and to collect the methane gas produced. In order for the bioreactor to produce methane, it must be held at a constant optimal temperature of about 35° Celsius, the pH must be neutral, and anaerobic conditions must be maintained. The following describes the process of methane production.

1. The first step in methane production in bioreactors involves the use of anaerobic bacteria to convert the manure or cellulose material into organic acids and alcohol.
2. The second step converts the organic acids and alcohol into acetate, hydrogen and carbon dioxide.
3. The final step involves the use of methanogenic bacteria either to convert acetate into methane and carbon dioxide or to convert carbon dioxide and hydrogen into methane and water.

Carbon dioxide + hydrogen = methane + water

Acetate = methane + carbon dioxide

Anaerobic fermenters allow for the large-scale production of biogases from organic material. Biogases can undergo combustion. Combustion then releases energy that is used as fuel for heating buildings, cooking, fuelling automobiles and energy needs in industrial processes.

TOPIC CONNECTIONS

- 1.5 Origin of cells
- 3.5 Genetic modification and biotechnology

Every living organism contains DNA that is made of the same four nitrogenous bases: adenine, thymine, guanine and cytosine. For this reason, any organism can read the DNA bases of any other organism. The enzymes used for protein synthesis in each organism will function for all organisms. This allows scientists to transfer DNA from one species to another without the donor DNA being rejected. The protein coded for by the donor DNA will appear the same when the recipient of the foreign DNA produces it.

Gene Transfer

Gene transfer allows scientists to transfer genes (DNA) from one species into the genome (DNA) of another species. In order to do this, the gene of interest must be cut from the donor DNA using restriction enzymes. Many different restriction enzymes are available. Each is cut at unique base sequence sites. Some restriction enzymes cut DNA and leave sticky ends. Others cut DNA and leave blunt ends. Restriction enzymes that cut DNA and leave sticky ends are used to obtain DNA fragments that have matching complementary bases and can easily form bonds joining the two pieces of DNA. Refer to Figure 3.20 in the core material.

A plasmid (small, circular piece of DNA) is obtained from a prokaryotic (bacterial) cell and cut with the same restriction enzyme that is used to cut the foreign DNA from its genome. Both the plasmid and foreign DNA must be cut with the same restriction enzyme to ensure complementary bases on the sticky ends of the DNA pieces.

The two cut pieces of DNA (plasmid and foreign) are joined together by the enzyme DNA ligase. The new DNA is referred to as *recombinant DNA (rDNA)* since it contains both the original and new foreign combinations of bases.

Bioinformatics

Bioinformatics is an interdisciplinary branch of science that allows the receiving, storing, sorting, analysing and retrieval of large amounts of biological data. Bioinformatics includes the working of biologists as well as computer scientists and computer engineers. Bioinformatics is used to store information about nucleic acid sequences, gene locations and protein structures. An important implication of bioinformatics is the role it plays in identifying genes of interest (target genes).

Transgenic Organisms

Recombinant DNA (rDNA) must become part of an organism's genome in order for the protein coded for by the gene to be produced. Organisms that contain rDNA are referred to as transgenic organisms. They produce proteins that are not normally part of their proteome (normally

> ### REMEMBER
>
> Restriction enzymes are produced only by prokaryotic cells (bacterial cells) but will cut any organism's DNA. This is possible because DNA is universal!

expressed proteins). Direct methods of transfer involving both chemical processes and physical processes can be used to introduce rDNA into a cell's or an organelle's (chloroplast) genome. Vectors such as viruses or bacterial cells can be used as indirect methods for the introduction of rDNA. Table B.2 outlines methods for the introduction of rDNA into a host cell.

Table B.2: Direct Methods for the Introduction of rDNA

Chemical Transfer	Physical Transfer
Calcium chloride: Used to produce small holes in the cell membrane of cells in order to allow rDNA to enter the host cell	Biolistics (gene gun): Heavy metal (gold or tungsten) particles are coated with the rDNA and "shot" with great force, penetrating the cell membrane and entering the host cell.
Liposomes: Artificially produced vesicle made from the same substances as those of the cell membrane; fuses with the cell membrane and deposits the rDNA into the host cell	Electrophoration (electrical current): An electrical field is used to increase the permeability of the cell membrane, facilitating the movement of rDNA into the host cell.
	Microinjection: The rDNA is injected into the nucleus of the host cell.

Open Reading Frames and Target Genes

Open reading frames (ORF) are sections of nucleotides that are located on DNA between a start and a stop codon. Open reading frames code for the production of a complete polypeptide. Open reading frames are transcribed without interruption, making them the preferred site for insertion of the target gene (gene of interest). The target gene is read during transcription and translated as a complete protein due to the placement of the gene into an ORF. The sequences of nucleotides that target genes are linked to control the expression of the target gene. An ORF is located on the section of DNA shown below between the start codon ATG and the stop codon TAA. Notice that no DNA stop codons (tga, taa, TAG) appear within the ORF.

5'atg**aagctgaatagcgtagaggggttttcatcatttgaggacgatgta**taa3'

Marker Genes

Marker genes are genes located in known locations (loci) on DNA that are linked to the expression of a specific trait. Marker genes are used in gene transfer since the expression of the trait coded for by the marker gene can be used as an indicator for successful gene transfer of the target gene. For example, a gene that codes for the resistance to an antibiotic could be the site of insertion of a target gene. A section of DNA containing the marker gene sequence along with the target gene sequence is transferred to the host organism. Successful transformation of the target gene is identified by the presence of antibiotic resistance in the host cells. The expression of the marker gene trait by the transgenic organism indicates successful uptake of the rDNA.

Genetically Modified Plants

Plants can be genetically modified with the introduction of rDNA into the genome of the entire plant, into parts of leaves (leaf discs) or into protoplasts. Protoplasts are plant cells that have chemically or enzymatically had their cell walls weakened or removed. In order to undergo transformation, the rDNA must become incorporated into the nuclear DNA (chromosome) or chloroplast DNA.

Genetically modified organisms (GMO) have been used to help produce crops that are resistant to pests, disease, extreme temperatures and salinity. The addition of genes such as those that code for resistance to salinity and frost into the genomes of plant species has allowed the harvesting of plants in environments in which they could not otherwise survive. The ability to grow crops in diverse environments increases a farmers' productivity and the ability of farming to meet the world's food demand. In addition, the use of recombinant DNA that provides plants with resistance to certain pesticides decreases the need to use pesticides in the environment. Less pesticide use helps to protect ecosystems from the release of potentially toxic substances.

Plants can be enhanced nutritionally by adding genes that code for substances such as the synthesis of beta-carotene (a precursor to vitamin A). Nutritionally enhanced plants can help in treating nutritional deficiencies that exist in many parts of the world. An excellent example is golden rice. Golden rice plants have been genetically modified to produce beta-carotene. Vitamin A deficiencies in children in poor countries (such as Africa) can lead to blindness. Golden rice has been vital in preventing the loss of sight due to the production of beta-carotene by the transgenic plant.

HERBICIDE RESISTANCE IN SOYBEAN CROPS

The bacterium *Agrobacterium tumefaciens* is used to produce soybean crops that express resistance to the herbicide glyphosate. A tumour-inducing (T_i) plasmid naturally produced by the bacterium is used to introduce the gene for resistance to glyphosate into the soybean plant cells. The glyphosate resistance gene is placed into the ORF of the gene for tumour formation. The tumour-inducing gene serves as the marker gene for recombination. Cells that successfully take up the rDNA can be easily identified by the presence of tumour formations.

HEPATITIS B VACCINE PRODUCTION BY TOBACCO PLANTS

Tobacco plant crops have been genetically modified to allow the production of a hepatitis B vaccine with the use of the tobacco mosaic virus. The tobacco mosaic virus is the most commonly used viral vector in the production of transgenic organisms. A gene coding for the production of a surface antigen for the hepatitis B virus is inserted into the genome of the tobacco mosaic virus. The tobacco mosaic virus infects the tobacco plant, delivering the rDNA-containing genes for the surface protein of hepatitis B. The tobacco plant produces the surface proteins for hepatitis B. The surface proteins are isolated and used to produce vaccines against the hepatitis B virus.

AMFLORA POTATO PRODUCTION

The amflora potato (*Solanum tuberosum)* has been genetically modified to produce pure amylopectin as its starch. Normally, potatoes produce both amylopectin and amylose as their main forms of carbohydrate storage (starches). The presence of amylose makes the use of potato starch difficult in the production of paper and adhesive products. Amylose tends to form jells that limit the paper's consistency and ability to display colours properly. The genes that code for the production of amylose have been genetically modified in the amflora potato in such a way that the genes for amylose are no longer expressed (genetically "turned off") in the potato. As a result, the amflora potato is the preferred source of starch in industrial processes.

Benefits and Risks of Genetically Modified Plants

Genetically modified plants possess both potential benefits and potential harmful effects. For example, a plant genetically modified to survive in a new environment may outcompete the natural plants found in the environment. (Refer to Table B.3.)

Table B.3: Benefits and Risks of Genetically Modified Plants

Benefits	Risks
■ Increased crop yield ■ Less land needed for crop growth ■ Increases global food supply ■ Less use of pesticides	■ Other organisms in the ecosystem could be negatively affected. ■ Genetically modified pollen could affect other plants in the ecosystem. ■ Individuals could be allergic to the foreign protein.

B.3 ENVIRONMENTAL PROTECTION

TOPIC CONNECTIONS

- Topic 1: Cell biology

Bioremediation

> **Remediate:** Fix a problem
> **Bio:** Life
> **Bioremediation:** The use of living organisms in the removal of waste

Bioremediation is the use of microscopic organisms to return an altered state of an environment back to its normal state. Bioremediation takes advantage of naturally occurring processes carried out by microorganisms. The disposal and cleanup of waste can be facilitated with bioremediation. Bioremediation is used for sewage treatment, cleaning up oil spills, degradation of toxic chemicals and the processing of chemical elements.

Biofilms

Biofilms are groups of many bacteria held together by adhesive polysaccharides (which have a gluelike structure) that allow Eubacteria to exist as large colonies. Eubacteria within biofilms use quorum sensing to respond to population density changes. High-density populations stimulate the Eubacteria to turn on specific genes that would otherwise remain silent. For this reason, biofilm aggregates exhibit characteristics not present in individual Eubacteria. Emergent properties that develop within biofilms can be either harmful or useful to humans.

> **Quorum sensing:** Responding to stimuli based on population density.

Biofilms can release toxic substances that individual Eubacteria would not normally produce. Breaking up biofilms is an important process in the treatment of many pathogenic diseases and in the prevention of the production of toxic substances by these colonized Eubacteria. The Eubacteria *Pseudomonas areuginosa* forms biofilms that are resistant to antimicrobial agents (such as antibiotics). Infection with *Pseudomonas areuginosa* is one of the main reasons for the death of patients with cystic fibrosis.

Pollution Control

Biofilms, along with other chemical and physical procedures, can be used for cleaning up oil spills. The salt-tolerant bacteria *Marinobacter hydrocarbonoclasticus* possess enzymes that degrade the hydrocarbon rings of benzene released into ocean waters during oil spills. *Pseudomonas* bacteria have also been used for bioremediation of oil spills. The *Pseudomonas* bacteria produce chemicals and surfactants that break up the oil spill and aid in degrading toxic compounds.

Methyl mercury is toxic to living organisms and can build up in marshes and wetlands. The toxic mercury enters the food chain and becomes more concentrated as it moves through trophic levels. The biomagnification of the toxic mercury can be lethal to top consumers and disrupt the entire food chain. Some strains of *Pseudomonas* avoid mercury poisoning by possessing the natural ability to degrade toxic methyl mercury into elemental mercury. Scientists take advantage of the mercury conversion abilities of the *Pseudomonas* and use the bacteria for bioremediation in mercury-contaminated environments.

Sewage Treatment

Bacteria and viruses are both used in the treatment of sewage waste. Bacteriophages are used in sewage treatment to help rid sewage of pathogenic bacteria. Bacteriophages infect bacterial cells, destroying the host cell and limiting population growth of the pathogenic cells. Biofilms in trickle filter beds can be used to treat sewage by removing organic matter from the waste. The sewage is filtered through stones containing biofilms of saprotrophic anaerobic and aerobic bacteria. The saprotrophic bacteria feed on the organic waste, removing it from the sewage. Table B.4 outlines the current uses of bioremediation.

Table B.4: Bioremediation

Bioremediation Use	Process
Removal of oil spills	<ul><li>Hydrocarbon-degrading bacteria naturally found in the environment (*Marinobacter* and *Pseudomonas*) oxidize the hydrocarbons that make up crude oil, removing it from the water.</li><li>Additional hydrocarbon-degrading bacteria can be added to speed up the process.</li><li>Fertilizers can be added to increase the reproductive processes of the hydrogen-feeding bacteria.</li></ul>
Mercury pollution	<ul><li>Bacteria produce elemental mercury from methyl mercury (*Pseudomonas*).</li><li>Elemental mercury is much less toxic than methyl mercury.</li><li>Bacteria that convert methyl mercury to elemental mercury can be added to methyl mercury–polluted environments to help in removal of the toxic substance.</li></ul>
Sewage treatment	<ul><li>Bacteriophages infect pathogenic bacteria, destroying the host cell and decreasing the population of bacteria in raw sewage.</li><li>Biofilms of saprotrophic bacteria (anaerobic and aerobic) are used in filter beds to feed on organic waste, removing it from the raw sewage.</li></ul>

NOTE: *B.4 and B.5 are for HL students only, and material from these two sections will not appear on the SL exam.*

B.4 MEDICINE

TOPIC CONNECTIONS

- 3.5 Genetic modification and biotechnology
- 6.3 Defence against infectious diseases
- 11.1 Antibody production and vaccination

Biotechnology has impacted the way in which diseases are detected and treated. Pathogenic diseases can be detected within an organism by screening for the presence of protein antigens or genetic material from the pathogenic agent in the blood or urine of the host. Genetic markers have been identified for many diseases. Testing for the presence of these genes can identify individuals who run the genetic risk of developing the disease.

DNA Microarrays

Microarrays are used in genetic testing to identify the presence of specific DNA gene sequences in a given genome. Microarrays test for the presence of specific genes by identifying the presence of mRNA transcripts. Microarrays are created with the use of polymerase chain reaction (PCR) and the enzyme reverse transcriptase. Reverse transcriptase is an enzyme produced by retroviruses (viruses that possess RNA as their sole genetic material). Reverse transcriptase can make DNA from single-stranded mRNA. The mRNA for the target gene is obtained, and reverse transcriptase is used to produce a complementary DNA strand from the existing mRNA strand. The RNA template used to create the double-stranded molecule is degraded, producing a single-stranded complementary DNA molecule. The single stranded DNA molecule is referred to as cDNA. The cDNA samples are fixed on microarray plates and are referred to as probes.

Microarrays are fixed on glass plates or silicon chips that contain specific cDNA probes at specific locations. Each fixed probe binds specific complementary cDNA target genes. The target cDNA is labelled with a fluorescent tag before being exposed to the probes. If the target genes are complementary to any of the probes, hybridization will occur. The microarray plate is washed after hybridization to remove any target cDNA that was not fully complementary to the probes. The presence of a target gene is detected by the presence of the fluorescent markers that can be seen prior to washing. The level of fluorescence seen on the microarray plate identifies the level of expression of the target gene. Figure B.2 shows the process of gene identification with the use of microarrays.

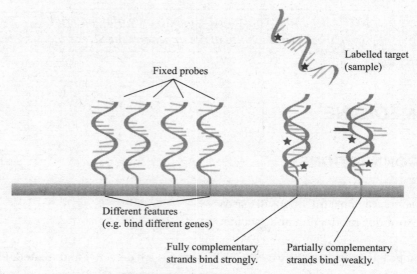

Figure B.2. Microarray process

Microarrays can be used to test for the presence of a gene that makes an individual predisposed to the development of a genetic disease or to identify the actual presence of the genetic disease. They can also be used to identify the presence of viral diseases such as different strains of the influenza virus.

The *BRCA1* gene is a tumour-suppressing gene that plays a role in regulating cellular division. Mutations in the *BRCA1* gene are linked to the development of breast and ovarian cancer. Detection of the mutated form of the gene using microarrays allows individuals to decide if they should take measures to reduce the risks of developing the cancer by the surgical removal of breast and ovarian tissues.

Treatment with Viral Vectors

Viruses infect cells by injecting their nucleic acid (DNA or RNA) into the host cell. Only the nucleic acid enters the cell, with the other viral parts remaining outside the host cell. Once the viral DNA is inside the host cell, it inserts into the host's genome. Viruses are the perfect choice for use as vectors for delivering genetic material to cells. Viruses can be used to deliver healthy genes to cells that have defective forms of a gene.

GENE THERAPY AND SCID

Gene therapy has been used to correct defective genes in children suffering from SCID—severe combined immunodeficiency. The defective gene does not produce the enzyme ADA (adenosine deanimase) that is vital for the production of T-cells. Without T-cells, SCID patients cannot fight off infections and are at risk of early death due to disease. By using gene therapy, SCID patients can develop healthy immune systems and live normal lives.

RISKS AND BENEFITS

Gene therapy has the benefit of treating genetic disorders in order to allow individuals with genetic diseases to live healthier lives. However, no procedure should be carried out without considering the risks that may be involved (see Table B.5).

Table B.5: Risks Associated with Gene Therapy

Possible Risk	Consequence of Risk
Unwanted immune response to the altered gene	Inflammation due to immune response can lead to tissue and organ damage. In extreme cases, it may cause death.
Development of cancerous cells (tumors)	■ The altered gene could be taken up by healthy cells, causing those cells to become cancerous. ■ The gene could be inserted into the wrong place in the target cell genome, causing the target cell to become cancerous.
Overproduction of the protein or hormone coded for by the gene	The levels of protein or hormone coded for by the gene could disrupt the homeostatic levels in the body and lead to physiological issues.
Long-term negative effects of the new gene	The long-term risks of gene therapy cannot be explained since it is such a new technology. Young children receiving the new gene could have side effects later in life that we can only speculate about today.

Biopharming

Biopharming involves the use of plants and animals to produce desired proteins for human therapeutic use (pharmaceuticals). Crops such as corn and tobacco can be genetically modified to produce drugs and vaccines. Bacterial cells are used to produce hormones such as insulin and growth hormone (GH). Crops that have been genetically modified for the production of protein for therapy uses are not allowed to enter the food market.

Antithrombin (AT) Production

Hereditary antithrombin deficiency (HAT) is an autosomal dominant disorder caused by the inheritance of a mutated gene coding for the production of antithrombin. Without the presence of the regulatory protein antithrombin, the clotting process proceeds uncontrolled. Hereditary antithrombin deficiency can lead to the development of clots within veins, a condition referred to as deep vein thrombosis (DVT). Clots that form within veins can travel to the lungs, resulting in pulmonary embolisms and death. Antithrombin is used in the medical field as an anticoagulant to regulate blood clotting in patients suffering from HAT. Biopharming with the use of goats has allowed for the mass production of human antithrombin. (Refer to Figure B.3.)

> **REMEMBER**
>
> More than 90,000 human blood donations would be needed to obtain the same amount of antithrombin that can be produced by just one genetically modified goat.

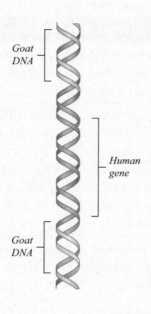

1. The human gene for the production of antithrombin is inserted into goat DNA.

Goat DNA

Human gene

Goat DNA

Egg

2. The recombinant DNA is injected into the nucleus of a fertilized goat egg.

3. The offspring of the transgenic goat are left to mature and then tested for presence of the protein in their milk.

4. Offspring that test positive for the production of antithrombin are bred to produce a herd of transgenic goats that produce vast amounts of antithrombin.

Figure B.3. Biopharming production of antithrombin

Tracking Tumour Cells

Tracking experiments are used to identify the location and interactions of specific proteins. Cancerous cells can be identified by tracking methods that identify high levels of production of cellular receptors for transferrin. Transferrin binds iron and attaches to transferrin receptors. Bound iron is then brought into the cell by endocytosis. Cells that are rapidly dividing possess an increased number of transferrin receptors compared with other tissues. Since cancer cells divide rapidly, they produce high levels of transferrin receptors. Cancer cells can be detected with the use of transferrin molecules that have attached luminescent probes. Cells taking in large amounts of the luminescent probes can be easily identified by the resulting luminescence of the cell. Transferrin can also be used to treat cancer by attaching a toxic agent to transferrin for targeted deliver to rapidly dividing cells. (Refer to Figure B.4.)

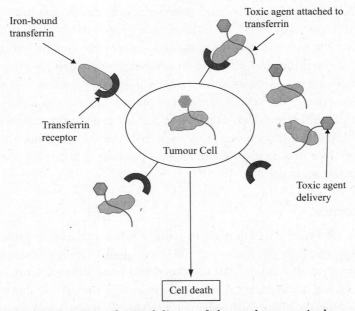

Figure B.4. Transferrin delivery of chemotherapeutic drug

B.5 BIOINFORMATICS

Databases can store increasingly exponential amounts of data. Scientists rely on databases for quick and easy access to stored genetic information. Software tools allow for comparisons of biological sequences found in proteins or nucleic acids. Software tools such as the Basic Local Alignment Search Tool (BLAST) electronically store known biological sequences of nucleic acid and protein structures.

- BLASTn—used for nucleotide sequence alignment and comparisons
- BLASTp—used for amino acid alignment and comparisons

BLAST databases can be used to compare nucleic acid sequences among members of the same species as well as among members of different species. Newly identified sequences can be run through BLAST software in order to identify any existing similar sequences. Multiple sequence analysis allows the comparison of three or more protein or nucleic acid sequences among different organisms. Multiple sequence analysis is used to predict possible evolutionary pathways and determine phylogeny. For example, if a new gene sequence is discovered in one mammal, it can be compared with known existing sequences of other mammals to see if the gene codes for a shared evolutionary trait. If genes are common among similar species, the function of the gene can be investigated using any of the organisms that share the common gene sequence. BLAST results are reported for similarities among sequences as well as exact nucleotide matches.

Expressed Sequence Tag

Expressed sequence tags (EST) are short pieces of cDNA that are complementary to mRNA. Over 72 million EST that recognize known gene nucleotide base sequences are available from databases. EST are used for gene discovery and to identify the presence of known gene sequences in DNA molecules based on the presence of gene transcripts (mRNA). Information gained from mRNA transcript levels can be used to compare gene expression among different tissues, compare gene expression in the same tissue over time and analyse gene expression for genetic disease identification. Changes in gene expression can be compared with phenotypic changes that occur along with the changes in gene expression. This allows scientists to evaluate the function of the gene. Expressed sequence tags are especially useful in cancer research to compare the expression of genes in normal cells to the expression of genes in cancerous cells. Expressed sequence tags are used in data mining and have led to the discovery of new genes. Data mining involves looking for patterns in large sets of data such as the sequences of nucleic acids in DNA.

Knockout Genes

Knockout genes are genes that are used to replace functional genes. Knockout genes are produced by identifying specific gene sequences for a gene of interest from a database such as BLAST. The creation of a similar but nonfunctional gene allows scientists to replace the working gene with the knockout gene. Knockout genes get their name due to their effect of knocking out the function of the gene. Scientists use knockout genes to discover the functions of working genes by looking for phenotypic changes in the organism carrying the knockout gene. Mice are often used for experimentation with knockout genes since they share many genes with humans.

Yeasts are unicellular organisms that respire anaerobically. Three populations of yeast were fermented for 45 minutes in different solutions of glucose, sucrose and pure water. An additional population was boiled prior to fermenting in a solution of sucrose. The gas released from the four populations of yeast was measured at 5-minute intervals for a total of 45 minutes. The results from the investigation are displayed on the graph below.

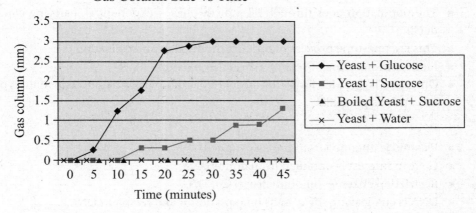

1. (a) Calculate the average amount of gas produced per 5-minute interval by the yeast in glucose over the 45-minute time period. (2)
   (b) State the name of the gas released by the yeast. (1)
   (c) Explain batch fermentation. (3)

2. Outline the process of gene transfer using a viral vector. (5)

3. Explain one use of biofilms in bioremediation. (4)

4. State 2 uses of DNA microarrays. (2) *HL only*

5. Explain the use of knockout gene technology. (3) *HL only*

BIOTECHNOLOGY/BIOINFORMATICS

ANSWERS EXPLAINED

1. (a) 0.25 + 1.50 + 1.75 + 2.75 + 2.90 + 3.00 + 3.00 + 3.00 + 3.00 = 21.15 (+/– 0.25) mm (One point is awarded for the correct answer, and one point is awarded for showing the calculations.)

   (b) Fermenting yeast release carbon dioxide.

   (c)
   - Batch fermentation occurs in a closed system.
   - Microorganisms and nutrients are combined and left/allowed to ferment until the desired level of product is produced.
   - The population goes through all growth phases/exponential, carrying capacity, decline.
   - This is a one-time process of product production and collection.
   - Limited monitoring of very few variables is required.
   - One point is awarded for listing any correct example of a batch fermentation product, such as penicillin.

2.
   - Plasmid is obtained from prokaryotic cell.
   - Gene of interest is obtained from donor DNA.
   - Restriction enzymes/endonucleases cut DNA
   - DNA is cut, leaving sticky ends on both donor and recipient DNA.
   - Same restriction enzyme must be used to cut both DNA molecules to ensure complementary bases.
   - Donor/foreign DNA is sealed with plasmid using DNA ligase.
   - Recombinant DNA is produced/results from donor and plasmid DNA binding.
   - Recombinant DNA/rDNA must be placed into a host to enable production of protein coded for by the gene.
   - Virus/bacteriophage is stripped of its nucleic acid/DNA.
   - rDNA plasmid is inserted into viral protein cap.
   - Viral vector delivers the rDNA to the host bacterial cell.
   - Transgenic organism is produced when the host takes up the rDNA/plasmid.
   - Bacterial cell produces foreign protein coded for by plasmid/rDNA.
   - Bacterial cells most commonly used/unicellular organism
   - Gene gun (biolistics)/calcium carbonate/liposomes or example of other method used for gene transfer to the host

3.
   - Biofilms are aggregates of bacterial colonies.
   - Bacterial cells within aggregates exhibit different gene expression than when in isolation/exhibit quorum sensing.
   - Biofilms can secrete toxic substances or substances that can degrade toxic substances.
   - Biofilms are used to clean up oil spills by degrading hydrocarbon rings/*Marinobacter* and *Pseudomonas*.
   - Sewage treatment with biofilms involves filtering the sewage as microoganisms feeding on dead organic material, thereby removing potentially toxic substances from the waste.

- *Pseudomonas* converts toxic methyl mercury into elemental mercury.
- The removal of toxic methyl mercury prevents biomagnification of the toxic substance.

4. DNA microarrays are used for:
   - Gene discovery
   - Disease identification and diagnosis (genetic)
   - Phylogeny/evolutionary relationships
   - Gene expression/level of gene expression at different times or in different tissues
   - Comparison of gene expression in cancer and normal cells

5.

   - Knockout genes are mutated forms of functional genes.
   - Knockout genes do not code for a functional protein.
   - The use of knockout genes allows the identification of a gene's function.
   - Looking at phenotypic changes due to the presence of a knockout gene can aid in determining gene function.
   - Knockout genes are similar to the functional genes with enough base changes to interfere with gene function.

Ecology and Conservation

CENTRAL IDEAS
- Community structure and interactions
- Energy flow and ecosystems
- Human impact on ecosystems
- Conserving biodiversity
- Population dynamics

TERMS
- Alien species
- In situ conservation
- Niche
- Ex situ conservation
- Keystone species
- Simpson index

INTRODUCTION

Organisms rarely live in isolation from members of their own species or from members of other populations. The biotic (living) world and abiotic (nonliving) world that organisms inhabit bring both benefits and challenges to species' existence. Competition for resources impacts ecosystems and determines which organisms will be able to survive. Different organisms have adapted to survive best in specific environmental niches. Changes in the habitats of organisms occur naturally and can be greatly affected by the action of humans. When the niches of populations of organisms are threatened, the organism's existence depends on its ability to adapt to the changing conditions. Human actions often involve the destruction of habitat and death of many species of organisms. However, humans also carry out actions to protect the loss of species and ensure the diversity of life is maintained.

NOS Connections
- **Models are used to represent the natural environment (tolerance graphs, species diversity, energy transfer and pyramids).**
- **Scientists must assess both the risks and benefits of research (biological control methods, agriculture and disturbances to nutrient cycles).**
- **Cooperation and collaboration among scientists enhances scientific knowledge and endeavours (international cooperation with preservation of species).**
- **Biases exist and need to be avoided in scientific research (data collection methods—sampling).**

ECOLOGY AND CONSERVATION

C.1 SPECIES AND COMMUNITIES

Species Distribution and Limiting Factors

Limiting factors influence the rate at which a population can grow. Limiting factors for plants include water, light, space and proper mineral nutrients. If any of these factors become limited in the environment, the plant population declines. Factors such as soil pH, temperature, light intensity, water availability and salinity influence a plant's ability to live in a particular area. Adequate space is necessary in order for plants to develop roots and branching shoots. A plant's ability to obtain needed nutrients, water, and sunlight determines the environment in which the plant can exist.

Animals must be able to obtain needed nutrients from the environment in which they live, find mates for reproduction, locate water for metabolic processes and find territory in which to survive. Limitations in obtaining any of these resources can greatly affect an animal's ability to survive and reproduce. Great variation exists among animals and their needs and adaptations for survival. Some animals are adapted to specific niches and rely on these niches for survival. Other animals can adapt to different niches should the need arise. Limiting factors in ecosystems are outlined in Table C.1.

Table C.1: Limiting Factors

Factors Limiting Plant Populations	Factors Limiting Animal Populations
■ Water for metabolic processes ■ Space for growth of roots and shoots ■ Mineral nutrients in soil or water ■ Sunlight for photosynthesis ■ Temperature (proper range for metabolic processes)	■ Water for metabolic processes ■ Territory to find mates and rear young ■ Nutrients from food ■ Temperature (proper range for metabolic processes)

Community Structure

Ecosystems consist of many different species of plants and animals. The population size of different species within ecosystems depends on various factors, including competition for limited available resources. Scientists investigate population sizes within communities in order to gain a better understanding of population interactions. Since counting every organism in a population is often difficult or impossible, scientists use random sampling to estimate population numbers.

Sampling Techniques

Random sampling involves selecting organisms from a larger population. Each member of the population must have an equal chance of being selected.

> **Random sample:** Selecting representative individuals from within a population. Each member of the larger population must have an equal chance of being selected.

QUADRAT SAMPLING

The quadrat method of population estimation involves marking off smaller areas within a large sample area in order to select marked areas randomly for counting population numbers. Each quadrat marked from the larger area is given an identification number for sample selection. Quadrats are randomly selected, and population numbers in each selected quadrat are counted. The average number of organisms found in each quadrat is calculated and used to estimate the entire population of the selected organism in the chosen area.

1. Select the region to be sampled.
2. Divide the area into smaller quadrats marked for ease of identification.
3. Randomly select numbers that identify each quadrat.
4. Count the members of the selected population that are found in each randomly selected quadrat.
5. Estimate population size.

$$\text{Population size} = \frac{\text{Average number per quadrat} \times \text{total quadrat area}}{\text{Area of each quadrat}}$$

Quadrat sampling is useful for comparing population sizes of different organisms within the same habitat or in different habitats. For example, random sampling using the quadrat method can be used to estimate the population of two different types of sea oat grasses inhabiting a dune area. The population estimates of the two species can be compared in order to make hypotheses concerning the interaction of the two species.

TRANSECT METHOD

Transect sampling is another method that is used for random sampling of populations. A transect is a line drawn across a chosen area of investigation in order to determine where sampling within the site should occur. Transects are useful when determining population numbers in areas that show variation in population numbers due to an abiotic factor. The effects of abiotic factors on the population of organisms can be investigated by comparing the abundance of organisms found at various locations along each transect. For example, the population numbers for areas near shorelines will vary as the amount of moisture in the soil changes based on changing tide flow.

1. Determine the area to be sampled.
2. Run the transect in a specific predetermined straight line through the area to be sampled.
3. Determine equal distances from one end to the opposite end of the transect that are to be sampled.
4. Determine the size of the sampling area for each sample site chosen for sampling along the transect line. Use quadrats at each sample site to identify the chosen area for population counting, and count the number of members of the population(s) of interest in each quadrat.
5. Measure the abiotic factor being tested (moisture, pH, etc.) in order to evaluate any possible correlation between the population size and the abiotic factor.

Niche Concept

The area an organism inhabits and how that organism interacts with living and nonliving components of an ecosystem is referred to as its niche. Each species has evolved to exist in distinct niches.

> **Niche:** The location and life activities of an organism. A niche includes an organism's spatial habitat, feeding activities and interactions with other species.

Niche concept deals with how organisms interact with each other, obtain food and interact with other species within the environment they inhabit. The niche of one species affects the niche of other species. No two species can share the exact same niche since competition within the niche will cause one species to be excluded from the niche. This phenomenon is known as the competitive exclusion principal.

> **Competitive exclusion principal:** No two species can occupy the exact same niche. One of the species will outcompete the other for survival in the niche. Hence, each species occupies its own distinct niche.

Niches that overlap one another can exist as long as the competition is limited to only one or very few resources. For example, two different species could share similar niches that overlap. The species could be competing for breeding space but not for food or other resources. The overlapping niches of various shore birds are shown in Figure C.1.

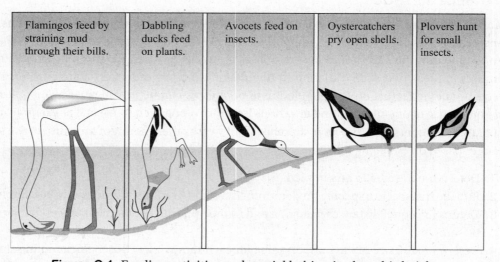

Flamingos feed by straining mud through their bills.

Dabbling ducks feed on plants.

Avocets feed on insects.

Oystercatchers pry open shells.

Plovers hunt for small insects.

Figure C.1. Feeding activities and spatial habitat in shore bird niches

Figure C.1 shows the feeding and spatial activities of various shore birds. Notice that each bird occupies a specific feeding and spatial niche on the shore. The birds may compete for breeding space. However, they do not compete for other resources in the environment, which allows each bird population to exist in its own unique niche.

Species Interactions

When species interact within ecosystems, many distinct types of interactions develop. Each type of interaction develops due to the need to obtain limited resources from the environment. Table C.2 outlines the main types of species interactions.

Table C.2: Species Interactions

Type of Interaction	Description	Examples
Competitive	Two species occupying the same territory compete for the same limited resource.	■ Cheetahs and lions sharing habitats compete for the same prey. ■ Different species of sea anemones in tide pools compete for space (territory).
Herbivory	Consumption of plant or primary producer material by herbivores or primary consumers	■ Plants fed on by herbivores have developed mechanisms such as thorns to ward off consumption by herbivores, such as the thorns of a cactus. ■ Some beetles have evolved to be immune to toxic substances produced by plants to ward off herbivory.
Predation	One consumer (predator) feeding off of another (prey)	■ Barn owls are predators of rodents. A change in either population affects the survival of the other. ■ The bottlenose dolphin feeds on fish whose numbers influence the ability of the dolphins to survive.
Parasitism	Parasites live on or in a host and take food resources from the host; the parasites benefit and the host is harmed in parasitic relationships.	■ Deer ticks feed on deer, taking blood meals containing nutrients from the host. ■ Butterfly larvae feed on plant tissue as they develop, taking nutrients from the plant.
Mutualism	Two species form a relationship in which each species benefits from the relationship.	■ Oxpecker birds eat ticks and other parasites off of rhinos and zebras. The birds gain food, and the rhinos and zebras gain pest control. ■ Flowers provide food for bees. Bees help the plant reproduce by dispersing pollen for the plant.

REMEMBER

Keystone species are named after the keystone brick in an archway that holds the entire structure together. If the keystone brick was removed, the stability of the entire structure would be affected.

Keystone Species

Members of an ecosystem that have a disproportionate impact on the diversity of species found within an ecosystem are referred to as keystone species. The abundance of a keystone species is small in relationship to the abundance of other populations within the ecosystem. If the keystone species was removed from the population, the dynamic interactions within the ecosystem would change. An example of a keystone species is the sea otter. The sea otter feeds on sea urchins that are damaging to kelp forests. If the sea otter was not feeding on the sea urchins, the sea urchin population would consume vast amounts of the kelp. A kelp forest provides breeding sites for organisms, serves as nurseries for young offspring, protects organisms from predation and supplies food for many members of the ecosystem. The loss of a kelp forest due to the absence of a keystone species would affect the balance of the entire ecosystem. Declines in sea otter populations have been caused by killer whales as well as human poaching of sea otters for fur.

Symbiotic *Zooxanthellae* and Coral

The algae *Zooxanthellae* have developed a mutualistic relationship with coral polyps. The photosynthetic algae provide oxygen and produce nutrients for the coral. The coral offers the *Zooxanthellae* protection as well and also provides the algae with carbon dioxide and other materials to aid in its metabolic processes. Both organisms benefit in this mutualistic, symbiotic relationship.

Competitive Exclusion

The principal of competitive exclusion states that two species cannot co-exist if they are competing for the same limited resources. One species will outcompete the other species for the limited resource. This will lead to the extinction of the outcompeted species or to the outcompeted species finding a new niche.

When *Paramecium aurelia* and *Paramecium caudatum* are grown in isolation from each other on the same growth medium, they both survive. However, when the two species of *Paramecium* are grown in the same habitat, *Paramecium aurelia* outcompete *Paramecium caudatum* for the limited resources and excludes it from the habitat.

Fundamental and Realized Niches

The actual niche that a species occupies is referred to as its realized niche. The realized niche is an organism's actual mode of existence. The realized niche evolves due to competition and interactions with other species. The fundamental niche is the potential mode of existence of a species. This refers to the total range of an area in which the organism could live with its given adaptations (see Figure C.2).

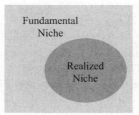

Figure C.2. Realized and fundamental niches

Species Tolerance and Zones of Stress

The range of an environment in which an organism can survive is known as the zone of tolerance. The optimal range of the zone of tolerance includes areas that the organism can inhabit without experiencing any loss of homeostatic ability or stress. The upper and lower limits of the zone of tolerance stress the organism and limit reproductive abilities (see Figure C.3).

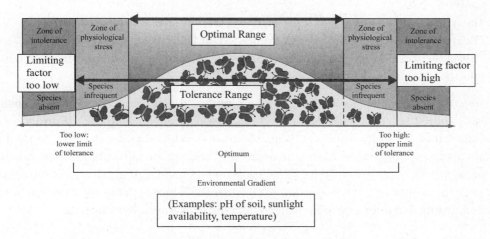

Figure C.3. Zones of tolerance and stress

Plants inhabiting intertidal zones have become adapted to survive in the harsh environment of fluctuating water exposure. Plants living near the intertidal zone are not adapted to the intertidal environmental conditions and cannot survive the harsh conditions.

C.2 COMMUNITIES AND ECOSYSTEMS

TOPIC CONNECTIONS

- 4.2 Energy flow

Biomass

One method used to evaluate the success of organisms in various niches is to measure the biomass of organisms in particular areas at particular times. Changes in biomass could indicate changes in the success of a particular organism to survive in its realized niche. Biomass can also be used to evaluate the number of organisms inhabiting different trophic levels within niches.

> **Biomass:** The total dry weight of organic material derived from individual organisms or from within ecosystems. Since obtaining dry weight destroys living organisms, small samples are taken from an area and used to estimate the biomass of the larger populations.

Trophic Levels and Biomass

In order to evaluate the biomass found in different trophic levels within an ecosystem, the following steps are used:

1. Samples are taken from within the ecosystem using random sampling methods such as quadrats.
2. The samples are sorted according to trophic level and dried in ovens to remove all water content.
3. The mass of organisms in each representative trophic level is measured.
4. The masses of each trophic level are compared for analysis of the ecosystem.
5. This process can be repeated at a later time with the same ecosystem to evaluate if and how the biomass is changing over time.

When collecting organisms for determining biomass, the ethical treatment of the living organisms should be considered. Samples should be kept small to limit the destruction of organisms found in the environment. Scientists have developed tables that can be used to estimate biomass of organisms based the organisms' size and height without destroying the organisms by determining their actual dry mass. When removing animals from ecosystems in order to estimate biomass, it is important to consider the ethics of doing so. Humane methods for capturing and returning organisms to sampled ecosystems must be used.

Gross and Net Production

Biomass productivity can be calculated in several ways. The total amount of organic material produced determines the gross production of the ecosystem. Subtracting the energy cost due to cellular respiration from the gross production determines the net production of the ecosystem.

- **Gross production:** The total amount of organic material (biomass or energy) produced by an organism or within an ecosystem over a period of time without consideration of any loss of energy through processes such as respiration.
- **Net production:** Gross production minus energy lost due to respiration. (gross production (GP) – respiration (R) = net production (NP) or GP – R = NP)

Table C.3 shows a sample calculation of net production.

Table C.3: Marsh Snail Data for Calculating Net Production

Mass of 6 Marsh Snails (+/– 0.1 g)	Grams of Food Consumed by 6 Marsh Snails (+/– 0.1 g)	Grams of Faeces Excreted by 6 Marsh Snails (+/– 0.1 g)
January 1: 320.2 g January 6: 340.4 g	200.6 g total for 6 days by all snails (January 1 to January 6)	6.8 g total for 6 days by all snails (January 1 to January 6)

In order to calculate the net production of the 6 marsh snails over the 6 days shown in Table C.3, you subtract the beginning mass of the snails from the ending mass of the snails and divide this number by 6.

1. Net production = (Change in the biomass) ÷ (Number of days)

        340.4 g – 320.2 g = 20.2 g (change in biomass) ÷ 6 days = NP = 3.4 g/day

2. Gross production = Net production – Loss from respiration

The actual amount of food material consumed and used to build organic material is calculated by subtracting the mass of faeces from the mass of food consumed. This is the best estimate of loss of energy due to processes such as respiration.

200.6 g of food consumed – 6.8 g of faeces/waste = 193.8 g ÷ 6 days = 32.3 g
Gross production per day = 32.3 g/day
Respiration = (GP – NP = R) 32.3 g – 3.4 g = 28.9 g/day

From the data shown in Table C.3, the following can be concluded:

- GP = 32.3 g
- NP = 3.4g
- R = 28.9g

Food Industry and Productivity

The food production industry relies on the ability of organisms to produce usable products efficiently. The input of food mass in relationship to the output of food product determines the conversion ratio for the productivity of farming practices. Industry uses a food conversion ratio to determine the productivity of food intake compared with product production (for example, meat products, eggs, milk). The cost of producing meat by farm industries must be considered in order to determine the net gain in profit from the sale of usable product. Homeotherms (animals that regulate their body temperature) are much less efficient when converting food to biomass then are poikilotherms (animals with variable body temperature). Homeotherms lose a lot of energy by producing heat in order to maintain body temperature.

Trophic Levels

It is often difficult to classify organisms into specific trophic levels due to the fact that some organisms can exist in more than one trophic level. For example, *Euglena* can feed as a primary consumer on algae and bacteria. *Euglena* can also produce their own food through photosynthesis. This allows them to occupy two different trophic levels. Another issue is that many heterotrophic organisms feed at more than one trophic level. The hawk can feed on both mice that are classified as primary consumers and on snakes that are classified as secondary consumers. This makes the hawk both a secondary and a tertiary consumer. In addition, filter feeders (clams and oysters, for example) feed on producers (plankton) and on consumers (zooplankton). Filter feeds also function as decomposers by consuming dead organic material. For these reasons, food webs are more appropriate to use when showing the complex feeding relationships that exist in ecosystems. Refer to Figure 4.1 in the core material (food web).

ENERGY TRANSFER BETWEEN TROPHIC LEVELS

Most of the energy from organic material travelling through trophic levels is lost as heat to the environment. Approximately only 10% of the energy available in each trophic level is passed to the next-higher level. The loss of available energy from each trophic level limits the amount of biomass that can be supported as energy is passed up to higher trophic levels. Cellular respiration is responsible for most of the biomass that is lost because water and carbon dioxide are released from the oxidation of glucose. In addition, energy is lost when organic material available in a lower trophic level is not completely consumed by organisms in higher trophic levels. Organic material is also lost from organisms as waste in faeces. Refer to Figure 4.2 in the core material.

ENERGY PYRAMIDS

Energy pyramids are used to represent the amount of energy available for each trophic level as energy passes from one trophic level to the next. Remember that only 10% of the energy available is passed from one level to the next (90% is lost). Make sure the size of each bar you use to represent energy passing through trophic levels reflects this by showing the size of the bar about 10% of the size of the bar beneath it. Mathematically, you need to make sure you multiply the energy available at one trophic level by 10% to calculate the amount that is available for the next trophic level. The amount of available energy in each trophic level is usually expressed in kilojoules (kJ) per area per unit of time (energy/area/time). The most commonly used area is square meters (m^2), and the most commonly used time is per year (yr).

$$kJ\,m^{-2}\,yr^{-1} = kilojoules/meter/year$$

The successive loss of energy between trophic levels limits the number of trophic levels that can exist. This loss also limits the number of organisms (amount of biomass) that can be supported at each successive level. Refer to Figure 4.3 in the core material.

Succession

Succession refers to the changes that occur in an environment over time. Succession in ecology follows predictable changes as organisms enter into environments and develop niches. There are two main types of ecological succession. Primary succession explains the pattern of development of ecosystems in areas in which living organisms did not previously exist. Secondary succession explains the pattern of development of ecosystems in areas where living systems once existed but have been drastically disrupted due to changes in the environment. Primary and secondary successions are outlined in Table C.4.

The predictable changes that occur during succession begin with the introduction of pioneer species. Pioneer species represent the first organisms to appear in an ecosystem. These species change the environment and make it suitable for additional species to enter the ecosystem. Pioneer species change the composition (depth, minerals, organic material) of soil, making it suitable for additional plant and animal life. Eventually, the ecosystem will become stable and fully mature, reaching what is called the climax community. (Refer to Figure C.4.)

Table C.4: Primary and Secondary Succession

Type of Succession	Pattern of Succession	Example
Primary succession	Ecosystems develop in previously uninhabited areas.	Newly formed islands due to volcanic action begin hosting life-forms and eventually establish mature ecosystems.
Secondary succession	Ecosystems develop after previously inhabited areas have undergone changes that greatly affected the presence of living organisms.	Life begins to return to an area that has undergone catastrophic changes in the environment due to events such as fire, hurricanes and so on.

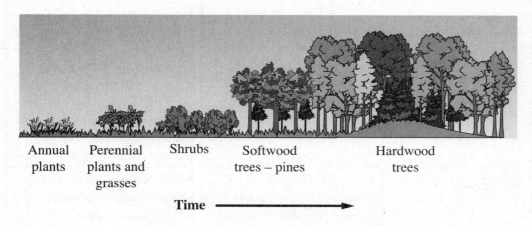

Annual plants   Perennial plants and grasses   Shrubs   Softwood trees – pines   Hardwood trees

Time ⟶

Figure C.4. Primary and Secondary Succession

The changes that occur during primary succession follow predictable pathways. The diversity of species in a developing ecosystem increases as the process of succession occurs. The primary succession involved in the development of a mature forest is outlined as follows:

- Pioneer species enter the environment and begin to break down the soil. They release organic material into the soil as they decompose.
  - □ Mosses start to grow on rocks and slowly break down the rocks, releasing minerals into the soil. As the mosses die, they add organic material to the developing soil.
- Larger plant life can be sustained by the developing soil in the environment.
  - □ Grasses move into the area, supporting small insect life. The grass roots further break down the soil and prevent erosion.
- Larger plant life can be sustained as the soil begins to deepen and increase in nutrient content (organic matter and minerals).
  - □ Shrubs grow and support more animal life such as birds and small rodents. Insects continue to enter the community as well.

- Mature trees develop in the ecosystem, replacing some of the smaller plant life as the dominant plants in the climax community.
  - Mature trees replace the smaller shrubs. These trees allow more animals to live in the forest and to feed on smaller animals in the ecosystem.

Living organisms influence the abiotic parts of the ecosystem during the process of succession. They impact the composition and development of soil as well as reduce the process of erosion. Biotic influences on abiotic features during succession are presented in Table C.5.

Table C.5: Biotic Influence on Abiotic Features During Succession

Biotic Influence	Process	Importance
Development of soil	Plant material produces structures that infiltrate rock and break down rocks, producing soil.	Thicker soil allows for larger plants to exist by allowing roots to spread deeper into the soil.
Mineral accumulation in soil	As plant material breaks down rocks, the minerals in rocks are released into the soil.	Plants require access to minerals to carry out life processes, and additional minerals support more plant life. Animals entering the ecosystems consume the plants and obtain essential minerals from the plants.
Reduction of erosion	Plant material growing below and above the soil prevents the loss of mineral nutrients due to erosion.	Minerals remain in the soil and are available as plant and animal nutrients to support the ecosystem.

Biomes

Ecosystems can be identified based on the characteristics of the climax communities that develop during the processes of succession. Mature ecosystems that share similar characteristics are grouped into biomes. All the biomes of the world are collectively referred to as the biosphere.

> **Biome:** A major type of ecosystem that is recognized globally based on shared characteristics (climate and life forms).
> **Biosphere:** All the overlapping and interacting biomes on Earth.

CLIMATE AND BIOMES

The type of biome that emerges in an area is dependent upon climate. The two main features of climate affecting the distribution of biomes are the annual rainfall and the average temperature of the region. Plant materials that exist in each of the world's biomes have adapted to the annual rainfall and average temperatures prevalent in those regions. The interaction between the average temperature and the annual amount of precipitation determines the type of biome that exists.

MAJOR BIOMES

There are six major types of biomes that exist within the biosphere. The major biomes are the desert, grassland, shrubland, temperate deciduous forest, tropical rain forest and tundra. The average temperature, annual rainfall amounts and vegetative characteristics of each of the six major biomes are outlined in Table C.6.

Table C.6: Major Biomes

Major Biome	Annual Rainfall and Average Temperature	Characteristic Vegetation
Desert	Very low levels of annual rainfall with very hot days and cold nights	Few plants adapted to reduced rainfall with water storage organs or rapid growth after rainfall. Cactus and tumbleweeds
Grassland	Low levels of annual rain with hot summers and cold winters	Grasses and herbaceous plants
Shrubland	Dry, hot summers followed by wet, cold winters; Fires are prevalent due to the dry, hot summer	Shrubs, grasses, aromatic herbs (oregano, thyme, rosemary)
Temperate deciduous forest	Moderate levels of annual precipitation with warm summers followed by moderately cold winters	Deciduous trees (shed leaves each year), shrubs, and herbaceous plants
Tropical rain forest	High levels of annual rainfall with warm or hot seasons year-round	Most diverse plant life of all biomes Canopy layer of tall trees, vines, herbs, shrubs, ferns
Tundra	Low levels of annual rainfall with very cold temperatures; Permafrost (frozen soil) exists, and most precipitation is in the form of snow or ice	Very little vegetation with only a few or no trees present Lichens, mosses, grasses, shrubs

Closed Ecosystems

Closed ecosystems exchange energy with the surrounding environment but do not exchange matter. Closed ecosystems are contained within sealed environments, preventing the exchange of matter with the surrounding environments. Other organisms inhabiting closed ecosystems must consume waste that is produced by organisms existing within the same closed ecosystem. If the waste is not consumed, it will build up and become toxic to the organisms within the closed ecosystem. Closed ecosystems must include photosynthetic organisms as primary producers of energy. Energy must enter the ecosystem in the form of light.

C.3 IMPACTS OF HUMANS ON ECOSYSTEMS

Alien Species

The introduction of a species that is not indigenous (normally found or native) to an ecosystem into an ecosystem can affect the ability of the entire ecosystem to survive. Those species that have been introduced and that were originally foreign to the local population are referred to as alien species. Human interactions, knowingly and unknowingly, have introduced alien species into many ecosystems. Examples of the introduction of alien species into ecosystems include kudzu, zebra mussels and cane toads. Table C.7 outlines several alien species that have greatly impacted native ecosystems.

Table C.7: Alien Species

Alien Species	Method of Invasion	Impacts of Alien Species
Kudzu (*Pueraria lobata*): a plant native to Japan	Brought to the United States as an ornamental plant and thought to help benefit controlling soil erosion	■ Kudzu grows very rapidly (up to 1 foot a day), smothering native plants with its leaves and stems. It can acquire nutrients from the soil very efficiently, making fewer nutrients available for the native plants. ■ Kudzu is responsible for an estimated $500,000,000 in annual financial losses.
Zebra mussels (*Dreissena polymorpha*): a bivalve native to Russia	Thought to have been transported from Russian waters to the United States in the ballast water of ocean ships	■ Zebra mussels reproduce rapidly and form sticky fibres that enable them to attach to many structures. They clog pipelines, disrupting the water pipes of many factories and water treatment facilities. ■ Zebra mussels filter algae and small organisms from the water at such rates that they impact the ability of native filter feeders to survive.
Cane toads (*Bufo marinus*): native to South America	Introduced to Australia in an attempt to control the cane grub that was rapidly destroying native sugar cane plants	■ The cane toad did not feed on cane grub as expected but instead fed on many other native species. ■ The cane toad has no native predators and produces a poisonous toxin that kills any organism that tries to eat the cane toad. ■ The cane toad reproduces very rapidly, allowing its numbers to grow quickly and cover large territories. ■ Cane toad creates a hallucinogenic toxin that has become a source of social issues in Australia.

IMPACTS OF ALIEN SPECIES

Alien species impact ecosystems by increasing interspecific competition for resources and by causing the extinction of native species. Alien species are often the source of new predators for native species. Invasive species lack natural predators, allowing their population size to grow. These invaders consume resources previously available for endemic species. The invasive species outcompetes the endemic species for resources, leading to competitive exclusion of the endemic species (see Table C.8).

Table C.8: Impacts of Alien Species

Impact	Outcome of Impact	Example
Interspecific competition	Invasive species outcompete the native species for resources.	Kudzu outcompetes native plants for light and nutrients.
Predation	The invasive species preys on a native species. The native species has not evolved with the predator, so the native species lacks behaviours to avoid the predator.	The invasive sea lamprey preys on and has devastated some fish populations in the Great Lakes.
Extinction	The invasive species exhibits some behaviour that causes the native species to become extinct. This could be due to outcompeting the native species for resources or due to preying on the native species to the extent that all members of the species are killed as prey.	The invasive brown tree snake has caused the extinction of 12 bird species in the Northern Mariana Islands.

Biomagnification

Ecosystems are negatively impacted by toxins that are introduced into the environment. These toxins become more concentrated in the tissues of organisms that occupy higher trophic levels.

> **Biomagnification:** An increase in the concentration of a substance as it moves from one organism to another organism higher in a food chain.

DDT AND MOSQUITOES

An excellent example of biomagnification involves the toxic chemical DDT (dichlorodiphenyltrichloroethane). DDT was used as a pesticide to control insect populations such as mosquitoes. DDT is a fat-soluble compound that accumulates in the fat cells of organisms that consume the toxic substance. The concentration of the toxin is amplified each time an organism in the food chain consumes an organism containing the toxin. The biomagnification of DDT causes it to reach toxic levels at higher trophic levels in the feeding chain. Predatory birds occupying the highest trophic levels suffered greatly from the bioaccumulation of DDT.

The toxin inhibited the deposition of calcium in egg shells, resulting in weak eggs that could not support developing chicks. A rapid loss in the numbers of predatory birds became associated with the use of DDT in the environment.

Plastic Pollution

Plastic debris entering marine environments from human activity has greatly affected organisms that rely on the ocean waters for survival. Marine mammals, sea birds, fish, coral reefs and many other marine organisms are negatively impacted by the presence of plastic pollution. Plastic pollution includes objects such as old toothbrushes, bottle caps, plastic bags, food wrappers and other plastic items that arrive in the ocean from land litter. Marine organisms can get trapped in macroplastic debris, leading to death. Organisms that ingest microplastic debris can die from blockages in their digestive tracts or from damage to their gastrointestinal tract caused by the indigestible material.

The Laysan albatross (*Phoebastria immutabilis*) migrates great distances across ocean waters. During the migratory process, the bird feeds on organisms that inhabit the ocean waters. Plastic debris floats on the surface of the water and is mistaken for food by the albatross. The ingested plastic is indigestible and remains in the bird's digestive tract. An adult bird regurgitates its food along with the plastic debris as it feeds the developing offspring. Young birds die due to the plastic filling up their small stomachs and preventing them from receiving nutrition or from the damage to their digestive tract from the plastic pollution.

> **REMEMBER**
>
> Close to 100 million tons of plastic pollution are estimated to be in the world's oceans.

The presence of the plastic pollution has contributed to the decline of many sea turtle populations, including the loggerheads, leatherbacks and green sea turtles. Sea turtles ingest plastic pollution that cannot be regurgitated due to the downwards-facing anatomy of their throats. The plastic builds up in their stomachs and prevents food from entering. Large plastic bags that are ingested trap gas produced by digestive processes. The presence of gas adds buoyance to the sea turtles and prevents the sea turtles from being able to descend in the water properly and locate food. Floating turtles are also unable to hide from predators and therefore become easy targets for prey.

C.4 CONSERVATION OF BIODIVERSITY

Simpson Diversity Index

Biodiversity is used to describe the variety of living organisms that are found on Earth. Biodiversity can be used to describe diversity at many levels within communities. This includes the diversity of genes within a population, the diversity of species in a community and the diversity of ecosystems on Earth. Human interactions with the environment have impacted biodiversity at all levels of the community.

Scientists evaluate ecosystems based on the biodiversity present. The number of different species inhabiting an area determines the richness of the area. One method used to calculate the biodiversity or richness of an area is the Simpson diversity index. The Simpson diversity index can be used to evaluate species diversity and to compare changes in diversity over time.

In addition, by comparing diversity indices among similar ecosystems, scientists can evaluate the relative health of ecosystems.

> D = Diversity index
>
> N = Total number of organisms present (including all species found)
>
> n = Number of organisms of a particular species
>
> Simpson diversity index: $D = \dfrac{N(N-1)}{\sum n(n-1)}$

The first step in calculating the Simpson diversity index is to do random sampling of an area to identify the total number of organisms of each representative species. For example, plant diversity can be compared between different locations, as in the following example.

Plant species present in a field 10 metres from a school and 20 metres from a school were randomly sampled in order to evaluate any changes in biodiversity based on the location of the sample in relation to the location of the school. The data collected from the two sites were analysed for diversity using the Simpson diversity index, as shown in the tables that follow.

Site A: 10 Metres from School

Species (N)	Number Identified (n)	$n(n-1)$ for Each Species
Dandelion	8	$8(7) = 56$
Daisy	6	$6(5) = 30$
Buttercup	5	$5(4) = 20$
$N(N-1) = 3(3-1) = 6$		$\sum n(n-1) = 106$

$D = 6/120 = 0.05$

Site B: 20 Metres from School

Species (N)	Number Identified (n)	$n(n-1)$ for Each Species
Dandelion	5	$5(4) = 20$
Daisy	4	$4(3) = 12$
Buttercup	6	$6(5) = 30$
White clover	7	$7(6) = 42$
Common nettle	4	$4(3) = 12$
$N(N-1) = 5(5-1) = 20$		$\sum n(n-1) = 116$

$D = 20/116 = 0.17$

According to the Simpson diversity index, site B is more diverse than site A. This makes sense since site B contains a larger variety of species. However, the distribution of species (numbers) within an area impacts the diversity index. For example, assume that instead of finding 5 dandelions in site B, we found 45. Recalculate the Simpson index for site B. Perhaps the change in number could be explained because the schoolchildren like to pick the dandelions that grow closer to the school, thereby artificially deflating the number of dandelions.

Species (*N*)	Number Identified (*n*)	$n(n-1)$ for Each Species
Dandelion	45	45(44) = 1980
Daisy	4	4(3) = 12
Buttercup	6	6(5) = 30
White clover	7	7(6) = 42
Common nettle	4	4(3) = 12
$N(N-1) = 5(5-1) = 20$		$\sum n(n-1) = 2076$

$D = 20/2076 = 0.01$

The new calculation suggests that site *A* is more diverse than site *B*. Simply having more of a species present does not mean that the community exhibits greater diversity. It could be argued that the dandelion population is limiting the diversity of the community.

Indicator Species

Biotic indices and indicator species are used to assess and monitor environmental conditions. Indicator species are sensitive to changes in the environment and can be used to assess the overall health of an ecosystem. The populations of indicator species change as environmental conditions that affect their ability to survive in ecosystems fluctuate. Indicator species are very sensitive to specific abiotic factors. The abundance of these species in an ecosystem can be used to determine if the abiotic factor is prevalent in amounts that are encouraging or inhibiting their growth and reproduction. Some of the most commonly used indicator species for monitoring the levels of an abiotic factor in ecosystems are outlined below. Table C.9 presents examples of indicator species.

Table C.9: Indicator Species

Indicator Species	Abiotic Feature	Effect on Ecosystem
Lichens	Sensitive to acid rain; pollutants in water such as sulphur dioxide from burning fossil fuels and volcanic eruptions	Produce sulphuric acid found in acid rain; the acid rain alters the pH of the environment, denaturing proteins needed by living organisms
Mayfly larvae	Sensitive to dissolved oxygen levels	Low oxygen levels in water (dissolved oxygen) inhibit the ability to carry out cellular respiration.
Tubifex worms	Sensitive to heavy metal deposits (lead, mercury, cadmium) in lakes and rivers	Heavy metals are toxic to tissues; tubifex worms can survive in environments with high levels of heavy metals; an overabundance of these worms indicates that the pollutant levels are rising; these worms can also survive in water low in oxygen.

Biotic indices are determined by using scales to show the relative number of indicator species present in an ecosystem. The biotic index can be calculated and used as a way of comparing changes in the health of ecosystems. Environments that contain high levels of tolerant species (species that are not as affected by pollution, oxygen levels or other abiotic changes) receive a low biotic index. Environments with high levels of less tolerant species receive higher biotic index scores. Changes in biotic indices allow scientists to evaluate changing abiotic characteristics in the environment.

Extinction

When conditions within ecosystems change in such a way that organisms cannot adapt for survival, the entire species is threatened by extinction. Many species have become extinct during the history of life on Earth.

> **Extinction:** The end to the existence of a species.

The dodo bird (*Raphus cucullatus*) is an excellent example of an extinct animal. The dodo bird went extinct in the late 17th century after having thought to have existed on Earth for less than 100 years. The dodo's extinction was due mainly to human actions that affected its niche in the ecosystem. The dodo bird inhabited the island of Madagascar in the Indian Ocean. Human development on the island took territory from the bird. Introduction of domesticated animals brought in new predators that further decreased the number of reproducing dodo birds. The extinction of entire species limits the biodiversity of the world and affects all organisms that exist in the extinct species habitat.

Conservation of Biodiversity

Nature reserves are an important component of the struggle to protect and maintain biodiversity. Nature reserves help to protect wildlife from being threatened by human actions. The size of the reserve, the environments neighbouring the reserves and the methods used to allow animals access to habitats on the reserve are all important factors influencing the success of the nature reserve (see Table C.10).

Active Management Techniques

Active management techniques are used to ensure the survival of threatened species. Nature reserves employ individuals to monitor the land and ensure the environment is promoting the repopulation of endangered species. The implementation of measures to help keep invasive species out of the reserve, the revegetation of any damaged land areas and monitoring human intervention in protected areas are vital for the success of the reserve in promoting biodiversity. Active management ensures mating and survival of endangered offspring by intervening when needed. For example, some reserves have built boxes that serve as nesting sites for birds. Some reserves use captive breeding or artificially populate the environment to ensure the survival of endangered populations.

Table C.10: Biogeographical Features of Nature Reserves

Biogeographical Feature	Impact	Example
Size of the reserve	■ Larger reserves allow for large species to roam freely. ■ Larger reserves decrease the effects of the environments neighbouring the reserve.	■ The Gebel Elba reserve in Egypt includes over 35,000 miles of territory. Gebel Elba is home to more species than anywhere else in Egypt.
Edge effect (the impact from neighbouring environments)	Larger reserves possess smaller perimeters, decreasing the effect of interactions from unprotected neighbouring environments.	■ The impact of the habitats adjacent to the new edges of the rain forests affects organisms inhabiting areas at the edge as well as deeper into the rain forest.
Wildlife corridors (roads, pathways connecting fragmented reserves)	Corridors allow wildlife to migrate among fragmented reserves, increasing the diversity of reproducing populations and giving organisms access to more land for foraging.	■ Building overpasses or tunnels under reserves that keep busy roadways from fragmenting reserves. ■ Building a strip of land that connects two fragmented reserves.

IN SITU CONSERVATION

In situ conservation refers to the protection of an endangered species in its natural habitat. Advantages of in situ conservation include:

■ Maintaining the population in the environment in which it survives best.
■ Encouraging the natural evolution of the organism as it adapts to changes in its natural environment.
■ The ability of protecting the population from the stresses that relocation could produce.
■ Greater genetic diversity due to the presence of a larger gene pool by keeping the larger population together.
■ The natural environment encourages behaviours within the population that promote species reproduction and survival.

EX SITU CONSERVATION

When in situ conservation fails to promote the survival of a species, survival can be encouraged by ex situ conservation measures. Ex situ conservation involves the protection and promotion of the repopulation of a species outside of its natural habitat. This method is often a last resort to save a species from extinction and is frequently done along with in situ conservation.

Seed banks allow the conservation of plant species by storing seeds for use when the population of a plant is threatened. Zoos and botanical gardens encourage the reproduction of threatened species as well as educate the public about the importance of the conservation of biodiversity. Ex situ conservation methods have several advantages:

- Greater control over the environmental conditions, such as food and shelter for the captive organisms.
- Greater control over reproductive success with the use of artificial insemination, IVF and embryo transfer techniques.
- Human care and intervention are ongoing in botanical gardens, seed banks, aquariums and zoological parks.
- Zoos, botanical gardens, aquariums and seed banks allow the conservation of many different species of threatened organisms.

Ex situ conservation methods have several disadvantages:

- The gene pool is limited to the organisms in the ex situ conservation site, decreasing genetic diversity within the population.
- The reintroduction of captive species back into their natural habitat is difficult.
- Ex situ conservation does not offer protection to the natural habitat of the threatened species.

California Condor

Poaching, habitat destruction and toxins in the environment depleted the total population of the California condor populations to less than 25 known living members of the species by the late 1980s. The San Diego Zoo and the Los Angeles Zoo began captive breeding and reinstatement of the California condor in 1987. Captive breeding efforts for the California condor have brought the population of California condors to their current population of over 400. More than half of the over 400 living California condors are currently found in their natural habitat.

C.5 POPULATION ECOLOGY

R-Strategies and K-Strategies

The life cycles and reproductive strategies of organisms exhibit patterns that are based on the environment of the population. Two main patterns of population growth are exhibited by organisms based on environmental conditions. Stable environments encourage *k*-strategies for population growth. Unstable environments encourage *r*-strategies for population growth. These two strategies represent the extremes in reproductive strategies. Many organisms exhibit intermediate strategies that are difficult to place into either category. In addition, some organisms can switch between strategies depending on the conditions in the environment. It is important to keep this in mind when discussing populations exhibiting *r*-strategies and those exhibiting *k*-strategies (see Table C.11).

Table C.11: *K*-Strategies and *R*-Strategies

Population Characteristic	*R*-Strategies	*K*-Strategies
Adult body size	Small	Large
Maturation	Early	Late
Rate of reproduction	Only once	Many times
Parental care	Little or none	Much (very common)
Number of offspring	Many	Few
Life span	Short	Long
Size of populations	Vary	Stable
Environment	Unstable (changing)	Stable (unchanging)
Types of organisms exhibiting strategy	Bacteria, rodents, insects	Whales, elephants, humans

Populations exhibit *r*-strategies in order to mature quickly and produce many offspring since the unstable environment does not favour their survival. Unstable environments are continually changing, making survival difficult since there is little time to adapt to the environment. Producing many offspring in short reproductive cycles helps to ensure that at least some of the population members will survive to continue repopulating the species. Since it is highly unlikely that the population in unstable environments will ever become too large to be supported by the environment, *r*-strategists are not density dependent. Environments under constant change, such as tidal pools or coastal areas, possess organisms that exhibit *r*-strategies. Because *r*-strategists are more likely to survive in unstable environments, they are the first organism to inhabit an area during the process of secondary succession.

Populations exhibit *k*-strategies in order to mature slowly, allowing for increased adult size for survival in a stable environment where competition is high. Possessing larger adult sizes and having parental care helps to ensure the population's ability to survive in the competitive environment. Competition in stable environments increases the effect of density-dependent factors on the survival of populations. The *k*-strategists will be prevalent late in secondary succession as the ecosystem nears the climax community.

Capture-Mark-Release-Recapture

One method of population estimation that scientists use in order to investigate population numbers when it is impossible to count every member of the population is the capture-mark-release-recapture method of population estimation. This method allows scientists to estimate population numbers in populations with members that move around or are difficult to find in a given habitat (see Figure C.5).

The following steps are used in the capture-mark-release-recapture method of population estimate:

- Capture as many of the organisms as possible in the selected area (habitat).
- Mark each captured individual with a distinct mark, making sure the mark does not decrease the ability of the organism to evade predators.
- Release the marked individuals back into the sampling habitat. Allow time for the marked individuals to reenter the population.
- Recapture as many of the organisms as possible from the selected area (habitat).
- Calculate the number of organisms in the recapture as well as the number of organisms that were marked in the first capture but were recaptured in the second capture.
- Use the Lincoln index to calculate the estimated size of the population inhabiting the sampling area.

  □ N(population size) = $(n_1 \times n_2)/n_3$, where n_1 = total number caught and marked in original capture, n_2 = total number caught in second capture, and n_3 = total number marked in second capture

---

Sample Lincoln Index Calculation

Random sampling resulted in the capture of 200 crickets that were marked and released. A month later, 200 crickets were captured from the same area. Of the 200 recaptured crickets, 20 were marked. The Lincoln index was used to estimate the population of the crickets in the given area. It was determined to be 2000.

$$\frac{200 \times 200}{20} = 2000$$

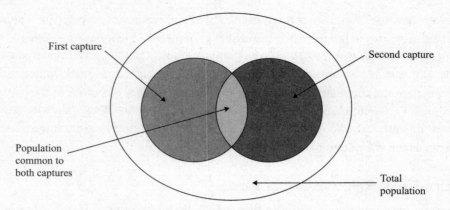

Labels on figure:
First capture
Second capture
Population common to both captures
Total population

Figure C.5. Capture-mark-release-recapture selections

Estimating the Size of Commercial Fish Stock

Fish are a major source of food for humans and for other animal populations. Fish populations need to be monitored in order to determine how many fish can safely be harvested from waters without causing a decline in fish population numbers. Overharvesting of fish could lead to the loss of populations of fish for future generations. It is difficult to estimate populations of fish since they can live in large bodies of water and can travel at various depths in the water. Methods to calculate fish populations include:

- In contained bodies of water (lakes), the fish population can be estimated with the capture-mark-release-recapture method of population estimation. This is not effective for large bodies of water such as oceans.
- Echo sounders can be used to measure fish populations based on density and location. Trawls are used to capture fish in the area where the echo sounders are used in order to identify the species of fish detected.
- Fishermen can share information about fish populations based on their catches and experiences in the harvesting environment (average size of fish, numbers caught, females with eggs and the conditions of the water environment).
- Research involving catching and tagging fish with transmitters can help calculate fish populations and motility.

The challenges in obtaining information to determine the exact populations of fish make it difficult to determine the maximum sustainable yields. Fishermen and conservationists argue about the values used to determine how many fish can safely be harvested from fish populations. The livelihood of the fishermen relies on the ability to harvest fish freely. The survival of the fish for future populations of humans relies on limiting the number that can be captured.

> **Maximum sustainable yield:** The maximum number of fish that can be harvested without causing a decline in fish stocks (subpopulations of fish required to maintain the population).

It is important to ensure that the harvesting of fish does not limit the ability of the fish population to be maintained. The maximum sustainable yield represents a balance between the number of fish captured and dying naturally compared with the rate of reproduction and growth of the fish. Not exceeding the maximum sustainable yield of a population will allow the population to continue to reproduce at levels that ensure its continual survival. The maximum sustainable yield is best held at about half the carrying capacity of the fish population since the most rapid reproduction of a population occurs at this point in the growth curve.

The continual survival of fish populations is as important to humans as it is to the fish. If the maximum sustainable yield is exceeded, the population could fail to recover and there will be no fish available for future harvesting. Laws have been put in place to ensure the maximum sustainable yield is enforced. It is difficult to monitor fish yields locally and even more difficult to do so internationally. In order to preserve the fish populations for current and future generations, it is vital that all human populations limit the number of fish captured from bodies of water.

Growth Curves

The rate of natality (birth rate) and mortality (death rate) in a population is one determining factor in how the population size will change. If no other factors are influencing the population size, the population will grow if natality exceeds mortality and will shrink if mortality exceeds natality. Immigration (individuals entering the population) and emigration (individuals leaving the population) also play a role in population size. If no other factors are influencing population size, the population will grow if immigration exceeds emigration and will shrink if emigration exceeds immigration. Natality and mortality rates as well as immigration and emigration must be considered when determining whether a population is growing or shrinking in size.

SIGMOID POPULATION GROWTH CURVE

The sigmoid population growth curve represents the normal changes that occur in population size. During the lag phase, there are only a few reproducing organisms and the population grows slowly. As more organisms become available to reproduce, the population enters the exponential growth phase. Many offspring survive during the exponential growth phase since enough resources are available to support the growing population in an ideal, limitless environment. The exponential growth of the population ultimately results in members of the population competing for limited resources. Competition for limited resources results in the population entering into the transitional phase until the carrying capacity (K) is reached. Once the carrying capacity is reached, the population levels off and enters the plateau phase. The plateau phase represents the carrying capacity of the environment and is based on the availability of limited resources (see Figure C.6).

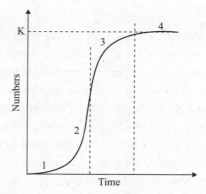

Figure C.6. Sigmoid (population) growth curve (1) lag phase; (2) exponential growth phase; (3) transitional phase; (4) plateau (carrying capacity—*K*)

Top-Down and Bottom-Up Limiting Factors

Two types of controls that affect population sizes are top-down and bottom-up limiting factors. Top-down limiting factors involve increases in the number of organisms at higher trophic levels in the food chain. Bottom-up limiting factors involve decreases in the number of organisms at the producer level of the food chain. The combinations of top-down and bottom-up limiting factors maintain a stable population at each trophic level (see Table C.12).

Table C.12: Bottom-Up and Top-Down Forces

	Bottom-Up	Top-Down
Algae blooms	Shortage of available nutrients decreases population growth rate.	Increase in herbivory on algae populations decreases growth rate.
Rabbit populations	Limited plant material for consumption decreases population growth rate.	Predation by higher-order consumers decreases population growth rate.

C.6 NITROGEN AND PHOSPHORUS CYCLES

Nitrogen Cycle

Nitrogen is an essential mineral for all living organisms. Nitrogen must be available in order for cells to produce proteins and nucleic acids. Most available nitrogen exists as an atmospheric gas. Microbes carry out vital processes in ecosystems, including fixing nitrogen, producing energy and decomposing organic matter. Several forms of microbes play a role in the nitrogen cycle. Nitrogen-fixing bacteria are responsible for converting atmospheric nitrogen into ammonia and other nitrogen-containing compounds. Microscopic algae and photosynthetic bacteria carry out photosynthesis, thereby supplying food for many ecosystems. Microscopic decomposers such as bacteria, fungi and protists decay dead organic matter, recycling nutrients through ecosystems. The main roles of microbes in ecosystems are outlined in Table C.13.

Table C.13: Roles of Microbes in Ecosystems

Microbes	Benefit to Ecosystem
Producers	Supply food for ecosystems by carrying out photosynthesis
Nitrogen fixers	Convert atmospheric nitrogen to ammonia and other usable nitrogen compounds
Decomposers	Break down dead organic matter, returning nutrients to ecosystems

Nitrogen-fixing bacteria are responsible for making nitrogen available to all living organisms. Nitrogen is essential to living organisms since it is a component of all proteins and nucleic acids. Plants have additional needs for nitrogen since nitrogen is a component of chlorophyll. Nitrogen enters ecosystems, cycles through ecosystems and is returned to the atmosphere. Plants actively pump nitrates into their roots since the nitrate demand of plants is higher than the nitrate concentration normally found in soil. The active transport of nitrate ions into the roots of plants is vital in supplying nitrogen to the nitrogen cycle. The nitrogen cycle is shown in Figure C.7. The roles of microbes in the nitrogen cycle are outlined in Table C.14.

Microbes are involved in each stage of the nitrogen cycle. The major steps of the nitrogen cycle and the microbes involved in each step are outlined below:

1. **Nitrogen fixation:** Nitrogen is taken from the atmosphere and incorporated into nitrogen compounds (such as ammonia) that can be used by living organisms. This is carried out by free-living *Azotobacter* and *Rhizobium* bacteria living in the soil and in the root nodules of plants. Nitrogen fixation can be accomplished by:

   - **Free-living bacteria:** Nitrogen-fixing bacteria found in the soil produce ammonia from atmospheric nitrogen.
   - **Mutualistic relationships:** *Rhizobium* bacteria live in the roots of plants and fix nitrogen into ammonia for the plant while receiving carbohydrates for food from the plant.
   - **Industrial nitrogen:** Industrial processes can produce ammonia from atmospheric nitrogen (Haber–Bosch process).
   - **Putrefaction (rotting organic matter):** Decaying organic material releases ammonia into ecosystems.

2. **Nitrification:** Ammonia is oxidized into nitrite followed by the oxidation of nitrite into nitrate. Ammonia is converted to nitrite by *Nitrosomonas* microbes. Nitrite is converted to nitrate by *Nitrobacter* microbes.

3. **Denitrification:** Nitrogen is removed from nitrates and returned to the atmosphere. This process is carried out by *Pseudomonas denitrificans*.

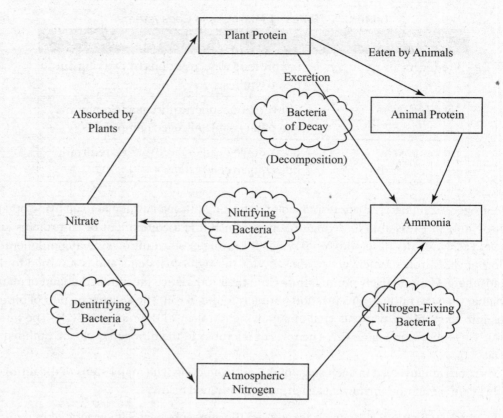

Figure C.7. The nitrogen cycle

Table C.14: Roles of Microbes in the Nitrogen Cycle

Microbe	Role in the Nitrogen Cycle
Rhizobium	Carry out nitrogen fixation in the roots of plant cells (mutualistic relationship), producing ammonia from atmospheric nitrogen
Azotobacter	Carry out nitrogen fixation in the soil, producing ammonia from atmospheric nitrogen
Nitrosomonas	Oxidizes ammonia into nitrite
Nitrobacter	Oxidizes nitrite into nitrate
Pseudomonas denitrificans	Denitrifies nitrate into nitrogen, releasing the nitrogen back into the atmosphere

The processes of denitrification and nitrification are favoured under different conditions. *Pseudomonas denitrificans* uses nitrates instead of oxygen as the final electron acceptor during the production of ATP. This process results in the release of nitrogen from nitrates. Since denitrification is carried out under anaerobic conditions by *Pseudomonas denitrificans,* low oxygen levels encourage this process. Nitrification occurs when the bacteria *Nitrosomonas* and *Nitrobacter* oxidize nitrite and nitrate, respectively. Oxygen must be available for these oxidative (gain of oxygen) processes to occur (see Table C.15).

**Table C.15: Conditions that Favour
Denitrification and Those that Favour Nitrification**

Conditions Favouring Denitrification	Conditions Favouring Nitrification
■ Oxygenated soil (aerobic—aerated soil) ■ Neutral pH (near 7) ■ Warmer temperatures	■ Deoxygenated soil (anaerobic—compact/flooded soil) ■ High nitrogen availability

WATERLOGGING AND THE NITROGEN CYCLE

Waterlogged soil is deprived of oxygen and is the perfect environment for anaerobic respiration. The anaerobic bacterium *Pseudomonas denitrificans* is prevalent in waterlogged soil. The metabolic processes of *Psuedomonas denitrificans* converts nitrate into nitrogen gas. The anaerobic respiration of *Pseudomonas denitrificans* depletes the soil of nitrogen. Without sufficient supplies of nitrogen, plant growth is greatly reduced.

Some plants make up for the low nitrogen levels available in the soil by obtaining nitrogen from other living organisms. Insectivorous plants possess modified leaves that trap insects and excrete enzymes that digest the organic material, thereby releasing nitrogen from the insect tissue. The pitcher plant and the Venus flytrap are examples of insectivorous plants.

Phosphorus Cycle

Plants require phosphate in order to assemble organic compounds such as ATP and other nucleic acids. The phosphorus cycle is much slower than the nitrogen cycle, making phosphate more limited in the soil. Harvesting of crops contributes to decreased levels of phosphorus in the soil since decomposition of the organic material does not occur. Farmers often supplement the phosphorus in the soil by using fertilizers that contain high levels of phosphate.

Phosphate is released into the soil from the erosion of rocks. Phosphate is slowly lost from the land as runoff into aquatic environments. Plants absorb phosphate and transfer it to humans through the consumption of plant tissues. Decomposition releases phosphate from organic material into the soil and water. Since phosphate has limited water solubility, it accumulates in the bottom sediment of oceans and eventually into sedimentary rock formations. Sedimentary rock is brought to Earth's surface again through the process of geologic uplifting. Processes of uplifting are very slow, leading to the amount of usable phosphate becoming limited in the environment.

Eutrophication

Eutrophication is due to increased algal growth in aquatic systems. Leaching from agricultural land carries mineral nutrients into rivers and other bodies of water. The release of mineral nutrients such as nitrogen and phosphate into aquatic environments gives algae the ability to increase in population size. High levels of algae can cause the death of plants living under the water as the plants are blocked from receiving sun. As the algae die, decomposers feed on them and deplete the oxygen in the water. The decomposition and decay of the algae leads to low oxygen levels in the water that can result in the death of aerobic aquatic organisms.

1. The graph below shows the relationship between species extinction and the human population.

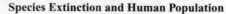

Species Extinction and Human Population

Graph source: USGS

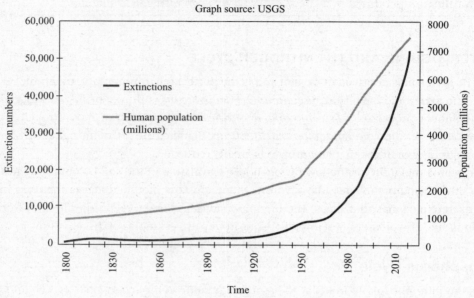

Time

(a) Explain the relationship between species extinction and the human population. (2)

(b) State the trend seen in the human population. (1)

(c) Suggest reasons for the similarities seen in the graph of extinctions and of human populations. (2)

(d) Calculate the percent increase in extinctions between 1920 and 2010. (2)

2. Outline the use of the capture-mark-release-recapture method. (4) *HL only*

3. Explain the use of indicator species. (2)

4. Discuss the advantages of in situ conservation methods. (3)

ECOLOGY AND CONSERVATION

ANSWERS EXPLAINED

1. (a)
   - There is a direct or positive correlation between the human population numbers and the number of extinctions.
   - Both show exponential growth starting around 1950 +/– 20 years or greatest increases occur after 1950.
   - Both reach their highest values after 2010.
   - Both were lowest in the 1800s.

   (b)
   - The human population is rapidly increasing.
   - The human population showed steady growth through 1950 and then grew exponentially.

   (c)
   - Humans hunting/harvesting food sources to the point of extinction
   - Human development of land, destroying habitats for species
   - Humans/industrialization contributing pollutants to the environment that affect species survival

   (d) 2500% (+/– 500%) 1920 = 2000 and 2010 = 52,000.

   (New value – Old value) ÷ (Old value) × 100 = % change

   (52,000 – 2000) ÷ 2000 × 100 = 2500%

2.
   - The capture-mark-release-recapture method is used to estimate population numbers.
   - Use a random sampling of area for the desired organism.
   - Count and mark captured organisms in a manner that does not affect their ability to evade predators.
   - Release the marked organisms back into the sampled area.
   - Allow the organisms time to reenter the population.
   - Randomly sample area again for the desired organism.
   - Count how many are captured and how many of the captured are marked.
   - Use the Lincoln index to estimate the population size.
   - Correctly state the Lincoln index formula: N(population size) $= n_1 \times n_2/n_3$, where n_1 = total number caught and marked in original capture, n_2 = total number caught in second capture, and n_3 = total number marked in second capture

3.
   - Indicator species are sensitive to abiotic features in the environment.
   - Indicator species can be used to analyse changes in the abiotic features in the environment.
   - Name an abiotic change monitored (pH, light, heavy metals, oxygen levels). One abiotic factor must be stated to get this point.
   - Some indicator species are very sensitive to small changes in abiotic factors, and others are very tolerant.
   - Large numbers of tolerant species or small numbers of sensitive species indicate a threat to biodiversity in the area
   - Name organism used as an indicator species (moss, Tubifex worms, mayfly larvae or other named indicator species).

4.
- In situ allows the conservation of endangered species in its native habitat.
- Organisms are adapted to their natural habitat.
- Behaviours in the native habitat help encourage mating and normal life activities of the organism.
- The animals can occupy their natural place in the food chain.
- Removal with ex situ may make the organism unable to survive when returned to the habitat.

Human Physiology

CENTRAL IDEAS
- Human health and nutrition
- Human digestion
- Liver functions
- Human heart
- Hormones and metabolism
- Nutritional issues in humans
- Respiratory gases

TERMS
- Body mass index (BMI)
- Egestion
- Malnutrition
- Minerals
- Normal weight
- Nutrient
- Obese
- Vitamins

INTRODUCTION

All living organisms must take in nutrition from the environment in order to survive. The human body must take in the appropriate amounts and types of nutrients in order to maintain life processes. The human body can build most of the needed nutrients from the digestion and assimilation of molecules. When the human diet lacks any needed type of nutrition, its physiological functions are impaired.

Cardiovascular, digestive and respiratory systems must function optimally in order to maintain human health. The body will show signs of distress when the homeostatic level of a body system is altered. The human body possesses natural physiological responses to counteract changes in homeostatic levels. If homeostatic levels cannot be maintained, human disease can develop. Human behaviours contribute to the progression of disease and, if not modified, could be life threatening.

NOS Connections

- Scientists are obligated to educate the public as to scientific discoveries and research findings (human health disease progression and prevention, nutritional recommendations).

- Collaboration and cooperation among scientists leads to greater scientific understanding and discoveries (disease cause and prevention, human anatomy and physiology).

- Scepticism exists within the scientific community and can lead to new theories replacing old ones (cause and treatment of disease, such as scurvy).

- Technology enhances scientific exploration and discovery (stethoscope, heart valves, ultrasound use, ECG, defibrillator).

- Ethical issues exist (animal experimentation).

D.1 HUMAN NUTRITION

TOPIC CONNECTIONS

- 6.1 Digestion and absorption

Essential Nutrients

> **Nutrient:** A chemical substance that is used by the human body and supplied by the consumption of foods.

Many substances are essential in the human diet. Essential substances consist of those that the human body cannot synthesize. Among the nutrients essential to the human diet are essential amino acids, essential fatty acids, minerals, vitamins and water. Of the 20 amino acids used by the body, 11 are considered to be nonessential. Nonessential amino acids can be synthesized in the body from other available nutrients. Carbohydrates are needed in the human diet; however, none are essential since the body can synthesize all needed carbohydrates from the breakdown of ingested precursors. Essential nutrients are outlined in Table D.1.

Table D.1: Essential Nutrients

Nutrient	How Obtained	Essential Types
Amino acids	Protein consumption	9 of the 20 amino acids are essential; the body can produce the other 11
Fatty acids	Lipid consumption	Polyunsaturated fatty acids
Minerals	Varied diet of animal and plant material (taken in from soil, rocks or seawater by living organisms)	All are essential and inorganic, such as sodium, potassium, calcium
Vitamins	Varied diet of animal and plant materials	All are essential and organic, such as vitamin C and vitamin A
Water	Eating, drinking and metabolic processes (metabolic water)	Water

Malnutrition

The lack of, excess of, or imbalance of any required nutrients in the body may lead to malnutrition. Protein deficiency malnutrition occurs when an individual fails to consume one or more of the 9 essential amino acids. The lack of any essential amino acid leads to health consequences that include oedema (fluid retention), abdominal swelling and a lack of needed proteins in the blood. Protein deficiency malnutrition can lead to both physical and mental impairments in children.

> **Deficiency:** A health problem caused by the lack of a needed nutrient in the diet.
>
> **Malnutrition:** An imbalance of the required amounts of one or more nutrients. Malnutrition can result from too much or too little of the nutrient.

Phenylketonuria (PKU)

Phenylketonuria is a genetic disease caused by a mutation in a gene coding for an enzyme that is responsible for converting the amino acid phenylalanine into tyrosine. The lack of the ability to carry out this conversion leads to high levels of phenylalanine in the blood of affected individuals. As the body tries to eliminate the excess phenylalanine, toxic by-products are produced. These toxic by-products (toxins) interfere with normal brain development and lead to severe mental problems. Individuals suffering from phenylketonuria must avoid food that is high in phenylalanine in order

> **REMEMBER**
>
> Kwashiorkor is a protein deficiency disease prevalent in nonindustrialized nations. This disease occurs due to the lack of resources to purchase the more expensive food items that contain protein with the essential amino acids.

to prevent the accumulation of the toxic by-products. In most developed countries, newborns are tested for phenylketonuria within the first week of birth. The test for phenylketonuria is done to ensure that those who carry the mutated gene follow a strict diet in order to avoid the mental problems associated with the genetic disease.

Fats and Disease Progression

All fats contain high levels of energy. Excess consumption of fats (both saturated and unsaturated) should be limited in the diet. Too much fat in the diet can cause obesity. Obesity is linked to many health issues, such as diabetes, coronary heart disease, hypertension, cancer and respiratory issues.

> **Types of lipoproteins:** Transporters of cholesterol in the blood. (Cholesterol cannot dissolve in the blood due to its hydrophobic nature.)
>
> - **LDL:** Causes plaque to build up as other substances bind with the cholesterol-forming plaque and block arteries. This leads to atherosclerosis (hardening of the arteries).
> - **HDL:** Carries cholesterol away from the arteries to the liver. There it can be broken down and removed from the blood, protecting the body from atherosclerosis.

Diets rich in saturated fatty acids have been linked to coronary heart disease. Saturated fatty acids form linear structures that can readily form plaque on coronary vessels. Saturated fatty acids are also linked to an increase in low-density lipoprotein (LDL) cholesterol levels. Unsaturated fatty acids are not associated with plaque buildup and increase levels of high-density lipoprotein (HDL), which is the good cholesterol. High levels of low-density cholesterol in the blood have been found to be associated with the development of coronary heart diseases.

High levels of cholesterol in the blood increase the likelihood of cholesterol attaching to the sides of arteries and clogging them. Arteries clogged with cholesterol can trap substances such as calcium. As a result, the arteries harden, which leads to a condition known as arteriosclerosis (hardening of the arteries). The clogged arteries lead to cardiovascular health problems such as high blood pressure.

Although unsaturated fatty acids are better for coronary health than saturated fatty acids, there are differences in health risks between the *cis* and *trans* form. Diets rich in *cis*-monounsaturated fatty acids are encouraged since these fats are linked to low incidences of coronary heart disease (CHD). Diets rich in *trans*-monounsaturated fats are linked to higher levels of CHD. Most of the arterial fat deposits that are found in individuals who suffer from atherosclerosis are from *trans* fats. The fat deposits (plaques) block blood flow and lead to hypertension (increased blood pressure). Hypertension increases the possibility of plaque deposits breaking free from the arterial wall. The travelling plaque deposit can block blood flow in vessels and lead to a stroke (blocked blood flow to the brain) or a heart attack (blocked blood flow to the heart).

Vitamins and Minerals

> **Vitamins:** A diverse group of carbon compounds that cannot be produced by the human body.

Vitamins and minerals are essential in the human diet. Diets lacking one or more of the essential vitamins and minerals will result in health issues. Examples of essential vitamins and minerals are outlined in Table D.2.

Table D.2: Vitamins and Minerals

Substance	Chemical Nature	Examples
Vitamins	Organic molecules—contain carbon and are made by living organisms	Vitamin C: $C_6H_8O_6$ Vitamin D: $C_{27}H_{44}O$
Minerals	Inorganic molecules in ionic form—do not contain carbon and are not made by living organisms	Sodium: (Na^+) Potassium: (K^+) Calcium: (Ca^{2+})

VITAMIN C (ASCORBIC ACID)

All vitamins must be taken into the body since they cannot be synthesized by body cells. In other words, all vitamins are essential. Although some mammals and some birds can produce vitamin C, humans must obtain vitamin C through their diet. Vitamin C (ascorbic acid) is needed in the diet since it plays a role in the formation of collagen, aids in the proper functioning of the immune system and functions as an antioxidant as it helps to protect cells from damage. Lack of vitamin C in the diet can lead to scurvy, a life-threatening disease. In order to prevent scurvy and promote overall general health, the recommended daily amount of this essential vitamin in the diet has been determined to be between 30 mg and 60 mg per day. Although this amount of vitamin C consumption will prevent the development of scurvy, many scientists recommend consuming higher doses in order to boost the immune system and prevent respiratory tract infections.

> **REMEMBER**
>
> In Latin, *scorbutus* means "scurvy". Hence vitamin C is known as a*scorb*ic acid.

> Scurvy was common in the 16th through 18th centuries in sailors who went for long voyages without sources of food that supply vitamin C (fresh fruit and vegetables). The disease results in bleeding gums, swollen joints, rapid breathing, anaemia and a feeling of paralysis. Many sailors perished at sea from scurvy.

VITAMIN D

Vitamin D can be synthesized in the skin when skin is exposed to UV light from the sun. This conversion will not occur without exposure to the sunlight. When the skin is exposed to the light and vitamin D is synthesized, excess amounts of the vitamin will be stored in the liver. The liver will release the stored vitamin D as needed. If an individual goes too long without sunlight exposure, the stored vitamin D will be depleted and a deficiency may result. Wearing sunscreen prevents the UV light from reaching the skin and blocks the production of vitamin D. However, too much exposure to the sun can result in an increased risk of malignant melanoma. For these reasons, it is recommended that exposure to the UV light of the sun be limited in order to prevent cancer from developing. Exposure to sun in small amounts (equivalent to about 15 minutes twice a week) has been shown to be enough exposure to produce sufficient vitamin D without greatly increasing the risk of developing melanoma. Taking supplements of vitamin D and consuming foods high in vitamin D can make up for the loss of the production of vitamin D due to limited or lack of exposure to UV light.

> Lack of vitamin D or lack of calcium affects mineralization in the bones.
> - Rickets (lack of vitamin D) causes a softening and weakening of the bones. Drinking milk fortified with vitamin D has virtually eliminated rickets in developed countries.
> - Lack of calcium in the body can lead to osteoporosis (osteomalacia), a condition causing weak "porous" bones.

Appetite Control

The appetite control centre of the brain is located in the hypothalamus. It responds to three chemicals within the body. The first is the hormone insulin, which is secreted by the pancreas when blood sugar levels are high. The second is the chemical PYY (peptide tyrosine tyrosine), which is secreted by the small intestine when food is present. The third is the chemical leptin, which is secreted by adipose (fat) tissue when fat is being stored. The combination of these chemical signals from the body cause the body to feel satiated (feel like it has eaten enough) after consuming food.

REMEMBER

Anorexia was originally named for *an* = "negation" and for *orexe* = "appetite".

The word *anorexia* means "loss of appetite" although this is not exactly true. Instead, anorexia is really the ability to ignore the appetite, not the lack of an appetite.

Anorexia Nervosa

Anorexia nervosa is a disorder in which an individual has an abnormal obsession with being thin even though the person is already underweight. The distorted body image of feeling overweight when the body is extremely underweight and malnourished leads to health issues and can ultimately lead to death. Individuals suffering from anorexia nervosa limit the amount of food consumed each day to levels too low to support even the basic bodily functions. Starvation of the body for nutrients will lead to the body breaking down its own tissues.

The consequences of anorexia nervosa include:

- The use of protein for energy increases since carbohydrates and fats become depleted. The muscles lose mass and become weaker.
- Energy levels fall due to the lack of muscle mass and the ability to carry out cellular respiration to produce ATP.
- An imbalance of electrolytes can cause the nervous system to malfunction, leading to seizures or body tremors.
- Heart disease and arrhythmias result when the heart muscle atrophies (loses mass) and an electrolyte imbalance interferes with the ability of the heart to maintain normal rhythm.
- Hormone levels cannot be maintained, causing reproductive hormone levels to drop. Many anorexic females will become infertile due to the loss of a menstrual cycle. In addition, growth hormones cannot be produced in levels that maintain normal growth. So anorexics may have stunted growth levels.
- Hair and nails may become very thin and brittle due to the lack of nutrition required to maintain them.
- The immune system will not be able to function well, resulting in an increase in infections and the loss of the ability to repair or replace damaged tissue.

D.2 DIGESTION

TOPIC CONNECTIONS

- 1.2 Ultrastructure of cells
- 6.5 Neurons and synapses

Digestive Juices

Food must be digested in order for the body to extract nutritional components (nutrients) from the food and for blood capillaries to be able to absorb the extracted nutrients. Enzymes are required to break down food into components that can be absorbed and transported throughout the body. Digestive components are excreted into the alimentary canal by glands located throughout the digestive system. The saliva, stomach gastric juices, pancreatic juices and juices from the lining of the intestines all contribute to the digestive process. The nervous system and the endocrine system (hormones) are both involved in controlling the volume, content of and secretion of digestive juices. Table D.3 outlines the sources of the main digestive juices.

Table D.3: Digestive Juices

Digestive Juice	Secreted By	Secreted Into
Saliva	Salivary glands	Oral cavity
Gastric juice	Gastric glands of the stomach wall	Stomach
Pancreatic juice	Pancreatic glands	Duodenum of small intestine
Intestinal juice	Cell wall of the intestine	Intestinal space (lumen)

Exocrine Glands

Exocrine glands release their products into vessels that transport the products to the gut lumen or to the surface layers of the body. The vessel delivering the products produced by the glands is referred to as a duct. The cells that secrete the products into the gland are only 1 cell thick, allowing for easy transfer of the products to the duct. The product is released from the secretory cells in secretory vesicles through the process of exocytosis into the duct. The duct contains many small branches that merge into the main duct transporting the products. The branches provide a large surface area for production and excretion of the products. Each branch surrounded by cells producing product is referred to as an acinus (see Figure D.1).

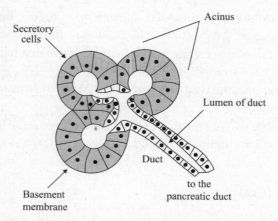

Figure D.1. Pancreatic duct and secretory cells

Exocrine gland cells contain prominent nucleoli for ribosomal formation, extensive rough endoplasmic reticulum for protein synthesis, many Golgi for processing proteins and numerous transport vesicles for exocytosis of products. Exocrine glands possess a large number of mitochondria to supply energy for the cells' production of product (protein synthesis) and excretory processes (exocytosis). Figure D.2 shows the main structures of exocrine gland cells, both as a drawing and as seen in an electron micrograph.

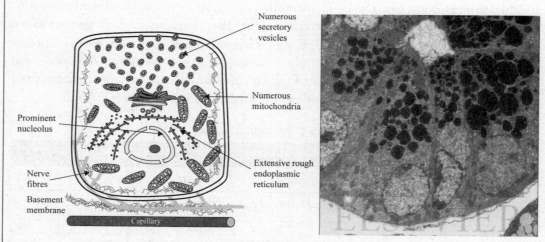

Figure D.2. (a) Structure of digestive exocrine gland cell and (b) electron micrograph of digestive exocrine gland cell

Control of Digestive Juices

The body avoids wasting energy by producing and excreting digestive juices only when they are needed. The release of digestive juices is under the control of the nervous system (nerves) and the endocrine system (hormones). Before food is consumed, the smell or sight of food triggers neurons in the brain to send messages to the cells in the wall of the stomach to start secreting digestive juices. Food entering the stomach triggers more gastric juice to be released. The presence of food activates chemical receptors, pressure receptors and stretch receptors in the stomach wall that send messages to the brain to stimulate the release of even more gastric juices. The presence of food in the digestive system stimulates the release of the hormone gastrin from cells in the stomach wall and cells in the duodenum. Gastrin stimulates the production of hydrochloric acid that activates the formation of pepsin from its precursor pepsinogen as well as stimulates the release of more gastric juices. (Refer to Figure D.3.)

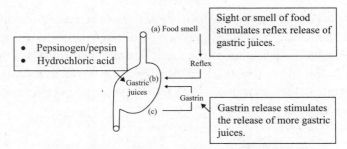

Figure D.3. Control of gastric juices

Helicobacter pylori and Stomach Ulcers

The pH of the stomach is extremely acidic. This acidic environment aids in the digestion (hydrolysis) of food and prevents many microbes from being able to survive in the hostile environment. Normally, the mucous lining of the stomach protects the stomach lining from damage due to the acidic environment. Stomach ulcers and stomach cancers develop when tissues in the stomach undergo changes. Stomach ulcers develop when the stomach mucosa (mucous lining) is damaged and the acidic juices reach the sensitive layers of the stomach lining that are usually protected by the mucosa layer.

> Proton pump inhibitors (PPI) are a group of drugs that reduce the acidity of the stomach by decreasing the production of gastric acid. PPI are used to treat heartburn, stomach ulcers and excess stomach acid. Examples of PPI include Nexium, Prilosec and Prevacid.

The bacteria *Helicobacter pylori* have been discovered to be responsible for much of the damage that occurs to the stomach lining. *Helicobacter pylori* can survive in the acidic environment by producing the enzyme urease that converts urea into ammonia. The ammonia neutralizes the stomach acid in the regions of infection, allowing the bacteria to survive and reproduce in the less hostile environment.

Helicobacter pylori inhabit the stomach wall layers beneath the mucosa lining, infecting these cells and causing the loss of the stomach lining. Damage to the cells lining the stomach wall brings the acidic digestive juices into direct contact with the cells that form the wall of the stomach. These stomach wall cells are destroyed, leading to ulcers that can reach deep into the stomach lining. Persistent long-term infection with *Helicobacter pylori* can result in cellular changes in the cells of the stomach wall that increase the risk of developing stomach cancers. Antibiotic treatment is the best course of action for treating ulcers caused by *Helicobacter pylori* and is recommended in order to avoid stomach ulcers and the increased risk of stomach cancers.

> **Stomach ulcer:** An open area in the mucous lining of the stomach wall that allows the acidic gastric juices to reach and damage cells of the stomach wall.

The Small Intestine

The structure of the cells lining the digestive system allow for maximum surface area for the absorption of nutrients. The small intestine consists of three parts: duodenum, jejunum and ileum. The ileum is the final section of the small intestine and functions mainly to absorb vitamins, bile salts and any nutrients left in the digested material as it passes by the cells of the ileum wall. Figure D.4 shows two views of a transverse section of the ileum: a drawing and a light microscope view.

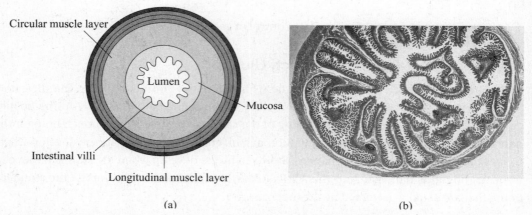

(a)   (b)

Figure D.4. Transverse section of ileum (a) drawing and (b) light microscope view

EPITHELIUM OF THE VILLI

The intestinal lining contains many villi that function to increase the surface area of the intestinal lining. The cells making up the lining of the intestine also possess folds in order to increase the surface area further. The folds on the cells of the villi are referred to as microvilli. Material can enter into the villi cells through membrane channels (pores) or by the process of endocytosis. Mitochondria in the cells produce the energy needed to pump digested nutritional material across the villi through membrane pores in order to be absorbed by capillaries. Pinocytotic vesicles form as material enters the villi through the process of endocytosis. Tight junctions form between the epithelial cells in order to ensure the cells of the villi are directly connected to each other. This prevents the movement of undigested material among the cells.

It also ensures absorption of nutrients into the villi cells and, ultimately, into the capillaries that reach into each villus. Table D.4 outlines the main structural features of the villi epithelial cells. Figure D.5 shows an electron microscope view of the villi structure.

Table D.4: Structural Features of Villi Epithelial Cells

Structural Feature	Function
Microvilli	Increase surface area for absorption of nutrients by simple diffusion, facilitated diffusion and active transport.
Mitochondria	Provide energy (produce ATP) for the active transport of nutrients such as glucose, amino acids and minerals.
Pinocytic vesicles	Formed during the process of endocytosis and allows for the bulk transport of material into the villi cells. This allows larger amounts of fluid with digested material to be transported into the cell.
Tight junctions	Attach adjacent villi cell membranes to each other to block material from moving among the epithelial cells, which would prevent material from being absorbed.

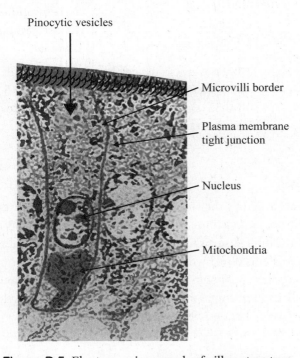

Figure D.5. Electron micrograph of villus structure

ABSORPTION OF FOOD

Material can be absorbed and transported into cells by facilitated diffusion, active transport and endocytosis. Methods of transport are outlined in Table D.5.

Table D.5: Absorption and Transport Mechanisms

Method of Absorption and Transport	Processes Involved	Examples
Simple diffusion	Material enters passively through the cell membrane	Small hydrophobic molecules such as lipids
Facilitated diffusion	Material enters passively through protein channels in the cell membrane	Small, charged molecules such as glucose, minerals and vitamins moving with their concentration gradient
Active transport	Material enters actively through protein channels that pump the material into the cell	Speeds up the absorption of molecules moving with their concentration gradient and pumps molecules against their concentration gradient, including glucose, amino acids and minerals
Endocytosis	Material enters the cell by the formation of vesicles from invaginations of the cell membrane	Large volumes of liquid containing nonspecific dissolved nutrients enter the cell

Egested Material

Material that is not absorbed by the small intestine will be egested (excreted) from the body. Undigested material (including fibre) leaves the small intestine and enters into the large intestine. The large intestine absorbs water from the digested material and forms faeces. The amount of fibre in the diet is directly related to the rate material will travel through the large intestine. If there is not enough fibre in the diet, food will travel slowly and too much water will be removed. This will result in constipation. If food moves too quickly, less water is absorbed by the large intestine and diarrhoea will result.

Cholera

Infections in the intestines can destroy epithelial cells, disrupting the tight junctions found in the tissue. This will cause fluid to leak across the intestinal lining and into the lumen (space) of the intestines. The result is too much water being lost from the body, leading to severe diarrhoea and dehydration. Extreme dehydration is life threatening. The bacterium *Vibrio cholerae* is the causative agent of cholera. It is a pathogen that releases toxins that destroy the lining of the intestinal walls.

D.3 FUNCTIONS OF THE LIVER

The liver is a large organ that sits on the right side of the body and weighs close to 3 pounds. The liver plays a role in every body system except for reproduction. Roles of the liver include producing bile for digestion, aiding in removing toxins from the body, producing plasma proteins, destroying old red blood cells (erythrocytes), helping to regulate cholesterol levels, storing and releasing nutrients and producing many chemicals needed by the body.

Two main vessels bring blood to the liver, and one main vessel takes blood away from the liver. The capillaries that reach into the villi merge together as they leave the intestines and collectively form the hepatic portal vein. The hepatic portal vein delivers material absorbed by the intestines to the liver for processing. The hepatic artery brings blood from the aorta to the liver to supply the liver with oxygenated blood. Blood leaves the liver through the hepatic vein that returns blood back to the heart, where it will cycle through the body once again. The main vessels of the liver can be seen in Figure D.6.

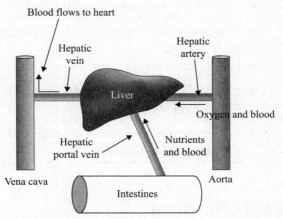

Figure D.6. Main vessels of the liver

Hepatocytes, which are cells in the liver, remove substances from the blood and add material to the blood. The exchange of material between hepatocytes and blood entering the liver from both the hepatic portal vein (bringing nutrients) and the hepatic artery (bringing oxygen) occurs in the sinusoids. Sinusoids are capillaries in the liver that consist of one layer of fenestrated endothelial cells. The word *fenestrated* means that larger-than-typical spaces occur between the cells. The fenestrations allow the hepatocytes located beneath the endothelial cells to be in direct contact with the components of the blood entering the sinusoids. Kupffer cells act as fixed macrophages that line the walls of the endothelial cells. Kupffer cells function to destroy foreign proteins, destroy foreign bacteria and break down old red blood cells by phagocytic processes. Sinusoids empty into central veins that ultimately merge to form the hepatic vein that takes material away from the liver.

Nutrient Storage

The hepatic portal vein delivers nutrients absorbed by the small intestines to the liver. The liver stores nutrients that are high in concentration in the blood in order to be able to release these nutrients when blood levels of them become too low. The liver plays a major role in regulating the concentrations of substances such as glucose, vitamin A (retinol), vitamin D (calciferol) and iron. When needed, iron is transported to the bone marrow. There it is incorporated into haemoglobin during the formation of new red blood cells.

The hormones insulin (lowers blood sugar levels) and glucagon (raises blood sugar levels) are released by the pancreas when blood sugar becomes too high or too low. Insulin stimulates hepatocytes to take up glucose and store it as glycogen, thereby lowering blood sugar levels. Glucagon stimulates hepatocytes to convert stored glycogen into glucose and to release the glucose into the bloodstream, increasing blood sugar levels. Nutrient storage in hepatocytes is outlined in Table D.6.

Table D.6: Nutrient Storage in Hepatocytes

Stored Nutrient	Function
Carbohydrates (glucose and glycogen)	■ Maintaining homeostatic blood sugar levels by storing glucose as glycogen and releasing glucose from glycogen stored in hepatic cells ■ Ensures glucose supply to body cells for cellular respiration
Vitamin D (calciferol)	■ Involved in regulating calcium and phosphorus levels in the body ■ Lack of vitamin D can cause rickets
Vitamin A (retinol)	■ Needed for production of visual pigments to ensure proper functioning of the eyes
Iron	■ Essential component in haemoglobin, which transports oxygen throughout the body

Plasma Proteins and Cholesterol

The liver is involved in the synthesis of plasma proteins and cholesterol. Hepatocytes in the liver are packed with rough endoplasmic reticulum and Golgi in order to produce and export plasma proteins such as fibrinogen and albumin. Although some cholesterol is absorbed by the intestines, most of it is synthesized by the liver. The liver will convert excess cholesterol into bile that will be stored in the gallbladder until needed.

Plasma Proteins
- Fibrinogen: needed for blood clotting
- Albumin: regulates body fluid osmotic pressure

Cholesterol
- Essential for cell membrane formation
- Production of bile
- Production of vitamin D and some steroid hormones (oestrogen and testosterone)

Detoxification

The material absorbed by the intestines and sent to the liver through the hepatic portal vessel may consist of toxic substances. The liver plays a major role in the detoxification of harmful substances, preventing them from entering the circulatory system. Toxic substances that may enter the body include ethanol from alcoholic drink consumption, toxic products from drug use, toxins from food consumption (pesticides and herbicides used on food) and other dangerous molecules. Since the liver is the first organ to receive blood that could contain toxins, it is at highest risk for damage when toxins are taken into the body.

Erythrocyte and Haemoglobin Breakdown in the Liver

Erythrocytes (red blood cells) lack nuclei, making them incapable of undergoing mitosis to produce new erythrocytes. For this reason, the life span of erythrocytes is limited to about four months. Every four months, each erythrocyte must be replaced by the formation of new erythrocytes in the bone marrow. Old red blood cells must be broken down and removed from the circulatory system. The liver breaks down old red blood cells, releasing their components for recycling or for excretion from the body. The Kupffer cells in the sinusoids of the liver take in haemoglobin that has been released from the rupturing of old erythrocytes. The haemoglobin is taken in by phagocytosis since it is a large protein molecule. Once in the Kupffer cells, the haemoglobin is broken down into its main components. Haemoglobin is a globular protein consisting of four polypeptides (two alpha and two beta chains) with iron as a prosthetic group in the centre core of the molecule. Refer to Table 2.8 in the core material for haemoglobin structure.

The polypeptides that make up haemoglobin are digested to their monomers (amino acids). These monomers are then released into the blood to be used by cells for protein synthesis. The iron is either stored in the liver or sent to the bone marrow for use in the formation of haemoglobin when producing new erythrocytes. The last remaining structure from the breakdown of haemoglobin is bilirubin (bile pigment). Bilirubin is absorbed by hepatocytes and used in the synthesis of bile.

Alcohol and the Liver

Excessive alcohol consumption can lead to damage and, ultimately, cirrhosis of the liver. The liver receives dissolved alcohol before any other organ in the body does. If too much alcohol is in the blood, it is released into the hepatic vein and circulates throughout the body. The alcohol will damage other tissues, but it will also return to the liver unchanged. The liver will keep receiving the alcohol until it can efficiently detoxify the blood alcohol and remove it from the body. High levels of alcohol consumption over a long period of time can cause fatty deposits in the liver, inflammation of the liver and cirrhosis of the liver. Death can result when damage to the liver is so severe that the damage prohibits the liver from functioning (see Table D.7).

Table D.7: Liver Damage from Alcohol Abuse

Liver Damage	Consequences
Fatty deposits	Fatty deposits enlarge the liver and replace healthy liver cells, limiting the ability of the liver to function properly.
Inflammation	Hepatitis results from the inflammation caused by the presence of damaged tissue.
Cirrhosis	Scar tissue develops from damaged tissue and replaces healthy tissue with dead tissue.

Jaundice

The presence of jaundice is an indicator of liver or of gallbladder disease. Jaundice is a condition resulting from the accumulation of bilirubin in the blood. Bilirubin is produced during the destruction of old red blood cells. The haemoglobin released from destroyed red blood cells is broken down, releasing the haem group from haemoglobin. The haem group is converted into bilirubin. Bilirubin is normally combined with bile from the gallbladder and removed from the body as waste. Excess bilirubin in the blood is a sign that the liver or gallbladder may not be functioning properly. Gallstones can cause jaundice by preventing the bile stored in the gallbladder from being released.

D.4 THE HEART

Cardiac Muscle Cells

The heart is a muscle composed of single nucleated involuntary striated cardiac muscle cells. The structure of cardiac muscle cells allows for the rapid propagation of action potentials (stimuli) through the wall of the heart. Striations result from the presence of light and dark bands containing different ratios of actin and myosin. Intercalated discs allow a direct connection among cardiac muscle cells that allows the uninterrupted flow of electrical impulses from one cell to another. Intercalated discs allow the cardiac muscle cells in the walls of the heart to contract in unison, producing strong smooth contractions. The cardiac muscle cells branch out in order to allow more surface area for stronger attachment to each other (see Figure D.7).

The Cardiac Cycle

The cardiac cycle consists of the series of events that occur in the heart, beginning when a collection of blood first enters the heart and ending when that same collection of blood leaves the heart. The cardiac cycle consists of two phases that involve contraction and relaxation of the heart muscle. The two atria contract at the same time, and the two ventricles contract at the same time. The contraction and relaxation of heart muscle is referred to as the systole and diastole stage of the cardiac cycle.

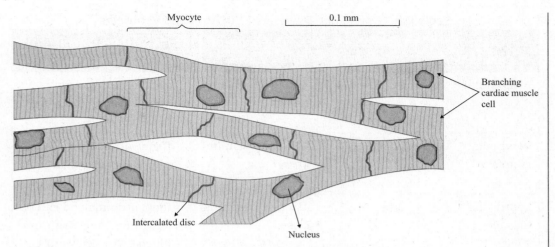

Myocyte   0.1 mm

Branching cardiac muscle cell

Intercalated disc

Nucleus

Figure D.7. Cardiac muscle cells

Blood returns to the heart from the body by the superior vena cava and inferior vena cava. The blood then enters into the right atrium. The right atrium contracts and pumps blood through the atrioventricular valve into the right ventricle. Next, the right ventricle contracts and pumps blood through the semilunar valve and out of the heart via the pulmonary artery. The blood then travels from the heart to the lungs, where the deoxygenated blood is oxygenated. The blood returns to the heart from the lungs through the pulmonary veins and enters into the left atrium. The left atrium contracts and pumps blood through the atrioventricular valve and into the left ventricle. The left ventricle contracts and pumps the oxygenated blood through the semilunar valve and out of the heart to the body through the aorta. Refer to Figure 6.9 in the core material for heart structure/blood flow.

> **Systole:** Occurs when chambers of the heart contract.
>
> **Diastole:** Occurs when chambers of the heart relax.

When all four chambers of the heart are in diastole (relaxed), blood passively moves into the atria from the vena cava and from the pulmonary artery. The blood also passively moves into the ventricles through the open atrioventricular (AV) valves. Pressure from the moving blood keeps the semilunar valves closed. Both atria and both ventricles go through systole and diastole stages together.

The atria undergo contraction (systole) to further empty the blood from the atria into the ventricles. During contraction of the atria, the ventricles are relaxed. The ventricles undergo contraction (systole) to force blood into the pulmonary arteries toward the lungs (from the right ventricle) or into the aorta (from the left ventricle). During contraction of the ventricles, the atria are relaxed. When the ventricles are contracting, the atria are relaxing.

The left ventricle pumps with the greatest force since it must pump blood out the aorta to all the body cells. The right ventricle forces blood to the lungs, which are much closer to the heart than all the body cells. Since the distance blood has to travel to reach the lungs is not far, less force is needed for blood to reach the lung tissue from the right ventricle. The systole and diastole phases of the cardiac cycle are divided between the actions of the atria and the actions of the ventricles (see Table D.8).

Table D.8: Systole and Diastole of the Atria and Ventricles

State of Contraction	Chambers Contracting	Condition of Valves	Direction of Blood Flow
Atrial systole	Atria	AV open; semilunar closed	Right atrium to right ventricle and left atrium to left ventricle
Ventricular systole	Ventricles	AV closed; semilunar open	Right ventricle to pulmonary artery and left ventricle to aorta
Ventricular diastole	None	AV open; semilunar closed	Right atrium blood enters left atrium and left atrium blood enters left ventricle due to volume changes, not pressure from contractions

The sequence of the cardiac cycle can be seen in an electrocardiogram and heard on a phonocardiogram. The pressure changes in the chambers and main vessels of the heart can be recorded on an electrocardiogram. The sounds generated by the beating heart can be picked up with a phonocardiogram. The sounds created by the heart are due to the force of blood hitting the valves, causing the valves to close. These sounds are heard as and referred to as "lub" (AV valve closes) and "dub" (semilunar valve closes).

> **"Lub"**: Heart sound caused by the ventricles contracting, forcing the atrioventricular valves to close.
>
> **"Dub"**: Heart sound caused by the atria contracting and forcing the semilunar valves to close.

When the ventricles contract, the semilunar valves open. Aortic pressure rises as blood is forced out of the left ventricle. The atrioventricular valves open when the atria contract as the ventricles relax, deceasing aortic and ventricular pressure. The highest spikes in the electrocardiogram are present due to the force of the electrical depolarization of the muscle in the ventricles.

Control of the Heart

The heart is myogenic, which means it beats on its own without stimulation from the central nervous system. There are two nodes (bundles of neurons and muscle fibres) that are both located in the right atrium wall and control the rate at which the heart beats. The sinoatrial node (SA) node is referred to as the pacemaker of the heart since it controls when the atria and ventricles will contract. The SA node initiates impulses to trigger contraction of the atria and to stimulate the AV node to initiate contraction of the ventricles. The AV node delays impulses that initiate contraction of the ventricles for about 0.1 second after the atria have contracted.

This delay ensures that the atria fully empty and fill the ventricles with blood before the ventricles contract. The smooth, simultaneous contraction of the ventricles is achieved by the presence of conducting fibres running down the septum and through the walls of the ventricles. The depolarization of the SA node and the AV node can be seen in electrocardiograms that pick up electrical activity in the heart.

1. The SA node initiates electrical impulses that trigger contraction of the atria. Blood is forced from the atria into the ventricles.
2. The SA node initiates electrical impulses that trigger the AV node to initiate contraction of the ventricles 0.1 second after the atria contract.
3. The AV node sends electrical impulses through conducting fibres in the septum and walls of the ventricles.
4. The ventricles contract, forcing blood into the pulmonary artery (from the right ventricle) and to the aorta (from the left ventricle).

The contractions of the ventricles are very powerful since more muscle fibres are triggered to depolarize. The contraction of the left ventricle is the most powerful since it contains the most muscle mass and needs to pump blood to the entire body. The forceful contraction of the left ventricle results in the largest spike seen in an electrocardiogram. The depolarization of the atria is not as powerful and sends about the same electrical signal level as does the repolarization of the ventricles. The atria do not need to contract as forcefully since they are sending blood a short distance to the ventricles. Table D.9 outlines the myogenic control of the heart.

Table D.9: Myogenic Control of the Heart

Control Mechanism	Location	Function
SA node	In the sinus of the right atrium (hence the name)	Triggers contraction of the atria and controls the AV node
AV node	In the right atrium near the AV valve (hence the name)	Triggers contraction of the ventricles, sending impulses through the conducting fibres
Conducting fibres	In the septum and muscular walls of the ventricles	Triggers contraction of the muscle fibres in the ventricles

Hypertension and Thrombosis

Hypertension (high blood pressure) and thrombosis (blood clots that form inside of vessels) are both life-threatening conditions. The causes and consequences of both are outlined in Table D.10.

Table D.10: Hypertension and Thrombosis

	Hypertension (High Blood Pressure)	Thrombosis (Blood Clots Inside of Vessels)
Causes	■ Genetic ■ Diet (high salt intake) ■ Stress	■ Changes in blood composition leading to clotting ■ Damage to vessels
Consequences	■ Kidney damage ■ Heart and artery damage ■ CNS damage—stroke	■ Reduced blood flow to tissues and organs ■ Stroke ■ Heart attack

D.5 HORMONES AND METABOLISM

Endocrine Glands

Endocrine glands secrete products into the bloodstream to be transported throughout the body. The human body possesses many endocrine glands that produce diverse hormones. Chemical messengers secreted by endocrine glands into the bloodstream are known as hormones. Hormones travel in the blood to all body cells and are recognized by the cells of target tissue. Cells making up the target tissue for a hormone possess receptors for the hormone. The recognition of a hormone by a target cell (formation of a receptor-hormone complex) initiates responses in the target cell.

> **Target tissue:** Tissue made up of cells that possess receptors for and respond to specific hormones.

Hormone Action

The structure of hormones determines their mode of action in the body. Hormones can be assembled from steroids or from proteins. The mode of action of a particular hormone depends on whether the hormone can enter into the cell or if it has to bind to an external receptor. The hydrophobic properties of the cell membrane determine the mode of action of hormones (see Table D.11).

Table D.11: Mode of Action of Hormones

Type of Hormone	Mode of Action	Example
Steroid	■ Easily passes into the cell through the membrane due to the hormone's hydrophobic properties (because the hormone is synthesized from cholesterol lipids) ■ Alters protein synthesis once inside the cell (bind to receptor proteins within the cytoplasm of the target cell)	■ Testosterone and oestrogen ■ Progesterone
Protein (peptide)	■ Does not enter into cells due to the hormone's hydrophilic properties; attaches to receptors on target cells, signalling a change on the inside of the cell without entering the cell ■ Attachment to target receptors activates second messengers inside of cells that alter actions of the cells	■ Insulin ■ ADH (antidiuretic hormone)

Hypothalamus and Pituitary Gland

The hypothalamus controls the release of hormones from the pituitary gland. Hormones released from the pituitary gland control reproduction, developmental changes, growth and homeostasis. The pituitary gland hangs from the hypothalamus and responds to hypothalamic-releasing hormones. The pituitary is divided into two lobes, each containing distinct hormones. The posterior lobe contains hormones produced by the hypothalamus and stored in the posterior lobe of the pituitary. The anterior lobe produces its own hormones that are released when releasing hormones are received from the hypothalamus. Each releasing hormone triggers the release of a specific anterior pituitary hormone. For example, gonadotropin-releasing hormone (GnRH) triggers the release of follicle-stimulating hormone (FSH) and luteinizing hormone (LH) from the anterior pituitary. The pituitary hormones are outlined in Table D.12.

Table D.12: Pituitary Hormones

Pituitary Lobe	Production and Control of Hormones	Names and Target Tissue of Hormones
Anterior lobe	Produces its own hormones that are released when releasing hormones are sent from the hypothalamus. Inhibiting hormones from the hypothalamus prevent their release.	■ Thyroid-stimulating hormone (TSH) targets the thyroid gland. ■ Follicle-stimulating hormone (FSH) and luteinizing hormone (LH) target the gonads. ■ Growth hormone (GH) targets the entire body. ■ Prolactin targets the mammary gland for milk production.
Posterior	Stores hormones produced by the hypothalamus and that are released when neurons from the hypothalamus send messages (action potentials), signalling the release of the hormones.	■ Antidiuretic hormone (ADH) targets the kidney (collecting duct of the nephron). ■ Oxytocin targets the uterus to stimulate uterine contractions and the release of milk from the mammary glands.

Examples of Artificial Use of Pituitary Hormones

■ Oxytocin controls uterine contractions. Synthetic oxytocin (Pitocin) is given to women to induce labour contractions.
■ Growth hormones can be taken to increase muscle mass and athletic performance.

Breastfeeding could not be accomplished without the combined effects of prolactin and oxytocin. Prolactin is responsible for the production of breast milk by the mammary glands. Hence, a breastfeeding woman is considered to be lactating. Oxytocin is responsible for the release of the milk from the mammary glands.

D.6 TRANSPORT OF RESPIRATORY GASES

TOPIC CONNECTIONS

- 6.4 Gas exchange

Partial Pressure

The cardiovascular system and respiratory system work together to carry out gas exchange. Inhaled air is composed mainly of nitrogen, oxygen and carbon dioxide. Together the combined partial pressures of gases in air make up air pressure.

> **Partial pressure:** The pressure exerted from an individual gas when mixed with other gases.

Haemoglobin

Haemoglobin is a globular protein that consists of 4 polypeptide chains with four central iron prosthetic groups. Haemoglobin can bind up to 4 oxygen molecules and 1 carbon dioxide molecule. Erythrocytes are packed with haemoglobin and transport gases to the alveoli of the lungs for gas exchange. The attraction of oxygen to haemoglobin increases as the number of oxygen molecules being carried by haemoglobin increases. A haemoglobin molecule lacking attached oxygen molecules has low affinity for attracting oxygen. However, when haemoglobin binds an oxygen molecule, it develops a greater attraction for another oxygen molecule. The increased attraction of haemoglobin for oxygen is due to the change in shape of the molecule caused by the first oxygen molecule binding to haemoglobin. This attraction of oxygen molecules increases as haemoglobin binds more oxygen molecules. Haemoglobin that is carrying the maximum number of oxygen molecules (4) no longer attracts oxygen since it cannot bind any further oxygen. Since oxygen is a diatomic molecule (always present as 2 bonded oxygen atoms), every time haemoglobin binds oxygen it is binding 2 oxygen atoms at once.

Myoglobin

Myoglobin is a protein that stores oxygen in muscle cells. It has a greater attraction for oxygen than does haemoglobin. Myoglobin consists of a single globin protein and 1 iron haem group. Myoglobin attracts oxygen at low concentrations and releases it when oxygen levels are depleted. This release of oxygen by myoglobin is used by the muscle when oxygen supply is depleted and helps prolong the onset of anaerobic respiration.

Foetal Haemoglobin

Foetal haemoglobin differs from adult haemoglobin by a few amino acid sequences. The change in amino acid sequence results in foetal haemoglobin having a greater attraction than adult haemoglobin for oxygen. Foetal haemoglobin needs to have a greater attraction for oxygen since foetal gas exchange occurs in the placenta where maternal haemoglobin releases its oxygen. The foetal haemoglobin readily binds the oxygen, ensuring the oxygen will be available for release in foetal circulation.

Oxygen Dissociation Curves

The oxygen dissociation curves for adult haemoglobin, foetal haemoglobin and myoglobin are shown in Figure D.8.

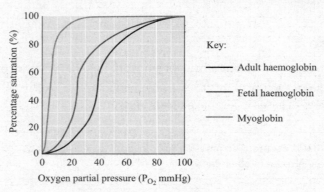

Figure D.8. Dissociation curves for adult haemoglobin, foetal haemoglobin and myoglobin

Oxygen dissociation curves display the ability of oxygen-binding molecules to attract oxygen at different partial pressures. Oxygen absorption by adult haemoglobin occurs most efficiently when oxygen partial pressure is very high. As shown in Figure D.8, once oxygen binds to haemoglobin, the haemoglobin will have a greater affinity for binding more oxygen molecules until it is fully saturated and holding oxygen molecules. The oxygen dissociation curve for foetal haemoglobin is very similar to that of adult haemoglobin except that foetal haemoglobin can absorb oxygen more efficiently at lower partial pressure since it has a greater affinity for oxygen. Myoglobin is the most efficient oxygen-absorbing molecule and can absorb high percentages of oxygen even when oxygen partial pressure is low. Table D.13 outlines the partial pressure and oxygen absorption of adult haemoglobin, foetal haemoglobin and myoglobin.

Transport of Carbon Dioxide

Cellular respiration results in the production of carbon dioxide that quickly diffuses into the capillaries surrounding cells. The carbon dioxide is transported to the lungs to be exhaled during gas exchange at the alveoli by one of the following processes:

- Small amounts of carbon dioxide are transported as dissolved gases in the blood.
- Some reversibly binds to haemoglobin molecules in erythrocytes.
- Most of the carbon dioxide is converted to hydrogen bicarbonate ions and transported in erythrocytes or blood plasma to the lungs.

The carbon dioxide that enters erythrocytes may be directly taken up by (bonded to) haemoglobin. Alternatively, the enzyme carbonic anhydrase may catalyse the formation of carbonic acid (H_2CO_3) from carbon dioxide by joining the carbon dioxide with water.

Carbonic Anhydrase

$$CO_2 + H_2O \rightarrow H_2CO_3$$

The carbonic acid quickly dissociates into a hydrogen carbonate ion and a hydrogen ion.

$$H_2CO_3 \rightarrow H^+ + HCO_3^-$$

Table D.13: Partial Pressure and Oxygen Absorption

Oxygen-Absorbing Molecule	Partial Pressure Curve	Role of Molecule in Human Systems
Adult haemoglobin	S-shaped; shows low affinity at low partial pressure and greater affinity as the partial pressure increases until saturation is reached	Carries oxygen throughout the body, becomes saturated at high partial pressures (within the lungs) and disassociates at lower partial pressures (within the tissues)
Foetal haemoglobin	S-shaped with a shift to the left of adult haemoglobin when graphed; shows increased affinity for oxygen at all partial pressures until saturated	Transports oxygen from the placenta to foetal tissues
Myoglobin	Much further shift to the left of foetal haemoglobin when graphed due to having the highest affinity for oxygen until becoming saturated	Stores oxygen in muscle cells for use when oxygen is limited; has a high affinity for oxygen even at low partial pressures that allows it to bind oxygen even when partial pressure is very low

Protein channels facilitate the movement of hydrogen carbonate ions out of the erythrocyte as chloride ions move in. The exchange of the negative ions is known as the chloride shift. This shift helps to keep the membrane potential (charge between the interior and exterior of the cell) at homeostatic levels. Some of the hydrogen ions leave the erythrocytes and can alter the pH of blood to dangerous levels. Plasma proteins serve as buffers in the blood by picking up the hydrogen ions, thereby preventing the pH from becoming too acidic. Hydrogen ions in the erythrocytes bind to haemoglobin, buffering the pH of the erythrocyte.

The Bohr Shift

The Bohr shift explains the changing affinity of haemoglobin for oxygen versus carbon dioxide. The partial pressure of carbon dioxide increases as cells carry out cellular respiration and release more carbon dioxide into the body fluids. The increase in carbon dioxide partial pressure decreases the affinity of haemoglobin for oxygen. This causes haemoglobin to release oxygen from erythrocytes and pick up carbon dioxide from the respiring tissues. Lung tissue has a high partial pressure for oxygen and a low partial pressure of carbon dioxide, causing haemoglobin to release carbon dioxide and pick up oxygen. In addition, lower pH levels of surrounding environments increase the affinity of haemoglobin for binding carbon dioxide and releasing oxygen molecules. Carbon dioxide combines with water to form carbonic acid, lowering the pH and increasing the partial pressure of carbon dioxide. The higher partial pressure of carbon dioxide stimulates haemoglobin to pick up carbon dioxide and release oxygen. The Bohr shift is displayed in Figure D.9.

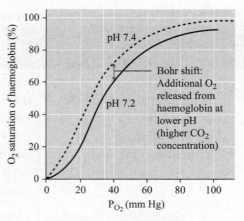

Figure D.9. Bohr shift

Ventilation Rates and Exercise

During exercise, the need for energy increases the rate of cellular respiration and results in the release of larger amounts of carbon dioxide. The change in pH caused by the oxidation of carbonic acid and the resulting release of hydrogen ions causes the pH of the blood to drop. The normal pH of blood is between 7.35 and 7.45. The increase in carbon dioxide production results in a drop in pH to near 7.0. Notice this value is still not acidic, but it does represent a decrease in the normally alkaline nature of the blood. This change in pH occurs because plasma proteins and haemoglobin cannot efficiently absorb the increased production of hydrogen ions that results from increased cellular respiration. Although the change in pH is slight, it is detected by chemoreceptors in the carotid arteries, aorta and medulla oblongata. The medulla receives messages from receptors as well as detects the change itself. When the drop in pH is detected, the medulla oblongata sends messages (neural transmission) to increase the heart rate and breathing. The diaphragm and intercostal muscles contract harder and more frequently to bring in more fresh air and to remove stale air. Increased ventilation rates facilitate the removal of carbon dioxide and increase the supply of oxygen to the body. The increase in heart rate brings more blood to the lungs in order to carry out gas exchange at the alveoli. When the pH of the blood returns to normal, the medulla stops sending messages to increase heart rate and breathing. In fact, both heart rate and breathing return both to normal levels. Table D.14 outlines the regulatory methods of ventilation.

Table D.14: Regulation of Ventilation

Body Part Involved	Regulation Method
Medulla oblongata	■ Sets ventilation rates ■ Monitors pH level of blood as well as partial pressures of carbon dioxide and oxygen to determine ventilation needs
Aorta and carotid arteries	■ Monitors pH level of blood as well as partial pressures of carbon dioxide and oxygen ■ Sends messages to medulla to increase breathing rates when pH and partial pressures of carbon dioxide and oxygen change from homeostatic levels

Emphysema

Emphysema is a condition in which the alveoli in the lungs become damaged and cannot efficiently carry out gas exchange. The fragile alveoli overinflate with trapped air and lose their elasticity. Without elasticity, gas exchange becomes difficult between the alveoli and capillaries. Scar tissue builds up in the damaged alveoli, further limiting gas exchange. This type of damage can be caused by cigarette smoking and exposure to air pollutants (asbestos, secondhand smoke). Rarely, the damage is caused by a genetic defect. Treatment of emphysema includes cessation of smoking, avoiding air pollutants, the use of chemical therapies and the delivery of low levels of oxygen directly to the lungs.

Gas Exchange at High Altitudes

At high altitudes, the amount of oxygen present in the air is reduced. This is due to the decreased pressure, allowing the gases to spread further apart and cover a larger volume of air. The partial pressure of oxygen in the atmosphere is therefore reduced in a given volume of air compared with that at sea level. The reduction in partial pressure decreases haemoglobin's affinity for oxygen. The reduction in the saturation level of oxygen as it carries out gas exchange at the alveoli reduces the amount of oxygen that can be absorbed by haemoglobin. The result is less oxygen in body tissues to be used for cellular respiration. Individuals travelling to high-altitude regions who are not accustomed to the pressure changes can experience altitude sickness. Altitude sickness can cause fatigue, headaches, nausea, breathlessness and dizziness. Severe altitude sickness can lead to psychotic behaviour, loss of coordination, coma and, ultimately, death. Individuals that live at high altitudes become physiologically acclimated to the lower availability of oxygen (see Table D.15).

Table D.15: Acclimation to High Altitudes

Acclimation	Result
Increase in erythrocyte production	More erythrocytes are available to transport more haemoglobin molecules.
Higher erythrocyte haemoglobin content	More haemoglobin is available to transport oxygen to offset the decrease in oxygen-carrying ability of haemoglobin.
Increased tidal volume	More air is exchanged during ventilation, bringing in more oxygen to the lungs.
Increased capillary beds in lungs	More capillaries transport blood to the alveoli to pick up more oxygen during gas exchange.
Increased myoglobin production	Muscle cells produce more myoglobin to store and supply oxygen when the partial pressure becomes low.

PRACTICE QUESTIONS

1. The graph below shows the relationship between weight, height and body mass. A healthy body mass index ranges from 18.5 to 24.9 and is calculated based on an individual's height and weight.

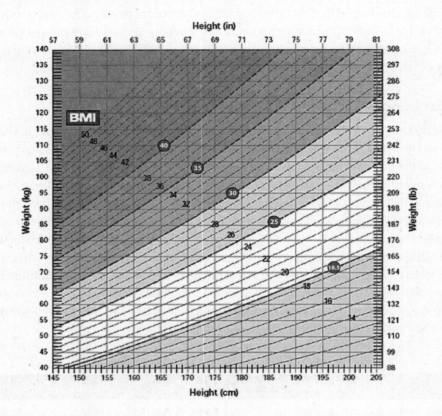

(a) State the body mass index of an individual who is 1.8 metres tall and weighs 80 kilograms. (1)

(b) State the weight range for a healthy body mass index (BMI) for an individual who is 73 inches tall. (1)

(c) Outline two health risks associated with a high BMI. (4)

2. Explain how the heart controls its own contractions. (4)

3. Outline the benefits of eating a diet high in fibre. (3)

4. State 2 roles of the acidic environment of the stomach. (2)

5. Explain the mode of action of steroid hormones. (2) *HL only*

ANSWERS EXPLAINED

1. (a) BMI of 25 (+/−2)—1.8 m, or 180 cm (remember to pay careful attention to units on axes and be prepared to make conversions).

   (b) The weight range for maintaining a healthy body mass index (BMI) for an individual who is 73 inches tall ranges between 138 and 187 lbs. (+/−5 lbs.).

   (c)

   - Individuals with a high BMI are overweight (possess excess body fat).
   - Hypertension can result from high BMI levels, leading to damage to the heart (coronary heart disease).
   - Coronary heart disease decreases the ability of the heart to efficiently deliver blood throughout the body.
   - The decreased ability of blood to be efficiently delivered throughout the body as a result of hypertension can lead to damage of all body tissues.
   - Hypertension can result in aneurysms that can bulge and burst, leading to life-threatening internal bleeding.
   - Hypertension can lead to stroke as a result of damage to and weakening of the blood vessels in the brain.
   - High BMI levels are associated with strokes that can block blood flow to various parts of the brain.
   - Blocked blood flow to the brain can lead to tissue loss in the brain, resulting in loss of physiological functions.
   - Type II diabetes can result from high BMI levels, causing insulin receptors to become less receptive due to the excess body fat (adipose tissue).
   - Type II diabetes can lead to blindness due to the high sugar concentration in the blood stream.
   - Type II diabetes can lead to the loss of sensation in extremities (neuropathy).

2.

   - The heart is myogenic/beats on its own without nervous system control.
   - The nervous system/medulla can influence heart rate.
   - Hormones/adrenaline can increase heart rate.
   - The sinoatrial node/SA node is the pacemaker of the heart.
   - The SA node initiates contraction of the atria.
   - The SA node sends a signal to the atrioventricular node (AV node) to initiate contraction of the atria.
   - There is a 0.1 second/short delay between the SA node initiating contraction of the atria and the stimulation of the AV node to initiate contraction of the ventricles.
   - The atria contract before the ventricles.
   - Conducting fibres transmit impulses through the septum and walls of the ventricles for ventricular contractions.
   - The atrioventricular valves (AV valves) open when the atria contract, and the semilunar valves open when the ventricles contract.
   - The left ventricle contracts with the most force/has the thickest muscular wall.

3.

- High-fibre diets provide bulk to food to help individuals feel full/satiated.
- Fibre keeps food moving and prevents digestion issues/constipation.
- Fibre decreases the amount of sugar absorbed from food, helping to prevent obesity/ diabetes.
- Fibre may help prevent cancer/haemorrhoids.

4.

- The low pH favours some hydrolytic/enzymatic reactions
- The low pH favours the conversion of pepsinogen to pepsin
- The low pH destroys some pathogens that may have been ingested

5.

- Steroid hormones easily pass into the cell through the cell membrane.
- Steroid hormones are hydrophobic in nature.
- Steroid hormones are synthesized from cholesterol lipids.
- Steroid hormones bind to receptors within the cytoplasm of the target cell.
- Receptor-hormone complex forms within the target cell cytoplasm.
- Receptor-hormone complex initiates transcription of select genes.
- Named example of steroid hormone (androgens, oestrogens, progestrogens, mineralocorticoids, glucocorticoids)

PRACTICE
TESTS

Practice Test 1

The practice test that follows is designed to mimic the actual IB Biology Exam Papers 1, 2 and 3. To get the most benefit from this practice exam, you should follow the IB guidelines.

- Adhere to the allotted time for each practice paper. It is a good idea to set a timer to ensure that you do not exceed the time limit.
- Respond to all questions presented. You will not lose points for incorrect responses.
- For Papers 2 and 3, make sure you include *at least* the number of responses required for full credit, as indicated by the point values located in brackets next to each question.
- Take Papers 1 and 2 on the same day, giving yourself only a 15-minute break between the two exams. You should not look at or review any notes during the break.
- Take Paper 3 on the day after your take Papers 1 and 2.

PAPER 1

Time: 45 minutes (SL)
Time: 1 hour (HL)

Paper 1 consists of 40 multiple-choice questions for higher-level (HL) students and 30 multiple-choice questions for standard-level (SL) students. No points will be deducted for incorrect responses, so you should choose a response for all questions presented. You will have 1 hour to respond to 40 questions (HL) or 45 minutes to respond to 30 questions (SL).

1. Which of the following are features of both prokaryotic and eukaryotic cells?

    I.   Cell membrane
    II. Mitochondria
    III. DNA

    A. I only
    B. I and II only
    C. I and III only
    D. I, II and III

2. Standard deviation is used to:

    A. Summarize the spread of values around the mean.
    B. Show the mode in a set of data.
    C. Show the range of a set of data.
    D. Show correlation and causation.

3. Which of the following are placed in correct order from smallest to largest in size?

    A. Virus, molecule, cell membrane, bacteria
    B. Molecule, cell membrane, virus, bacteria
    C. Bacteria, cell membrane, virus, molecule
    D. Cell membrane, virus, bacteria, molecule

Questions 4 and 5 refer to the following diagram.

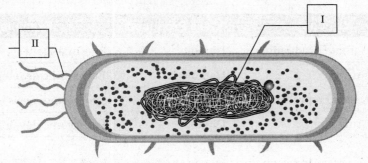

4. Identify the structures labelled below:

    A. I—nucleus; II—pili
    B. I—DNA; II—cilia
    C. I—DNA; II—flagella
    D. I—nucleus; II—flagella

5. The structure labelled II in the figure presented in question #4 functions for:

    A. Translation
    B. Movement
    C. Replication
    D. Storage

---

6. Which of the following processes are passive?

no energy

    I. Facilitated diffusion
    II. Osmosis
    III. Protein pumps

    A. I only
    B. I and II only
    C. II and III only
    D. I, II and III

7. Which parts of the cell membrane are hydrophilic?

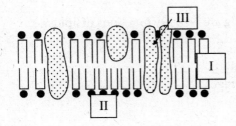

    A. I only
    B. I and II only
    C. II and III only
    D. I, II and III

8. Which of the following correctly depicts mitosis?

	Function	Cells Produced	Example
A.	Gamete production	Haploid	Spermatogenesis
B.	Growth	Haploid	Embryonic development
C.	Asexual reproduction	Diploid	Spermatogenesis
D.	Growth	Diploid	Embryonic development

9. What stages of cell division are shown in the figures below?

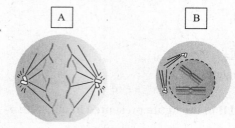

A
B

A. (A)—anaphase I; (B)—prophase II
B. (A)—anaphase II; (B)—prophase I
C. (A)—metaphase I; (B)—anaphase I
D. (A)—anaphase II; (B)—metaphase I

10. The most frequently occurring elements found in living organisms include:

A. Carbon, hydrogen and oxygen
B. Carbon, hydrogen and sulphur
C. Carbon, hydrogen, oxygen and nitrogen
D. Carbon, hydrogen and calcium

11. Which of the following are monosaccharides?

A. Glucose and galactose
B. Glucose and sucrose
C. Sucrose and fructose
D. Amylose and maltose

12. Which of the following are primary functions of lipids?

I.   Energy storage
II. Thermal insulation
III. Information storage

A. I only
B. I and II only
C. II only
D. I, II and III

13. Denaturation refers to:

    A. The breaking of peptide bonds

    B. The site of action of an enzyme.

    C. The loss of protein structure.

    D. The ability of a protein to catalyse a reaction.

14. What terms correctly identify the parts of the figures shown below?

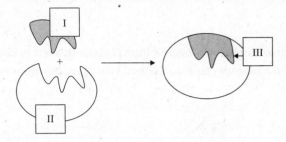

    A. I—enzyme; II—substrate; III—active site

    B. I—substrate; II—enzyme; III—allosteric site

    C. I—enzyme; II—active site; III—allosteric site

    D. I—substrate; II—enzyme; III—active site

15. What colours are most efficiently absorbed by chlorophyll?

    A. Red and blue

    B. Red and green

    C. Blue and green

    D. Red, blue and green

16. Where are the enzymes for the light-independent reactions located?

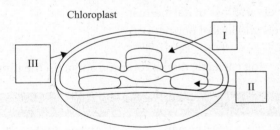

Chloroplast

    A. I only

    B. II only

    C. III only

    D. I and II only

17. Homologous chromosomes are best described as:

    A. Chromosomes that code for the same traits and have identical alleles.

    B. Chromosomes that code for the same traits and have different alleles.

    C. Chromosomes that code for different traits and have different alleles.

    D. Chromosomes that code for the same traits and may or may not have identical alleles.

18. Which of the following are sex-linked traits?

  I. Colour-blindness
  II. Haemophilia
  III. ABO blood types

  A. I only
  B. I and II only
  C. II and III only
  D. I, II and III

19. Black (*B*) is dominant to brown (*b*), and long (*L*) is dominant to short (*l*). A testcross was performed on a heterozygous organism. What possible genotypes could result from this cross?

  A. *BBll* and *BbLl*
  B. *BbLl* and *BBll*
  C. *BbLl* and *bbll*
  D. *bbLL* and *bbll*

  (handwritten annotations:)
	B l	b L
b	Bbll	bbLl
l		
b	Bbll	bbLl
l		

20. A community is correctly defined as:

  A. A group of members of the same species living in the same area.
  B. A group of members of different organisms living in different areas.
  C. A group of members of the same species living in different areas.
  D. A group of member of many different species interacting in the same area.

21. Which of the following would result from the increased release of greenhouse gases?

  A. Lower sea levels and changes in weather patterns
  B. Higher sea levels and no effect to weather patterns
  C. Higher sea levels and loss of permafrost
  D. Lower sea levels and loss of permafrost

22. Which of the following correctly depicts enzyme action in the human body?

	Enzyme	Location	Products
A.	Protease	Oral cavity	Amino acids
B.	Lipase	Small intestine	Glycerol and fatty acids
C.	Amylase	Oral cavity	Amino acids
D.	Protease	Small intestine	Monosaccharides

23. Which of the following is the correct sequence of classification from most inclusive to least inclusive?

   A. Kingdom, phylum, species, order
   B. Kingdom, class, phylum, species
   C. Kingdom, phylum, class, order
   D. Kingdom, order, class, genus

KPCOFGS

24. Arteries, veins and capillaries are correctly represented as:

	Arteries	Veins	Capillaries
A.	Carry oxygenated blood	Carry blood away from the heart	Allow exchange of material within the body
B.	Carry blood away from the heart	Possess the thickest muscular layer	Carry blood back to the heart
C.	Possess the thickest muscular wall	Possess the widest lumen	Allow exchange of material within the body
D.	Carry deoxygenated blood	Possess the widest lumen	Allow exchange of material within the body

25. Why are antibiotics ineffective against viral infections?

   A. Viruses lack a metabolism.
   B. Viruses do not contain DNA.
   C. Viruses possess plasmids that protect them from the effects of antibiotics.
   D. Viruses have a cell wall that blocks antibiotic entrance.

26. Identify the structures below:

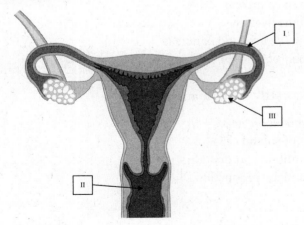

   A. I—oviduct; II—uterus; III—ovary
   B. I—uterus; II—vagina; III—ovary
   C. I—oviduct; II—vagina; III—ovary
   D. I—oviduct: II—vagina; III—uterus

27. Which of the following are roles of testosterone in the human body?

    I.   Development of male genitalia
   II.   Maintenance of sex drive
  III.   Growth of facial hair

   A. I only
   B. I and II only
   C. II and III only
   D. I, II and III

28. Identify the structure and function of the figure shown below:

	Structure	Function
A.	Nucleotide	Monomer of DNA
B.	Nucleosome	Packaging unit of DNA
C.	Nucleosome	Packaging unit of RNA
D.	Nucleotide	Monomer of RNA

29. Type 1 diabetes results from the loss of:
       ↳ cant produce
   A. Beta cells in the pancreas
   B. Beta cells in the liver
   C. Alpha cells in the liver and pancreas
   D. Alpha and beta cells in the liver

30. What is an effect of HIV on the human body?

   A. Decreased red blood cells
   B. Increased white blood cells
   C. Decreased antibody production
   D. Decreased antigen production

The next 10 questions are for HL students only.

31. Vaccinations (immunizations) are examples of:

    A. Active artificial immunity
    B. Passive artificial immunity
    C. Active natural immunity
    D. Passive natural immunity

32. Chemiosmosis in cellular respiration occurs:

    A. From the intermembrane space to the cytosol of the cell by facilitated diffusion
    B. From the matrix to the intermembrane space by active transport
    C. From the intermembrane space to the matrix by facilitated diffusion
    D. From the matrix to the stroma by facilitated diffusion

33. The following aid in mineral ion absorption by plants:

    I. Active transport
    II. Root hairs
    III. Facilitated diffusion

    A. I only
    B. I and II only
    C. II and III only
    D. I, II and III

34. Identify the structures below:

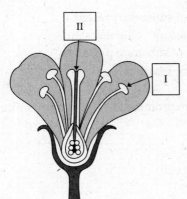

    A. I—anther; II—filament
    B. I—stigma; II—filament
    C. I—anther; II—style
    D. I—anther; II—stigma

35. Crossing-over occurs:

    A. Between sister chromatids during prophase I
    B. Between nonsister chromatids during prophase I
    C. Between sister chromatids during prophase II
    D. Between nonsister chromatids during prophase II

36. What is the correct order for seed germination?

    A. Water absorption, production of amylase, release of gibberellin (GA)
    B. Production of amylase, water absorption, release of gibberellin (GA)
    C. Water absorption, release of gibberellin (GA), production of amylase
    D. Release of gibberellin (GA), water absorption, production of amylase

37. What describes a polysome?
    A. The packaging unit of DNA
    B. Many ribosomes translating off of one mRNA molecule
    C. The monomer of nucleic acids
    D. Many chromosomes during metaphase

38. Which of the following are globular proteins?
    A. Haemoglobin and antibodies
    B. Keratin and haemoglobin
    C. Antibodies and myosin
    D. Antibodies and glycerol

39. What would increase the effects of a competitive inhibitor on an enzyme reaction?

    A. Increase the amount of substrate present
    B. Increasing the temperature of the reaction
    C. Decrease the amount of substrate present
    D. Changing the pH of the reaction

40. Which of the following are produced by the endomembrane system?

    I. Lysosomes
    II. Proteins for export
    III. Proteins for use in the cell

    A. I only
    B. I and II only
    C. II and III only
    D. I, II and III

PAPER 2

Time: 1 hour 15 minutes (SL)
Time: 2 hours 15 minutes (HL)

Paper 2 consists of two sections. Section A contains data analysis and short-response questions. Section B contains essay questions. Higher-level (HL) students will be presented with 3 essay questions and are required to answer 2. Standard-level (SL) students will be presented with 2 essay questions and are required to answer 1. Both SL and HL students must answer all questions presented in Section A.

Answer all questions presented in Section A.

Section A

1. Plants produce hormones that are involved in the regulation of physiological processes. The plant hormone auxin plays a role in the regulation of seed germination, phototropism and plant growth. An experiment was conducted to determine the effects of different levels of exposure to auxin on the rate of seed germination. Seeds from the **Dianthus caryophyllus** plant were soaked in auxin for 12 hours, 24 hours or 36 hours. After soaking the seeds for their respective times in the auxin solution (20 ppm), the seeds were removed and observed as they germinated. The data from the investigation are presented below.

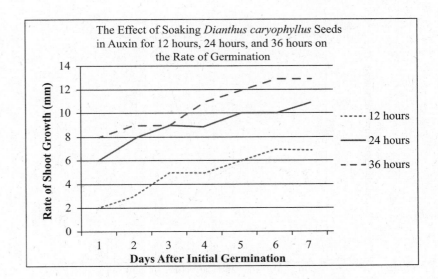

(a) Calculate the percent difference between the seeds soaked in auxin for 12 hours versus the seeds soaked in auxin for 36 hours on day 7. (2)

(b) Compare the rate of seed germination between the seeds soaked in auxin for 12 hours versus those soaked in auxin for 24 hours. (2)

The table below shows the results of data collected on seed moisture content of *Dianthus caryophyllus* seeds stored over a period of nine months. The seeds were removed from the fruit and stored in the same environment for a period of nine months. Stage 1 represents the amount of moisture found in *Dianthus caryophyllus* seeds removed from the fruit when the fruit was still green. Stage 2 represents the amount of moisture found in *Dianthus caryophyllus* seeds removed from the fruit when the fruit was ripe.

Storage Time (months)	Stage 1: Moisture Content (%) in Seeds Removed from Green *Dianthus caryophyllus* Fruit	Stage 2: Moisture Content (%) in Seeds Removed from Ripe *Dianthus caryophyllus* Fruit
0	37.20	38.30
3	40.10	41.30
6	42.10	43.60
9	43.40	44.70

(c) Deduce which seeds would most likely germinate the fastest. (3)

(d) Suggest reasons for the differences between the percent of moisture content in stage 1 and stage 2 seeds. (2)

The graph below shows the effect of the hormone auxin on the roots, buds and stems of the *Dianthus caryophyllus* plant.

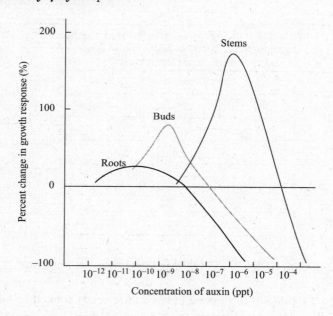

(e) Identify which part of the *Dianthus caryophyllus* plant is most sensitive to different auxin concentrations. Explain your reasoning. (3)

(f) Compare the effects of varying auxin concentrations on the percent change in growth response in the stems of *Dianthus caryophyllus*. (2)

2.

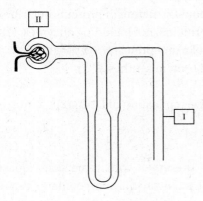

(a) State the name and function of each of the labelled parts in the structure shown above. (2)

(b) Annotate the structure to show the process of selective reabsorption. (2)

(c) Outline the process of ultrafiltration in the kidney. (3)

3. In order for cells to undergo mitosis, they must replicate their DNA. The process of DNA replication is shown in the diagram below.

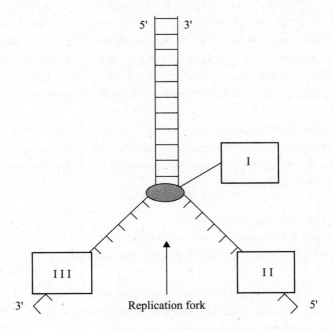

(a) Annotate the diagram above to show the direction of DNA replication on both open strands. (1)

(b) State the name of the enzyme labelled I. (1)

(c) Identify which strand is the leading strand. (1)

(d) Explain the process of DNA replication. (3)

(e) State the location in which this process occurs in eukaryotic cells. (1)

4. Proteins are vital to all living organisms. Although all proteins are composed of amino acids, the three-dimensional structure of proteins determines their function.
   (a) State 2 functions of proteins, not including enzymes. (2)
   (b) Draw the formation of a dipeptide. (3)
   (c) State the two parts of protein synthesis. (2)

5. Explain semi-conservative replication of DNA. (2)

Section B

Higher-level students will be presented with three essay questions. They are required to write essays for two of them. Each essay question contains three parts (a, b and c). Students must answer each part to receive full credit (18 possible points for each essay). Note that 2 additional points may be awarded for each essay. One of these additional points can be awarded if the essay shows clarity of expression. The other additional point can be awarded if the essay structure is of high quality. The extra clarity of expression point can be awarded when the essay response is easy to understand. The essay has to be read through only once in order to be understood, and the main concepts must be easily interpreted. The extra essay structure point is awarded when the essay flows naturally from the response to part (a) to that in part (b) and to that in part (c). These 2 points are in addition to the 18 points available for each essay, making the maximum possible score for each essay 20 points.

Standard-level students will be presented with two essay questions and are required to answer only one. For the sample essays in the following practice test, SL students may choose to write essays for questions 7, 8 or 9. Question 6 covers HL material that will not appear on SL exams. Higher-level students may choose from any of the questions presented (6–9). Only two choices will appear on the official IB Biology Exam for SL students.

6. (a) Explain the processes involved in the light-dependent reactions. (8)
   (b) Outline adaptations found in xerophytes. (4)
   (c) Describe the effect of high temperatures on the enzymes needed for the light-independent reactions. (6)

7. (a) Draw the structure of a neuron. (4)
   (b) Describe the divisions of the nervous system. (6)
   (c) Outline the processes involved in neural transmission. (8)

8. (a) Explain how the process of meiosis leads to genetic variation. (4)
   (b) Describe sex-linkage. Include a sample Punnett square to show possible outcomes of a cross involving sex-linkage. (6)
   (c) Outline the process of gel electrophoresis and its uses in biotechnology. (8)

9. (a) Draw a diagram of the cell membrane. (4)
   (b) Explain the endomembrane (vesicle transport) system. (8)
   (c) Outline the roles of extracellular structures. (6)

PAPER 3

Time: 1 hour (SL)
Time: 1 hour 15 minutes (HL)

Paper 3 consists of questions that will test knowledge on the optional areas of study. Paper 3 is administered the second day of testing. Questions for all 4 optional areas will be presented, and you will be required to respond to 2 of the 4 (A–D). The exams are similar for SL and HL students. However, HL exams include questions on the additional HL material presented in each option. The IB Paper 3 tests are created specifically for SL or HL, and you will be required to answer all questions presented. The Practice Paper 3 presented here combines SL and HL exams. Standard-level students must answer all questions except the last 2 in each optional area, which are identified as being reserved for higher-level students only. Higher-level students must answer all questions presented in each optional area.

Answer all questions in two of the options.

Option A

A1. Psychoactive drugs have the ability to alter the mind, changing the ways in which individuals normally respond to stimuli. The level of dependency on the drug influences the effect of the psychoactive drug on behaviour. Scientists collected data in order to analyse the correlation between psychoactive drug dependency and physical harm. The resulting data are displayed in the graph below.

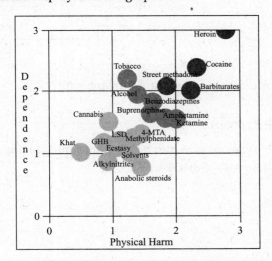

(a) Identify the psychoactive drug associated with the highest level of correlation between dependency and level of physical harm. (1)

(b) Identify the psychoactive drug associated with the lowest level of correlation between dependency and level of physical harm. (1)

(c) Analyse the relationship between the level of dependency and the level of physical harm. (2)

(d) Suggest reasons why higher levels of dependency are related to increased levels of physical harm. (2)

A2.

(a) Identify the function of Broca's area of the brain. (1)

(b) Explain the cause of spina bifida. (3)

(c) State 2 activities controlled by the medulla. (2)

(d) Identify the structures identified as *A, B* and *C* in the diagram of the ear. (3)

(e) Identify the function of each structure identified as *A, B* and *C* in the diagram of the ear. (3)

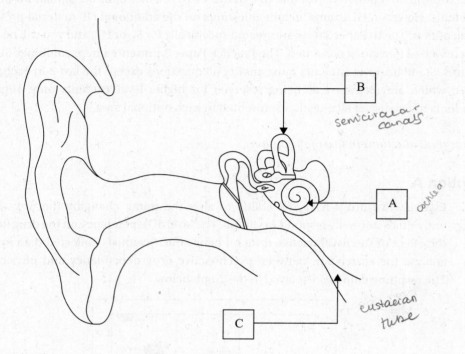

semicircular canals

B

cochlea

A

eustacian tube

C

A3. Explain how neurons develop from the neural tube. (6)

HL EXTENSION QUESTIONS

(a) Explain Pavlov's experiments into reflex conditioning in dogs. (6)

(b) State two inhibitory psychoactive drugs. (2)

Option B

B1. Gene transfer has allowed scientists to develop crops that are enriched with nutrients, can survive in previously uninhabited environments and are resistant to various chemicals. Popular crops and traits that are commonly genetically engineered are shown in the graphs below.

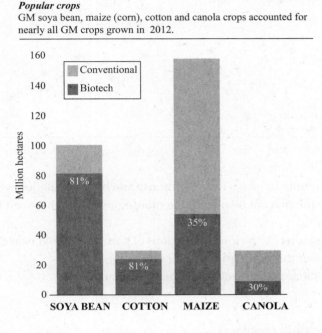

Popular crops
GM soya bean, maize (corn), cotton and canola crops accounted for nearly all GM crops grown in 2012.

(a) State the crop that experiences the highest level of gene modification. (1)
(b) Calculate the percentage of maize that is conventionally grown. (1)
(c) Explain the use of marker genes in gene transfer. (3)
(d) Outline the benefits of producing herbicide resistance in crops. (2)

B2.

(a) Define transgenic organism. (1)
(b) State two products produced by fermentation. (2)
(c) Explain why microorganisms are used in industry. (2)
(d) State one product produced with biotechnology in agriculture. (1)

B3. Outline the process of gene transfer using a named gene of interest. (6)

HL EXTENSION QUESTIONS

(a) Explain the use of databases in bioinformatics. (3)
(b) Outline the steps for PCR. (4)

Option C

C1. Predator and prey populations often fluctuate as the environment changes. The populations of moose and bear were calculated over a 5-year period, and the averages for each month are displayed in the graph below.

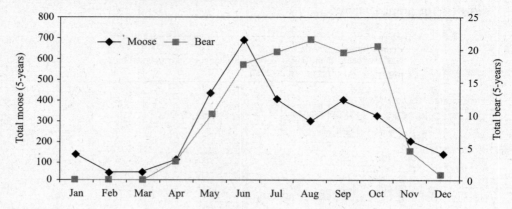

(a) State the month in which both the moose and bear population were the highest. (1)

(b) Calculate the percent change in the moose population between January and June. (2)

(c) Suggest reasons for the low populations of both moose and bear during the months of November through April. (2)

(d) Describe the trends in moose and bear populations. (3)

C2.

(a) Define indicator species. (1)

(b) Compare *in situ* and *ex situ* conservation. (2)

(c) Annotate the growth curve below. (3)

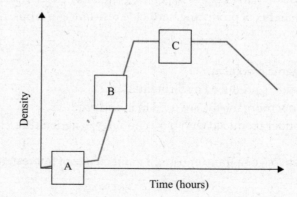

(d) Explain the process of biomagnification. (3)

C3. Explain why the number of trophic levels in a food chain is limited. (6)

HL EXTENSION QUESTIONS

    (a) Explain how populations can be estimated using the capture-mark-release-recapture method of population estimation. (6)

    (b) Explain why the turnover of phosphorus in ecosystems is slow. (2)

Option D

D1. High cholesterol levels have been correlated with increased risks of coronary heart disease. Low-density cholesterol molecules (LDL) can attach to the walls of blood vessels and lead to the development of atherosclerosis. The graph below shows the relationship between LDL levels and the relative risk of coronary heart disease.

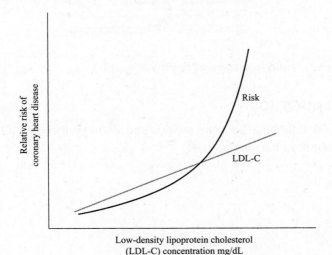

    (a) Describe the type of relationship shown between LDL levels and the relative risk of coronary heart disease. (1)

    (b) State the organ that is involved in regulating cholesterol levels. (1)

    (c) Outline two behaviours that could reduce cholesterol levels. (2)

    (d) State one role of cholesterol in the body. (1)

D2.

    (a) Define exocrine gland. (1)

    (b) Outline one cause of stomach ulcers. (4)

    (c) Explain why some amino acids are essential. (2)

(d) A diagram of a villus is shown below. Identify the structures labelled *A* and *B*. Annotate each labelled structure to identify its function. (2)

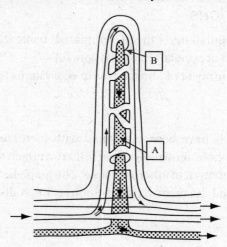

D3. Outline the causes and treatment of emphysema. (6)

HL EXTENSION QUESTIONS

(a) Explain the difference between peptide and steroid hormones. (6)

(b) State one role of haemoglobin. (1)

ANSWERS AND EXPLANATIONS

Paper 1

ANSWER KEY

1.	C	11.	A	21.	C	31.	A
2.	A	12.	B	22.	B	32.	C
3.	B	13.	C	23.	C	33.	D
4.	C	14.	D	24.	C	34.	D
5.	B	15.	A	25.	A	35.	B
6.	B	16.	A	26.	C	36.	C
7.	C	17.	D	27.	D	37.	B
8.	D	18.	B	28.	B	38.	A
9.	B	19.	C	29.	A	39.	C
10.	C	20.	D	30.	C	40.	B

ANSWERS EXPLAINED

1. **(C)** I and III only. Prokaryotic cells do not have membrane-bound organelles such as mitochondria.

2. **(A)** The standard deviation summarizes the spread of data around the mean.

3. **(B)** The correct order is molecule (1 nm), cell membrane (10 nm), virus (100 nm), bacteria (1 um). Remember to use MCV BOC (molecule, cell membrane, virus, bacteria, organelle, eukaryotic cell) to help you remember the order!

4. **(C)** I—DNA; II—flagella. Bacteria do not have cilia. Pili are short, hairlike extensions on bacterial cells.

5. **(B)** The flagella is a long, whiplike structure that functions for movement.

6. **(B)** I and II only. Protein pumps are active, which means they require ATP to function.

7. **(C)** II and III only. I refers to the hydrophobic tail region of the cell membrane.

8. **(D)** Mitosis produces diploid cells and functions for growth, embryonic development, asexual reproduction and tissue repair.

9. **(B)** Picture A shows anaphase II since homologous chromosomes are not present. Separation of chromatids has already occurred. Picture B shows prophase I since homologous chromosomes are present and the nuclear membrane is just beginning to break down.

10. **(C)** The most frequently occurring elements in living organisms are carbon, hydrogen, oxygen and nitrogen.

11. **(A)** Glucose, galactose and fructose are monosaccharides.

12. **(B)** Lipids do not store information. Nucleic acids store information.

13. **(C)** Denaturation is the loss of protein structure and does not necessarily include breaking peptide bonds.

14. **(D)** I—substrate; II—enzyme; III—active site

15. **(A)** Red and blue are most efficiently absorbed by chlorophyll, and green is reflected.

16. **(A)** The enzymes are located where the light-independent reactions occur, which is in the stroma.

17. **(D)** Homologous chromosomes code for the same traits (carry the same genes) but may or may not have the same alleles (forms of the gene).

18. **(B)** Colour-blindness and haemophilia are sex-linked traits. ABO blood type is an example of codominance and multiple alleles.

19. **(C)** *BbLl* and *bbll*. Remember that a test cross always involves the use of a homozygous recessive. The cross would be *BbLl* × *bbll*.

20. **(D)** A community is a group of populations (members of different species) interacting in a given area.

21. **(C)** The increased release of greenhouse gases would result in higher sea levels due to the melting of polar ice and the loss of permafrost from melting.

22. **(B)** Lipase is released into the small intestine (duodenum) from the pancreas and digests lipids into glycerol and fatty acids.

23. **(C)** Classification from most inclusive to least inclusive is kingdom, phylum, class and order. The list continues with family, genus and species.

24. **(C)** Arteries possess the thickest smooth muscle layer to handle the highest pressure of all blood vessels. Veins possess the widest lumen. Capillaries are the only vessels that allow for the exchange of material in the body because capillaries are only one cell thick.

25. **(A)** Antibiotics target and inhibit metabolic pathways. Since viruses have no metabolism (the sum of all reactions occurring inside of cells), antibiotics are ineffective against viruses.

26. **(C)** I—oviduct; II—vagina; III—ovary

27. **(D)** Testosterone is responsible for the development of male genitalia, the maintenance of sex drive and the growth of facial hair (a secondary sexual characteristic).

28. **(B)** The nucleosome is the packaging unit of DNA and helps regulate transcription. RNA is not packaged.

29. **(A)** Type 1 diabetes results from the loss of beta cells in the pancreas because beta cells produce insulin.

30. **(C)** HIV infects white blood cells and causes a decrease in the B-cells that produce antibodies.

31. **(A)** Vaccinations are examples of active immunity because they cause you to make your own antibodies. Vaccinations are also examples of artificial immunity because they are not from typical, day-to-day encounters.

32. **(C)** Chemiosmosis is a passive process (facilitated diffusion) that moves protons/hydrogen ions from the intermembrane space to the matrix through ATP synthase.

The protons/hydrogen ions are then pumped into the intermembrane space by proton pumps to start the process again.

33. **(D)** Mineral ions can be pumped into root hairs by active transport (against the concentration gradient) or can flow in with water (bulk transport) by facilitated diffusion (with their concentration gradient).

34. **(D)** I—anther; II—stigma

35. **(B)** Crossing-over occurs during prophase I of meiosis between nonsister chromatids (chromatids on different chromosomes of a homologous pair).

36. **(C)** Water absorption must occur prior to the embryo releasing gibberellin (GA). Gibberellin triggers the release of amylase by the aleurone layer.

37. **(B)** A polysome consists of many ribosomes translating off of one mRNA molecule.

38. **(A)** Haemoglobin and antibodies (immunoglobins) are both globular proteins. Keratin and myosin are fibrous proteins. Glycerol is not a protein. It is an alcohol.

39. **(C)** Decreasing the amount of substrate present allows for the competitive inhibitor to bind the active site more often. If the substrate binds less often, the rate of the reaction is slowed.

40. **(B)** The endomembrane system produces proteins for export and also includes lysosomes, which keep digestive enzymes packaged in the cell. Proteins for use by the cell are produced on free ribosomes, those not attached to the rough ER.

Paper 2

ANSWERS EXPLAINED

Section A

1. (a) 86% increase (+/– 4%)

    On day 7, seeds soaked in auxin for 12 hours had 7 mm of growth. On the same day, seeds soaked in auxin for 36 hours had 13 mm of growth.

    $$\text{Percent change} = \frac{\text{New value} - \text{Old value}}{\text{Old value}} \times 100$$

    $$\text{Percent change} = \frac{13 - 7}{7} \times 100 = 86\% \text{ increase}$$

    (b)

    - Both 12-hour and 24-hour seeds showed a steady increase in germination rate over time/showed positive linear growth.
    - 24-hour seeds grew more rapidly than did the 12-hour seeds. By day 7, 24-hour seeds had 13 mm in shoot growth while 12-hour seeds had only 7 mm of shoot growth.
    - The growth of both 12-hour and 24-hour seeds progressed at close to the same rate after day 1/show parallel relationships.
    - 12-hour shoot growth levelled off after day 6 while 24-hour shoot growth continued to increase.

- 12-hour shoot growth on day one was only 2mm while 24-hour shoot growth on day one was 6mm/24-hour shoot growth on day one was measured 4mm higher than day one 12-hour shoot growth.

(c)

- Stage 2 seeds stored for 9 months
- Stage 2 seeds with 44.70% moisture/highest moisture percentage
- Water absorption is necessary for seed germination.
- Stage 2 seeds/seeds with 44.70%/highest level of moisture do not need to absorb as much water to begin germination/break the seed coat.
- Absorption of water allows for metabolic processes to begin/allows the testa/ seed coat to break and allows the embryo to grow.

(d)

- Seeds removed from green fruit/stage 1 have not had as much time to absorb water/exist inside of the moist environment of the fruit.
- Seeds removed from ripe fruit/stage 2 have had more time to absorb moisture from the fruit.
- The ripe fruit/stage 2 seeds are exposed to more moisture because the fruit becomes moister as it matures.
- Seeds from ripe fruit may be larger since they are in the fruit longer and can absorb more moisture from the environment/have a larger surface area for absorption of moisture.

(e)

- The stems seem to be most sensitive to auxin concentrations since they exhibit the most growth at high levels of auxin.
- The stems are the most rapidly growing part of the plant. So they should be most sensitive to auxin levels since auxin plays a role in plant growth.
- Roots and buds seem to be the most sensitive in negative growth due to higher levels of auxin.
- Once buds have formed, they don't grow as rapidly as stems since the buds are waiting to flower.
- Once roots have been established, they grow at lower rates in a mature plant.
- Roots showed the least positive percentage of growth, and the stems showed the most positive percentage of growth.
- All parts declined in growth (exhibited negative growth) at a point in higher concentrations/all parts exhibited increased growth at the lower concentrations of auxin.
- Award a point for any correct comparison including exact values of auxin levels and growth percentages.
- The buds exhibit intermediate levels of response to the auxin concentrations when compared with the stem and roots.
- All parts showed positive percentages of growth at 10^{-9} ppt since all overlap a bit at that level of auxin.
- Different levels of auxin affect all plant parts differently.

(f)

- Greatest effect on growth of stems was seen between 10^{-5} and 10^{-6} ppt of auxin (200% change)
- No percent change seen in stem growth at levels below 10^{-9} ppt auxin.

- At concentrations greater than 10^{-4} ppt auxin percent change in growth response is negative/declines
- Stem growth was affected more than root or bud growth/stems showed a maximum 200% increase and roots and buds exhibited percent changes under 100% increase in growth response.

2. (a) Structure I is the collecting duct. Its function is secretion/water reabsorption. Structure II is the Bowman's capsule. Its function is ultrafiltration.

   (b)
   - The proximal convoluted tubule is the location of selective reabsorption.
   - Sodium/amino acids/glucose are actively removed/pumped out of the proximal convoluted tubule.
   - Water is passively removed from the proximal convoluted tubule/osmosis is occurring out of the tubule.
   - Convolutions/folding of the proximal convoluted tubule are drawn/identified. They increase surface area.
   - Stated protein pumps/channels are present.

   (c)
   - Ultrafiltration occurs between the glomerulus/capillaries and Bowman's capsule.
   - Capillaries are fenestrated/have larger spaces between them to allow more filtration.
   - Pressure increases the filtration/causes the afferent arteriole to become wider than the efferent arteriole and creates pressure.
   - Proteins and blood cells are too large to fit through the capillaries/fenestrations.
   - Small substances can pass (based on size).
   - Sodium/amino acids/glucose/urea pass. (A point is given for naming 2 of these substances.)
   - Water moves by osmosis into Bowman's capsule.

3.

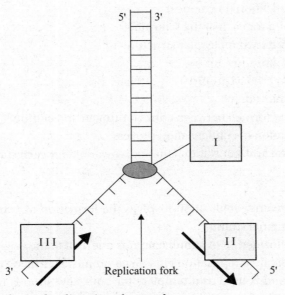

   (a) Arrows must be in the direction shown above.
   (b) Helicase

(c) Strand III is the leading strand since replication must occur in a 5′ to 3′ direction.

(d)

- Helicase opens up/unwinds/breaks hydrogen bonds in DNA.
- RNA primase lays down RNA primer to initiate replication/allow DNA polymerase III to recognize the location of the initiation of DNA replication.
- DNA replication occurs in a continuous fashion on the leading strand.
- DNA replication always occurs in the 5′ to 3′ direction.
- The lagging strand is discontinuous.
- New primers must be laid down on the lagging strand each time DNA polymerase III starts adding nucleotides.
- Nucleotides begin as nucleoside triphosphates.
- Nucleoside triphosphates lose 2 phosphates to become nucleotides.
- Phosphodiester bonds form between newly added and existing nucleotides on the DNA strand.
- Okazaki fragments are formed on the lagging strand.
- DNA ligase seals gaps/forms phosphodiester bonds between the gaps, thereby joining the Okazaki fragments together.
- DNA polymerase I replaces the nucleotides in the primer/RNA nucleotides with DNA nucleotides.
- DNA replication in eukaryotes is initiated at many points/many replication bubbles form
- Each new strand of DNA contains half of the original/semi-conservative
- Replication of DNA occurs during the S phase of the cell cycle

(e) DNA replication occurs in the nucleus of eukaryotic cells.

4. (a)

- Protection (antibodies)
- Transport (haemoglobin)
- Movement (muscles)
- Communication (hormones)

(b) Award 1 point for each of the following:

- Correctly drawn molecular structure
- Labelled amine group
- Labelled carboxyl group
- Water removed
- New bond formed between correct nitrogen and carbon
- Identification of condensation process

(c) Transcription and translation (1 point is awarded for each stated process)

5.

- Semi-conservative replication involves the formation of two new strands of DNA from one original strand
- Each newly formed DNA strand contains one-half of the original strand
- Semi-conservative replication cuts down on mistakes made during replication/An existing strand is used as a template for both new strands to help ensure correct complementary base pairing

Section B

6. (a)
- Photons/light are absorbed by photosystem II.
- Chlorophyll loses 2 electrons.
- Electrons go to the ETC/electron transport chain.
- ATP is produced from energy released from the electrons traveling through the ETC.
- Photophosphorylation occurs/ADP and P are bonded to form ATP.
- Chlorophyll replaces lost electrons by lysing/breaking open water/photolysis.
- Protons in the stroma are pumped into the thylakoid space.
- Proton pumps carry out the pumping of protons against their concentration gradient.
- ATP from the ETC fuels the proton/hydrogen pump.
- Protons flow/undergo facilitated diffusion through ATP synthase.
- Chemiosmosis describes the flow of protons through ATP synthase.
- ATP is produced by chemiosmosis/phosphorylation of ADP occurs.
- Photons/light are absorbed by photosystem I.
- Electrons are lost from photosystem I.
- Electrons lost from photosystem I are replaced by electrons in the ETC.
- Electrons are accepted by $NADP^+$ to become NADPH/NADP is reduced.
- Light-dependent reactions occur in the thylakoid membrane.
- The products of the light-dependent reactions are ATP and NADPH.
- ATP and NADPH will fuel/supply energy to the light-independent reactions.

(b) 1 point is awarded for each correctly outlined feature.
- Hairy leaves to trap water from the air
- Deep, branching roots to obtain water from a large surface area
- Few stomata to limit water loss/transpiration
- Reduced leaves to decrease surface area for water loss (e.g., spines on a cactus)
- Thick, waxy cuticle to act as a barrier to water loss
- Vertical growth to limit the surface area exposed to the direct sunlight in midday
- CAM photosynthesis/stomata open only at night to limit water loss

(c)
- As temperature increases, the rate of enzyme reactions increases.
- At too high of a temperature, the rate will drop drastically and stop.
- Enzymes denature/lose their structure at very high temperatures.
- At low temperatures, molecules do not collide as often and the rate slows.
- The active site of enzymes must have the correct structure to function/active site is where the enzyme will catalyze the reaction.
- The active site changes shape as an enzyme denatures. As a result, the site no longer fits the substrate.
- Denaturation is usually permanent.
- 1 point is awarded for a graph drawn with the correct shape.
- Without enzymes, the light-dependent reactions cannot occur.
- Enzymes for the light-dependent reactions are in the thylakoid/thylakoid space
- ATP synthase is an enzyme of the light-dependent reactions.
- Enzymes have optimal temperatures at which they function the best.

7. (a) 1 point is awarded for each structure correctly drawn and labelled.

- Dendrites
- Cell body
- Nucleus
- Axon
- Mylenated axon/Schwann cells
- Terminal branches
- Terminal buttons

(b)

- The nervous system is divided into the central nervous system (CNS) and the peripheral nervous system (PNS).
- The CNS consists of the brain and spinal cord.
- The peripheral nervous system consists of all the nerves branching off the CNS.
- The peripheral nervous system is divided into the somatic and autonomic nervous systems.
- The somatic nervous system consists of skeletal muscle/is voluntary.
- The autonomic nervous system is divided into the sympathetic and parasympathetic nervous systems.
- The sympathetic nervous system speeds up heart rate and breathing.
- The sympathetic nervous system causes the pupils to dilate and slows digestion.
- The parasympathetic nervous system slows heart rate and slows down breathing.
- The parasympathetic nervous system constricts the pupils and speeds up digestion.

(c)

- Neural transmission allows neurons to communicate/allows neurons to communicate with muscles.
- Neurons and muscle cells can send an action potential.
- Action potential arrives at the presynaptic membrane/cell.
- Calcium channels open in response to the action potential arriving.
- Calcium enters into the presynaptic button/cell.
- Calcium causes the exocytosis/release of the neurotransmitters.
- Neurotransmitters are stored in vesicles in the presynaptic cell/terminal buttons.
- Neurotransmitters attach to gates and open channels on the postsynaptic membrane.
- Excitatory neurotransmitters attach to and open sodium-gated ion channels.
- Sodium-gated ion channels open, and sodium diffuses into the postsynaptic cell.
- The postsynaptic cell depolarizes/reaches threshold/−50/−55 millivolts.
- An action potential is generated in a postsynaptic cell when threshold/depolarization to −50/−55 millivolts occurs.
- Inhibitory neurotransmitters attach to potassium channels and open potassium-gated ion channels.
- Potassium leaves the postsynaptic cell, and the cell becomes less negative/hyperpolarizes.
- The sodium-potassium pump returns the depolarized membrane back to its resting state/−70 millivolts.

- 1 point is awarded for correctly stating the name of a neurotransmitter (e.g., acetylcholine).
- 1 point is awarded for a correctly drawn and labelled diagram.

8. (a)

- Independent assortment
- Independent assortment in metaphase I distributes maternal and paternal chromosomes in random combinations to new cells.
- Crossing-over
- Crossing-over leads to the switching of genes between chromosomes.
- Crossing-over occurs during prophase I of meiosis.
- All cells resulting from meiosis are different from each other and from the parent cell.
- Sperm and egg/ova go through meiosis.
- Meiosis results in 4 genetically different haploid cells/cells with half the original DNA.

(b)

- Sex-linkage refers to genes on the sex chromosomes/X and Y chromosomes.
- Sex-linked traits are heterozygous in females, but males have only one allele for those traits.
- Males are more likely to show sex-linked traits.
- Males show sex-linked traits more often since they have only one X chromosome.
- The X chromosome contains many more genes than the Y chromosome, so most sex-linked traits are related to the X chromosome.
- Males with only one recessive allele will show the trait. Females who show the trait have two recessive alleles.
- A point is awarded for a correctly drawn Punnett square with X showing the allele and Y having no allele.
- A point is awarded for a correctly carried-out cross.
- A point is awarded for stating the resulting phenotypes of affected individuals.

(c)

- Gel electrophoresis is used to separate DNA.
- DNA is separated based on charge and size.
- The DNA is amplified by PCR/polymerase chain reaction.
- Gel plates are placed into a DNA chamber with buffer/solution.
- The DNA is placed into/dropped into the wells in the gel plates.
- Electrical currents are generated through the buffer.
- DNA migrates to the positive pole since DNA is negatively charged.
- DNA is negatively charged due to the phosphate groups.
- Molecules in the gel plate/agarose are the basis of separation since DNA molecules must migrate through and around the gel molecules.
- The gel plate is removed from the electrophoresis chamber when the DNA has travelled far enough to show separation on the gel plate/before it travels off the end of the gel plate.
- A loading dye is used to show how far the bands have travelled/loading dye travels slightly faster/directly in front of traveling DNA/loading dye ensures the

DNA is not allowed to travel off the gel plate/DNA bands cannot be seen as they travel, so loading dye is used to identify travel distance.

- Southern and northern blotting are different. Here the gel plate is removed from the chamber and placed in DNA staining dye in order for the bands to become visible.
- The plate is soaked in/subjected to dye that stains the DNA so the location of the DNA can be seen/produces the banding pattern to be seen.
- Banding patterns can be used for comparison to determine paternity/analyze DNA from a crime scene/forensics/evolutionary relationships/pedigree determination.
- Banding patterns are known as DNA fingerprints.

9. (a) Award a point for any of the correctly drawn and labelled structures:
- Phospholipids
- Phospholipid bilayer
- Integral/intrinsic protein/channel protein
- Peripheral protein/extrinsic protein
- Cholesterol
- Hydrophilic head
- Hydrophobic tails
- Glycoprotein

(b)
- The endomembrane system allows material to travel within and out of the cell.
- Exocytosis is accomplished by the endomembrane system.
- The endomembrane system consists of the rough ER, Golgi, vesicles and the cell membrane.
- Since all membranes are made of phospholipids, they can join together.
- The fluid nature of the membrane allows vesicles to pinch off and join membranes of the endomembrane system.
- When a vesicle pinches off, the natural hydrophobic nature of the tails attracts them quickly together to seal the opening
- Proteins for export are produced on the rough ER.
- The rough ER modifies the protein (changes its structure).
- Processed proteins from the rough ER are transported to the Golgi in a vesicle.
- The vesicle enters the Golgi (on the *cis* side), and the membranes fuse.
- The protein is further modified/packaged as it travels through the Golgi.
- The processed protein pinches off the Golgi (on the *trans* side).
- The vesicle moves toward the cell membrane.
- The vesicle fuses with the cell membrane.
- The vesicle contents are released outside the cell/exocytosis of protein.
- Vesicles containing digestive material can stay in the cell and are lysosomes. The endomembrane system produces lysosomes.

(c)
- Extracellular structures are found outside of the cell and can be attached to the exterior of the cell membrane.
- Extracellular structures are produced by the cell and exported to the exterior of the cell/outer face of cell membrane.

- The cell wall is an extracellular structure.
- The cell wall functions for support/turgor pressure/prevents cell from bursting/protection for cell.
- Plant cells/bacterial cells/fungi have cell walls.
- Glycoproteins are extracellular structures.
- Glycoproteins function for recognition/reception.
- Glycoproteins are a combination of proteins and carbohydrates.
- Glycoproteins can function for adhesion/form cell connections.

Paper 3

ANSWERS EXPLAINED

Option A

A1. (a) Heroin

(b) Khat

(c)

- As the level of dependency increases, so does the level of physical harm/positive correlation.
- Some psychoactive drugs show a higher correlation than others/example stated tobacco has high dependency and lower physical harm than barbiturates.
- Khat has the lowest physical harm, and anabolic steroids have the lowest dependency (one point is awarded for any valid comparison).
- All psychoactive drugs show some level of dependency and physical harm.

(d)

- Higher dependency makes the individual more irrational when the drug is not available.
- Dependency on the drug rewires the brain, making the drug necessary for normal functioning and thought processes.
- Reward pathways are activated when the drug is consumed and lacking without the drug, leading individuals to withdrawal and depression/feelings of wanting to hurt themselves.
- Dependency on drug leads to greater use, possibly causing hallucinations that could lead to harmful acts/drug abuser does not think rationally and may harm himself or herself.

A2. (a) Broca's area of the brain is involved in speech/language.

(b)

- Spina bifida is caused by an incomplete closure of the neural tube during early development of the foetus
- Severity varies due to level of incomplete closure of the neural tube
- Spinal nerves protrude through the spinal bones
- Involves the spinal cord and meninges
- Nerves located below the defect are affected
- Symptoms can be mild to severe/impaired nerve functions
- There is no treatment for spina bifida

(c) The medulla controls:
- Breathing
- Heart rate
- Digestion
- Swallowing[BL]

(d) *A*—cochlea; *B*—semicircular canals; *C*—eustachian tube

(e) *A*—converts sound waves into electrical impulses/perceives sound/sends messages to brain. *B*—functions for balance/perceives movements of the head. *C*—absorbs pressure waves from the cochlea, preventing them from moving backwards/relieves pressure from sound waves

A3.

- Three embryonic germ layers develop from the blastocyst: endoderm, mesoderm and ectoderm.
- The neural tube develops from the ectoderm tissue.
- The ectoderm folds inwards during development/gastrulation.
- The neural tube elongates into the brain and spinal cord/CNS.
- Neurons begin to develop within the neural tube.
- Chemical stimuli initiate the growth of axons from the developing neurons.
- Axons grow to reach great distances/reach outside the CNS.
- Neurons continue to develop and form connections with other neurons.
- Synapses continue to develop as a child develops.
- Neural pruning occurs/inactive synapses are not maintained.

HL Extension Questions

A3. (a)

- A reflex is an automatic response to a stimulus.
- Pavlov studied animal behaviour.
- Pavlov conducted experiments with dogs.
- Pavlov investigated classical conditioning.
- Pavlov presented dogs with food/unconditioned stimulus (US).
- US provoked salivation/unconditioned response (UR).
- Pavlov rang bell/made sound/conditioned stimulus (CS) when he presented the dogs with food.
- CS stimulus is paired with the US.
- Dogs salivated at the sound of the bell regardless of food being present/ conditioned response (CR).

(b) Benzodiazepines, alcohol or tetrahydrocannabinol (THC)

Option B

B1. (a) Soya bean

(b) $100 - 35 = 65\%$ conventionally grown

(c)

- Marker genes are genes of known phenotypic expression.
- An example of a marker gene is antibiotic resistance/other named example.
- The foreign gene is placed within the marker gene's transcription region.

- The transcription of the foreign gene occurs along with the marker gene's transcription.
- The presence of the marker gene trait ensures uptake of the foreign gene.

(d)

- Herbicide resistance increases crop yield.
- Increased crop yield can feed growing populations.
- Less use of herbicides in the environment decreases pollution.
- Saves the farmer money by not having to purchase/distribute herbicides.
- Keeps cost of food down since farmers can pass on savings to consumers.

B2. (a)

- Transgenic organisms are organisms that have foreign DNA/rDNA as part of their genome.
- Transgenic organisms produce proteins that they normally would not produce/foreign proteins.

(b)

- Citric acid
- Penicillin
- Any other correctly named product

(c)

- Microorganisms are small/unicellular, so they are easy to work with/manipulate.
- Microorganisms reproduce very rapidly, so metabolic processes occur quickly.
- Genetic engineering/manipulation of organisms is much easier when dealing with unicellular/small organisms.

(d)

- Amflora potato plants produce only amylopectin as the main starch.
- Tobacco plants produce hepatitis B vaccine.
- One point is awarded for naming any other biotechnology product that uses plants.

B3.

- Gene transfer is possible since DNA is universal/all organisms share the same genetic code.
- Gene transfer involves the insertion of a foreign gene into an organism's genome.
- Gene transfer uses restriction enzymes to cut the donor and recipient DNA.
- A plasmid is obtained from a prokaryotic cell.
- The donor gene is cut from donor DNA with the same restriction enzyme as that used to cut the recipient DNA.
- A restriction enzyme that leaves sticky ends must be used.
- The foreign gene/DNA is sealed into plasmid with DNA ligase.
- A plasmid with foreign DNA is known as recombinant DNA/rDNA.
- rDNA is placed into host organism/unicellular organism/bacteria.
- rDNA can be inserted using gene gun/biolistics; viral vector; calcium carbonate. environment; heat shock; microinjection; liposomes.
- An organism with rDNA is known as a transgenic organism.
- A transgenic organism produces a protein coded for by a foreign gene.
- Insulin production or other correctly named example.

HL Extension Questions

(a)

- Databases store vast amounts of biological information
- Amount of information stored is increasing exponentially
- Allows scientists to compare nucleotide sequences quickly/to compare amino acid sequences quickly
- BLAST is an example/basic local alignment search tool
- BLASTp allows protein alignment
- BLASTn allows nucleotide sequence alignment
- Used to compare sequences among different organisms

(b)

- PCR is polymerase chain reaction.
- PCR uses heating and cooling cycling/thermal cycling.
- Three steps are involved: denature, anneal and extend.
- The first step uses heat to denature DNA/break hydrogen bonds/denature.
- The second step is cooling and attachment of primers/anneal.
- The third stage is addition of nucleotides to growing DNA chains by DNA polymerase/extend.
- *Taq* polymerase/*Thermus aquaticus*/hot spring surviving bacterial polymerase is used due to heat-withstanding abilities/will not denature at high temperatures.
- The PCR cycle is repeated many times.
- PCR can produce vast amounts of DNA in a very short period of time/amplifies DNA.

Option C

C1. (a) June

(b) January moose population = 125 and June moose population = 700; 700 − 125 = 575/125 = 4.6 × 100 = 460% increase in moose population from January to June.

(c)

- Weather too cold for foraging/lack of food in environment
- Animals not as active and do not encounter each other as often
- Animals in hibernation/migrated elsewhere

(d)

- The moose and bear populations show similar trends.
- Both are low during the winter months/November to April.
- Highest in spring, summer and early fall/May–October.
- The greatest differences are in the months of July through October/most similar in February–June.
- The moose population is higher in all months except July–October/or vice versa.

C2. (a)

- An organism whose presence in the environment is determined by environmental conditions
- An organism that is sensitive to changes in the environment/its presence can be used to determine the health of the environment

(b)

- *In situ* conservation occurs in the organism's natural habitat, while *ex situ* occurs when the organism is removed from its natural habitat.
- *In situ* is on-site conservation, and *ex situ* is off-site conservation.
- Both *in situ* and *ex situ* conservation are involved in protecting endangered species.
- One point is awarded for including a correct example of each with named organisms.

(c) *A*—lag phase, population just beginning to grow; *B*—exponential growth, no limiting factors and growth is rapid; *C*—carrying capacity, limiting factors affect population growth/maximum number of organisms that can be sustained in the environment over time

(d)

- Biomagnification is the amplified accumulation of toxins as they move from lower to higher trophic levels.
- Pesticides and heavy metals become more concentrated in tissues as they are consumed by members of higher trophic levels.
- Substances leading to biomagnification are hard or very slow to metabolize.
- Biomagnification of DDT damaged bird eggs and greatly affected the bird population/one point is awarded for including any correct example.

C3.

- Energy transfer between trophic levels is very small/only 10%–20% of the available energy is transferred between trophic levels.
- Producers supply the initial source of energy in a food chain.
- Most energy is lost as heat.
- Energy is lost in faeces/waste/nonconsumed food material.
- The top trophic levels receive the smallest amount of available energy.
- Trophic levels are limited due to the loss of energy that occurs as energy moves up trophic levels.
- Energy is measured in kilojoules/$kJ/m^{-2}/y^{-1}$.
- One point is awarded for a correctly drawn energy pyramid with units of energy shown correctly.

HL Extension Questions

(a)

- The capture-mark-release-recapture method is used to estimate population sizes when each member cannot possibly be counted/organisms move around in a habitat.
- Members of the population that are seen are captured and counted.
- The captured organisms are marked for identification.
- The marking should be in such a way as to not decrease their ability to survive/hide from predators.
- Captured organisms are returned to the environment.
- At a later date/time, organisms are again captured from the environment.

- Organisms from the second capture are counted, and the number of marked organisms (from the first capture) in the second capture are recorded.
- The population size is estimated by the Lincoln index.
- Lincoln index/formula: estimated population size:

$$N = \frac{MC}{R}$$

N = Population estimate (number of organisms)

M = Number of animals in first capture (all will be marked)

C = Number of animals in second capture

R = Number of marked animals in the second capture

(b)

- Phosphorus is needed to build nucleic acids/organic compounds/DNA/RNA/ATP.
- Phosphorus is needed for cell membrane formation/phospholipids.
- Phosphorus is limited in the atmosphere/most available phosphorus is in soil and water.
- Most phosphorus is stored in marine sediment and rock formations.
- Phosphorus is not very soluble, so it settles out of water quickly/settles to the ocean floor.
- Decomposition of uplifted rock supplies phosphorus and is a very slow process.
- Phosphorus must be added to soils used for agriculture/fertilizers add phosphorus to the soil.

Option D

D1. (a) Positive correlation/as LDL levels increase, so does the risk of coronary heart disease

(b) The liver

(c)

- Change in diet to more plant based diet/reduce consumption of red meat
- Exercise
- Lose weight/reduce body mass
- Medications are available
- May be genetic and hard to control

(d)

- Essential for animal cell membrane structure (buffers the fluidity)
- Steroid hormone synthesis
- Bile formation

D2. (a) Exocrine glands secrete their products to the lumen of the digestive system or the skin's surface.

(b)

- Erosion of the lining of the stomach causes stomach ulcers.
- Stress can lead to ulcers.
- The bacterium *Helicobacter pylori* produces holes in the mucous lining of the stomach.
- The overuse of painkillers such as aspirin can cause ulcers.
- The overproduction of stomach acid can cause ulcers.
- Ulcers can become deep and become bleeding ulcers.

- Damage to the stomach lining allows acidic digestive juices to come into contact with the deeper stomach layers.
- *H. pylori* produces urease that can change the pH of the stomach lining, allowing the bacteria to survive and reproduce.

(c)

- Amino acids are the building blocks/monomers of proteins.
- Essential amino acids cannot be synthesized by the body/must be taken in with the diet.
- Lacking essential amino acids leads to protein deficiencies/inability to produce needed proteins.
- Point for named example of protein; for example, keratin, actin, myosin, named enzyme, haemoglobin, antibody/immunoglobin.

(d) *A*—lacteal, absorbs fats/glycerol and fatty acids, immune function/packed with white blood cells; *B*—capillary,– absorbs monomers and transports them to the liver for processing.

D3.

- Emphysema is a disease of the respiratory system.
- Emphysema is caused by damage to the lungs/alveoli.
- Alveoli are very fragile/only one cell thick.
- Gas exchange occurs between the alveoli and capillaries.
- Alveoli become enlarged with air and damaged/burst.
- Gas exchange cannot occur efficiently.
- Scar tissue builds up in alveoli, blocking the diffusion of gases.
- Cigarette smoking is the main cause/pollution in air can cause emphysema/coal dust.
- No cure is available, only treatments are available/oxygen tank delivers more oxygen to the lungs.
- A lung transplant may be needed.

HL Extension Questions

(a)

- Hormones are chemical messengers that travel in the blood.
- Hormones respond to cells with receptors/protein receptors.
- Peptide hormones do not enter into cells/attach to external receptor, and steroid hormones enter into the cells/pass through the cell membrane.
- Peptide hormones use second messengers, and steroid hormones do not.
- Steroid hormones have direct effects, and peptide hormones initiate a cascade of responses.
- Insulin is a peptide hormone, and testosterone is a steroid hormone/any correct example of each type.
- Peptide hormones are faster acting than steroid hormones/peptide hormones activate proteins already present, and steroid hormones must initiate production of proteins/transcription and translation.

(b)

- Transporting oxygen
- Transporting carbon dioxide
- Buffering the pH of the blood/keeping it from becoming too acidic
- Help to regulate blood pressure

Practice Test 2

The practice test that follows is designed to mimic the actual IB Biology Exam Papers 1, 2 and 3. To get the most benefit from this practice exam, you should follow the IB guidelines.

- Adhere to the allotted time for each practice paper. It is a good idea to set a timer to ensure that you do not exceed the time limit.
- Respond to all questions presented. You will not lose points for incorrect responses.
- For Papers 2 and 3, make sure you include *at least* the number of responses required for full credit as indicated by the point values located in brackets next to each question.
- Take Papers 1 and 2 on the same day, giving yourself only a 15-minute break between the two exams. You should not look at or review any notes during the break.
- Take Paper 3 on the day after your take Papers 1 and 2.

PAPER 1

Time: 45 minutes (SL)
Time: 1 hour (HL)

Paper 1 consists of 40 multiple-choice questions for higher-level (HL) students and 30 multiple-choice questions for standard-level (SL) students. No points will be deducted for incorrect responses, so you should choose a response for all questions presented. You will have 1 hour to respond to 40 questions (HL) or 45 minutes to respond to 30 questions (SL).

1. Plant and animal cells both possess:

   I.   Cell membranes
   II. Mitochondria
   III. Chloroplasts

   A. I only
   B. I and II only
   C. II and III only
   D. I, II and III

2. A student observes a structure under the microscope and draws it at a size of 9 mm. The actual structure is only 1 μm. What is the magnification of the structure?

   A. 0.9X
   B. 9X
   C. 900X
   D. 9,000X

Questions 3 and 4 refer to the section of a cell membrane shown below.

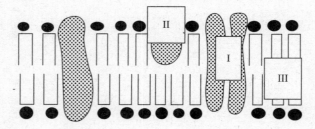

3. Identify the structures labelled in the section of the cell membrane shown above.

	Structure I	Structure II	Structure III
A.	Peripheral channel protein	Peripheral integral protein	Hydrophilic tails
B.	Peripheral protein	Integral channel protein	Hydrophobic tails
C.	Hydrophilic phosphate heads	Peripheral channel protein	Hydrophilic phosphate heads
D.	Integral channel protein	Peripheral protein	Hydrophobic tails

4. What could be a role of structure II in the cell membrane shown in question 3?

    I. Cell recognition
    II. Enzymatic activity
    III. Transport of material into the cell

    A. I only
    B. I and II only
    C. II and III only
    D. I, II and III

5. Which answer choice correctly identifies the structures labelled below?

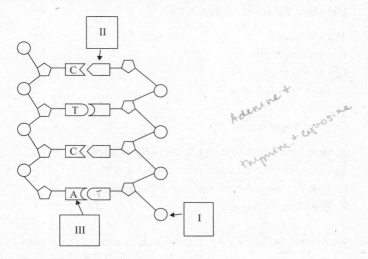

Adenine + Thymine + cytosine

	Structure I	Structure II	Structure III
A.	Deoxyribose	Purine	Pyrimidine
B.	Phosphate	Purine	Pyrimidine
C.	Ribose	Pyrimidine	Purine
D.	Phosphate	Purine	Purine

6. If a codon contains the triplet base of AAU, what will the tRNA complementary anticodon consist of?

    A. UUA
    B. TTA
    C. UUG
    D. AAU

7. Aerobic and anaerobic respiration both produce:

   I.   Pyruvate
   II. ATP
   III. Lactate

   A. I only
   B. I and II only
   C. II and III only
   D. I, II and III

8. What term best describes organisms that break down dead organic matter by external digestion followed by ingestion?

   A. Detritivore
   B. Autotroph
   C. Heterotroph
   D. Saprotroph

9. Oogenesis produces:

   A. One functional egg and three polar bodies
   B. Three functional eggs and one polar body
   C. Two functional eggs and two polar bodies
   D. Four functional eggs

10. What structures are important for absorption of nutrients in the small intestine?

   A. Alveoli
   B. Villi
   C. Smooth muscle
   D. Veins

11. Disease-causing organisms are known as:

   A. Antibodies
   B. Bacteria
   C. Pathogens
   D. Viruses

12. Identify the structures shown below:

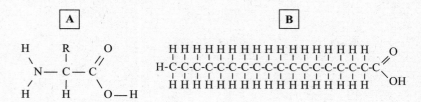

A. A—amino acid; B—fatty acid
B. A—fatty acid; B—glycerol
C. A—amino acid; B—glycerol
D. A—glucose; B—fatty acid

13. What are the products of aerobic cellular respiration?

   I. Carbon dioxide
   II. Water
   III. ATP

A. I only
B. I and II only
C. II and III only
D. I, II and III

14. The karyotype below indicates that:

A. Nondisjunction has not occurred and the individual is a male.
B. Nondisjunction has not occurred and the individual is a female.
C. Nondisjunction has occurred and the individual is a male.
D. Nondisjunction has occurred and the individual is a female.

15. In the human digestive system, most water reabsorption occurs in the:

A. Stomach
B. Small intestine
C. Large intestine
D. Rectum

16. Which graph correctly represents the effect of temperature on enzyme activity?

A.

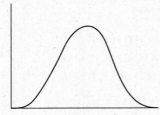

B.

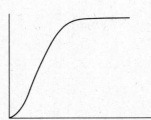

C.

 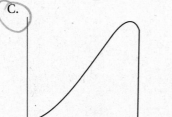

D.

17. Blood transports:

   I. Heat

  II. Antibodies

 III. Gametes

  A. I only

  B. I and II only

  C. II and III only

  D. I, II and III

18. What disease results from lack of beta cells in the pancreas?

 A. Type I diabetes
 B. Type II diabetes
 C. Atherosclerosis
 D. Haemophilia

19. Identify the structures shown below:

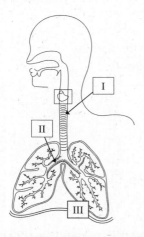

	Structure I	Structure II	Structure III
A.	Oesophagus	Trachea	Diaphragm
B.	Trachea	Bronchus	Diaphragm
C.	Oesophagus	Bronchus	Diaphragm
D.	Trachea	Alveoli	Diaphragm

20. What feature(s) of alveoli make(s) them perfect for gas exchange?

 I. Moist cells
 II. Flattened cells
 III. Close to capillaries

 A. I only
 B. I and II only
 C. II and III only
 D. I, II and III

21. Which of the following are components of the central nervous system?

 A. Peripheral nerves and central nerves
 B. Peripheral nerves and spinal cord
 C. Brain and spinal cord
 D. Brain and peripheral nerves

22. What describes a role of the hormone thyroxin in the body?

    A. Energy storage
    B. Thermoregulation
    C. Digestion
    D. Excretion

23. The correct order of the steps for IVF (in vitro fertilization) is:

    I. Collection of mature eggs
    II. The use of hormones
    III. Transfer of blastocyst to the uterus
    IV. Fertilization in the laboratory

    A. I, II, III and IV
    B. II, I, IV and III
    C. I, III, IV and II
    D. IV, I, II and III

24. Fertilisation should occur in the:

    A. Uterus
    B. Oviduct
    C. Vagina
    D. Ovary

25. What part of the mitochondrion contains the electron transport chain (ETC)?

    A. The matrix
    B. The outer membrane
    C. The intermembrane space
    D. The inner membrane

26. How many ATP are produced as a result of glycolysis?

    A. 2
    B. 4
    C. 32
    D. 38

27. What phylum of plants lack vascular tissue and reproduce by spores?

    A. Filicinophytes
    B. Bryophytes
    C. Angiospermophytes
    D. Coniferophytes

28. Which scientific name is correctly displayed?

    A. *Homo sapiens*
    B. Homo *Sapiens*
    C. homo sapiens
    D. Homo sapiens

29. What can be determined by the following scientific names?

Zea mays
Teosintes mays

A. The organisms are members of the same species.
B. The organisms are members of the same genus.
C. The organisms are members of the same genus and species.
D. The organisms are not members of the same genus or species.

30. Which of the following are greenhouse gases?

I. Methane
II. Carbon dioxide
III. Oxides of nitrogen

A. I only
B. I and II only
C. II and III only
D. I, II and III

The next 10 questions are for HL students only.

31. What enzyme replaces the RNA primers during replication?

A. Helicase
B. DNA polymerase I
C. DNA polymerase III
D. DNA ligase

32. Identify the structures shown below:

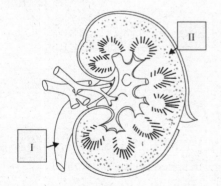

	Structure I	Structure II
A.	Urethra	Cortex
B.	Ureter	Pelvis
C.	Ureter	Cortex
D.	Urethra	Pelvis

33. On what part of the nephron does the hormone ADH (antidiuretic hormone) target?

   A. Proximal convoluted tubule
   B. Distal convoluted tubule
   C. Loop of Henle
   D. Collecting duct

34. What happens to the concentration of progesterone and oxytocin when labor is initiated?

	Oxytocin	Progesterone
A.	Increases	Decreases
B.	Increases	Increases
C.	Decreases	Increases
D.	Decreases	Decreases

35. Which chamber of the heart contains the thickest wall?

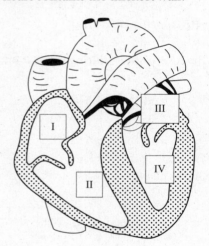

   A. I
   B. II
   C. III
   D. IV

36. Which of the following can speed up the heart rate?

   I. Adrenaline
   II. Exercise
   III. Parasympathetic nervous system

   A. I only
   B. I and II only
   C. II and III only
   D. I, II and III

37. How many amino acids does an mRNA strand containing 30 nucleotides code for?

   A. 10
   B. 15
   C. 30
   D. 90

38. Ligaments connect:

   A. Muscle to bone
   B. Muscle to muscle
   C. Cartilage to bone
   D. Bone to bone

Questions 39 and 40 refer to the following diagram.

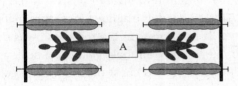

39. What structure is shown in the diagram above?

   A. Nucleosome
   B. Polysome
   C. Sarcomere
   D. Sarcoplasmic reticulum

40. What structure is labelled A in the diagram above?

   A. Actin filament
   B. Myosin filament
   C. Z line
   D. Keratin

PAPER 2

Time: 1 hour 15 minutes (SL)
Time: 2 hours 15 minutes (HL)

Paper 2 consists of two sections. Section A contains data analysis and short-response questions. Section B contains essay questions. Higher-level (HL) students will be presented with 3 essay questions and are required to answer 2. Standard-level (SL) students will be presented with 2 essay questions and are required to answer 1. Both SL and HL students must answer all questions presented in Section A.

Answer all questions presented in Section A.

Section A

1. Male gelada monkeys (*Theropithecus gelada*) possess a red patch of bare skin on the center of their chest. The red patch is brightest in mature males and fades as the gelada monkey ages. The level of brightness of the red patch was determined by comparing the brightness of the red patch of all male geladas. A scale was created based on the range of brightness levels of the red chest patches in the gelada monkey population. The scale was designed with 1 representing the lowest level of brightness observed and 2 representing the highest level of brightness observed. Error bars represent the standard deviations of the brightness levels of the red chest patches for each age group. The graph below shows the relationship between age of male gelada and level of brightness of red chest patch.

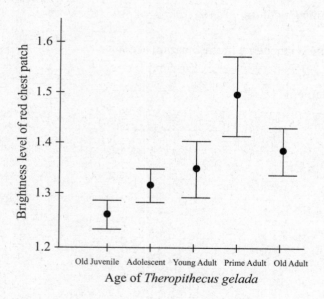

(a) State the age at which the male red chest patch is the brightest. (1)

(b) Describe the variation in red chest patch brightness in different ages of gelada monkeys. (2)

(c) Suggest reasons for male gelada monkeys having different levels of brightness of red chest patches. (2)

PRACTICE TEST 2

Researchers believe that the level of brightness of the red male chest red patch is involved in mate selection by female geladas. Mate selection by female gelada monkeys was observed over a period of two years. The data collected are displayed in the graph below.

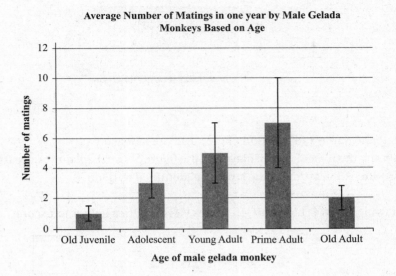

Average Number of Matings in one year by Male Gelada Monkeys Based on Age

2. (a) Explain the relationship between the levels of brightness of male red chest patches and the number of successful mating events achieved by male geladas. (2)
   (b) Calculate the percent increase in the average number of successful mating events for old juvenile gelada monkeys and prime adult gelada monkeys. (2)

Fifty mature female gelada monkeys were followed after successful mating, and the numbers of living offspring produced from the mating events were recorded. The table below displays the data collected.

	Old Juvenile	Adolescent	Young Adult	Prime Adult	Old Adult
Mating choice	2	2	10	32	3
Number of offspring	1	2	8	28	3
Number of living offspring	0	2	7	28	2

   (c) Identify which age of male gelada monkeys was the most successful in finding mates. Explain your reasoning. (2)
   (d) Suggest reasons, besides the brightness of the red chest patch, for the old juvenile not being chosen as often for mating as the older gelada monkeys. (2)
   (e) Some of the offspring did not survive. Describe the theory of natural selection in relationship to the survival of the gelada monkey offspring. (3)

3. All cells possess organelles that carry out specific functions needed to maintain homeo-static levels within the cell. A typical cell is shown below.

   (a) State the name and function of the structure labelled I. (2)
   (b) Identify the type of cell displayed in the figure above. Explain your reasoning. (2)
   (c) List three (3) features of the structure identified as II. (3)

4. The drawing below is based on structures viewed with a light microscope.

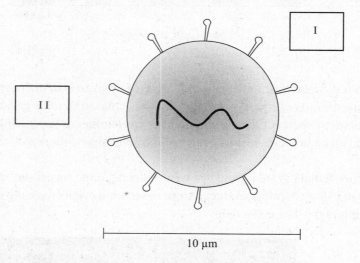

10 μm

   (a) State the names of the structures identified as I and II. (2)
   (b) Calculate the magnification of the image. (2)
   (c) State what the structure drawn represents. Explain your reasoning. (2)
   (d) Compare DNA found in prokaryotic cells with DNA found in eukaryotic cells. (3)
   (e) Explain how you would identify a drawing of a cell as that of a plant cell. (2)

5. All organisms must regulate an internal environment that is separate from the external environment. Organelles assist in the processes involved in the movement of material into and out of cells.

   (a) Annotate the parts of the section of the cell membrane shown below. Add annotated structures that could be present as well. (4)

   (b) State three methods by which material can enter into a cell. (3)
   (c) Explain the need for protein channels. (2)

Section B

Higher-level students will be presented with three essay questions and are required to write essays for two of them. Each essay question contains three parts (a, b and c). Students must answer each part to receive full credit (18 possible points for each essay). Note that 2 additional points may be awarded for each essay. One of these additional points can be awarded if the essay shows clarity of expression. The other additional point can be awarded if the essay structure is of high quality. The extra clarity of expression point can be awarded when the essay response is easy to understand. The essay has to be read through only once in order to be understood, and the main concepts must be easily interpreted. The extra essay structure point is awarded when the essay flows naturally from the response to part (a) to that in part (b) and to that in part (c). These 2 points are in addition to the 18 points available for each essay, making the maximum possible score for each essay 20 points.

Standard-level students will be presented with two essay questions and are required to answer only one. For the sample essays in the following practice test SL students may choose to write essays for questions 7, 8 or 9. Question 6 covers material required for HL students only. Only two choices will appear on the official IB Biology Exam for SL students.

6. (a) Draw and label a diagram of the female reproductive system. (4)
   (b) Explain the role of human chorionic gonadotropin (HCG) in pregnancy. (6)
   (c) Outline the processes involved in oogenesis. (8)

7. (a) Draw and label the structure of a mitochondrion. (4)
   (b) Outline the main events involved in the Krebs cycle. (8)
   (c) Compare anaerobic and aerobic respiration. (6)

8. (a) Explain the need for digestion of macromolecules. (4)
   (b) Describe the structure and function of the villi. (6)
   (c) Explain how changes in temperature affect enzyme activity. (8)

9. (a) Explain how bacteria develop resistance to antibiotics. (6)
   (b) Describe how natural selection impacts the evolution of organisms. (4)
   (c) Explain how nucleotide base substitution can lead to mutations in proteins with reference to sickle-cell anaemia. (8)

PAPER 3

Time: 1 hour (SL)
Time: 1 hour 15 minutes (HL)

Paper 3 consists of questions that will test knowledge on the optional areas of study. Paper 3 is administered the second day of testing. Questions for all 4 optional areas will be presented, and you will be required to respond to 2 of the 4 (A–D). The exams are similar for SL and HL students. However, HL exams include questions on the additional HL material presented in each option. The IB Paper 3 tests are created specifically for SL or HL, and you will be required to answer all questions presented. The Practice Paper III presented here combines SL and HL exams. Standard-level students must answer all questions except the last two in each optional area, which are identified as being reserved for higher-level students only. Higher-level students must answer all questions presented in each optional area.

Option A

A1. Many animals exhibit specific behaviours based on the time of the day. A group of cattle were observed form 7 a.m. until 7 p.m. to determine grazing behaviour. The results are shown below.

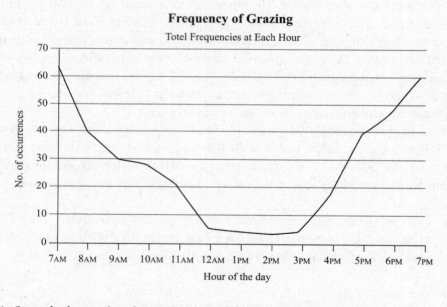

Frequency of Grazing

Totel Frequencies at Each Hour

(a) State the hours that the cattle engaged in the least amount of grazing. (1)
(b) Speculate as to why the frequency of grazing fluctuates within the time period shown. (2)
(c) Outline the trends seen in the graph. (2)
(d) Explain the plasticity of the nervous system. (2)

A2. (a) State the role of photoreceptors. (1)

(b) Identify the parts of the brain labelled *A–D*. (4)

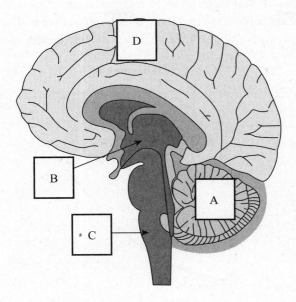

(c) Compare rods and cones. (2)

(d) State a function of the cerebral hemisphere. (1)

A3. Outline the process of neural transmission at the synapse. (6)

HL EXTENSION QUESTIONS

(a) Explain why the pupil reflex can be used to evaluate brain stem death. (4)

(b) Define ethology. (1)

Option B

B1. The process of fermentation converts sugars to products that can be used by living organisms. Fermentation takes place in anaerobic conditions. Many microorganisms rely on fermentation for all their energy needs. Yeasts were allowed to ferment for a period of 1600 seconds, and the level of carbon dioxide production was recorded.

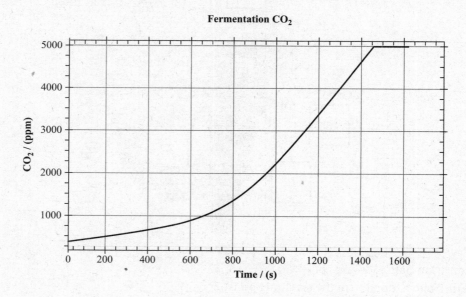

Fermentation CO_2

(a) State the time at which carbon dioxide production reached the maximum level. (1)

(b) Outline the trends seen in the production of carbon dioxide over the 1600-second time period. (3)

(c) State two products of fermentation. (2)

(d) Suggest a reason for the plateau phase of the graph. (1)

B2. (a) State one group of organisms that are used in the production of biogas. (1)

(b) Outline methods of introducing recombinant DNA into a host organism. (3)

(c) Suggest ways in which an ocean oil spill can be cleaned up. (3)

(d) State two environmental problems associated with biofilms. (2)

B3. Outline the process of continuous culture fermentation. (6)

HL EXTENSION QUESTIONS

(a) State one way in which the presence of a pathogen can be detected in a human host. (1)

(b) Outline the production of microarrays. (6)

Option C

C1. Net primary productivity refers to the rate at which photosynthetic organisms incorporate carbon from the atmosphere into their tissues. The net primary productivity of a freshwater pond was investigated at various water depths.

Primary Productivity (Net) in a Freshwater
Pond During the Month of May

Graph: y-axis "Net primary productivity" with 0 marked, x-axis "Depth of water (metres)" showing 10, 20, 30, 40.

(a) What depths of water show loss in net primary productivity? (1)

(b) Outline the trends seen in net primary productivity at the depths ranging from 10 to 40 metres. (2)

(c) Suggest reasons for the drop in net primary productivity after depths of 10 feet. (2)

(d) State 2 organisms that could be contributing to the net primary productivity in the freshwater pond. (2)

C2. (a) State two limiting factors in an ecosystem. (2)

(b) Fill in the energy amounts that will be transferred to each trophic level in the energy pyramid shown below. (3)

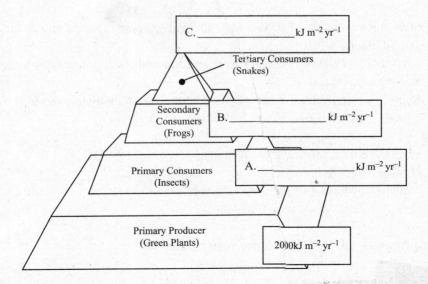

C. _____ kJ m^{-2} yr^{-1}

Tertiary Consumers
(Snakes)

Secondary
Consumers
(Frogs)

B. _____ kJ m^{-2} yr^{-1}

Primary Consumers
(Insects)

A. _____ kJ m^{-2} yr^{-1}

Primary Producer
(Green Plants)

2000kJ m^{-2} yr^{-1}

(c) Outline the effects of microplastic debris on marine environments. (3)

(d) Explain how captive breeding and reintroduction of a species can aid in preventing extinction (3)

C3. Explain how a named introduced alien species can affect an ecosystem. (6)

HL EXTENSION QUESTIONS

(a) Compare top-down and bottom-up limiting factors. (2)

(b) Outline the nitrogen cycle. (6)

Option D

Animals must acquire nutrition from their diet in the appropriate amounts to support metabolic processes. Minerals are inorganic compounds that cannot be made by living things and are essential in the diet. Mineral content varies among food sources. The graph below shows the mineral contents of different foods for cobalt, copper, iron, manganese and boron.

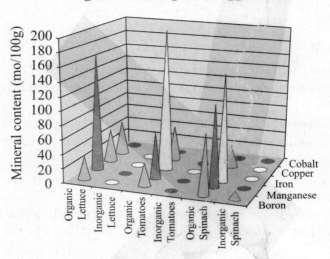

(a) State the food source that provides the highest total mineral content for all represented minerals. (1)

(b) Calculate the percent increase in boron mineral content between inorganic and organic lettuce. (2)

(c) Identify which mineral content is the highest overall in organic foods. (1)

(d) Explain the consequences of a lack of vitamin D. (2)

D2. (a) State the part of the brain involved in appetite control. (1)

(b) Outline two consequences of anorexia nervosa. (2)

(c) Explain why fibre is important in the diet. (2)

(d) State 2 roles of gastric secretions. (2)

D3. Explain how the villi aid in the absorption of food. (6)

HL EXTENSION QUESTIONS

(a) State 2 roles of pituitary hormones. (2)

(b) Explain the control of milk secretion from the mammary glands. (6)

ANSWERS AND EXPLANATIONS

Paper 1

ANSWER KEY

1.	B	11.	C	21.	C	31.	B
2.	D	12.	A	22.	B	32.	C
3.	D	13.	D	23.	B	33.	D
4.	B	14.	D	24.	B	34.	A
5.	D	15.	C	25.	D	35.	D
6.	A	16.	C	26.	A	36.	B
7.	B	17.	B	27.	B	37.	A
8.	D	18.	A	28.	A	38.	D
9.	A	19.	B	29.	D	39.	C
10.	B	20.	D	30.	D	40.	B

ANSWERS EXPLAINED

1. **(B)** Only plant cells possess chloroplasts. All eukaryotic cells possess mitochondria.

2. **(D)** 1,000 μm equals 1 mm. So 9,000 μm equals 9 mm. Since the magnified size is 9 mm, the actual structure, with a size of 1 μm, is being magnified 9,000 times (9,000X).

3. **(D)** Structure I, an integral (intrinsic) channel protein, spans the membrane with a pore. Structure II, a peripheral (extrinsic) protein, exists on the outside of the cell membrane. Structure III shows the hydrophobic tails.

4. **(B)** Peripheral proteins have several functions. They are involved in recognition of hormones and the immune system. They also function as enzymes. Only channel (integral) proteins can transport material into and out of the cell.

5. **(D)** I—phosphate; II—purine; III—purine. Adenine (A) and thymine (T) are pyrimidines. Guanine (G) and cytosine (C) are purines. Purines always form complementary bases with pyrimidines.

6. **(A)** UUA is the complementary codon that will match AAU at the ribosome during translation.

7. **(B)** Lactate is produced only by anaerobic respiration.

8. **(D)** Saprotrophs, such as fungi, digest and then ingest food. In contrast, detritivores, such as earthworms, ingest and then digest food. Both are decomposers.

9. **(A)** Oogenesis produces one functional egg and three polar bodies.

10. **(B)** The villi function for absorption of nutrients in the small intestine by increasing surface area.

11. **(C)** Pathogens are disease-causing organisms. Not all bacteria and viruses cause disease.

12. **(A)** A—amino acid; B—fatty acid

13. **(D)** Aerobic respiration produces carbon dioxide, water and ATP.

14. **(D)** Nondisjunction is the reason for the extra 21st chromosome (Down syndrome/trisomy 21). The presence of two X chromosomes makes the individual a female.

15. **(C)** The main role of the large intestine (colon) is water reabsorption.

16. **(C)** With increasing temperature, the rate of enzyme reactions increases up to an optimal level and then drops rapidly as the temperature becomes too high (denaturation occurs).

17. **(B)** Blood does not transfer gametes (sperm and egg).

18. **(A)** Type I diabetes is due to the loss of the beta cells (islet cells) from the pancreas.

19. **(B)** I—trachea; II—bronchus (plural bronchi); III—diaphragm

20. **(D)** Moisture aids in the diffusion of gases. Flattened cells leave less distance for gases to travel. Capillaries close by can bring and pick up gases to facilitate gas exchange.

21. **(C)** The central nervous system (CNS) is composed of the brain and spinal cord.

22. **(B)** A role of thyroxin in the body is to help control body temperature (temperature regulation). Thyroxin is produced by the thyroid gland.

23. **(B)** The female must take hormones to induce multiple follicles to develop and prepare the uterus for implantation. Then the mature eggs are collected. Next the eggs are fertilized in the laboratory. Finally the fertilized eggs (blastocyst stage) are transferred to the uterus.

24. **(B)** Fertilisation should occur in the oviduct (fallopian tube) to ensure time for the zygote to divide into a blastocyst (hollow ball of multiple cells) before implanting in the uterus.

25. **(D)** The ETC is found in the inner membrane of the mitochondrion.

26. **(A)** Glycolysis results in the production of 2 ATP. (Although 4 ATP are actually produced, 2 ATP are required for phosphorylation, resulting in a net gain of 2 ATP.)

27. **(B)** Bryophytes are the only plant phylum without vascular tissue (xylem and phloem). Bryophytes reproduce by spores.

28. **(A)** Genus is always capitalized, and species is always lowercase. Both genus and species are either underlined or italicized.

29. **(D)** You cannot be a member of the same species if you are not a member of the same genus. Be careful when taking the IB exam. Do not think that just because the species name is the same the organisms are of the same species. Look at all levels given.

30. **(D)** Methane (CH_4), carbon dioxide and oxides of nitrogen are all greenhouse gases. The IB sometimes gives only the molecular formula, not the name. So you should know the molecular formulas of substances that commonly appear on the exam.

31. **(B)** DNA polymerase I replaces the RNA primers during replication. DNA polymerase III adds nucleotides to growing strands of DNA.

32. **(C)** I—ureter; II—cortex. The urethra exits the body from the bladder.

33. **(D)** ADH (antidiuretic hormone) acts on the collecting duct, making the duct more permeable to water.

34. **(A)** When labour is initiated, oxytocin increases to induce uterine contractions. At the same time, progesterone decreases to shed the endometrium. (Progesterone functions to maintain the endometrium.)

35. **(D)** The structure labelled IV (left ventricle) has the thickest wall since it has to pump with the most force to pump blood to the aorta to be distributed to all the body cells.

36. **(B)** Adrenaline (epinephrine) and exercise both increase heart rate. The parasympathetic nervous system slows down heart rate (rest and digest).

37. **(A)** Three nucleotides make one codon (three nitrogenous bases), and each codon codes for one amino acid. 30/3 = 10 amino acids (as long as one triplet does not code for the stop codon).

38. **(D)** Ligaments connect bone to bone. Tendons connect muscle to bone.

39. **(C)** The structure shown is a sarcomere. The sarcomere is the contractile unit of muscle cells.

40. **(B)** Structure A is myosin—the thick filament. Actin molecules are attached to each Z line and are thin.

Paper 2

ANSWERS EXPLAINED

Section A

1.

   (a) Prime adult

   (b)
   - The brightness of the red chest patch increases as the gelada monkey ages. Then the brightness decreases when the monkey is an old adult.
   - There is a positive correlation between red chest patch brightness from old juvenile to prime adult ages.
   - Variation ranges from around 1.25 (+/– 0.2) to about 1.45 (+/– 0.2) for mean red patch brightness levels.
   - The red chest patch is brightest in the prime adult and the least bright in the old juvenile.
   - The red chest patch brightness level is higher in the old adult than in the old juvenile, the adolescent and the young adult.
   - The brightness of the red chest patch in prime adult gelada monkeys shows the greatest variability/has the greatest standard deviation.
   - The variation/standard deviation of the brightness of the red chest patch in the prime adult is more than twice the level of variation/standard deviation of brightness in the old juvenile.

   (c)
   - Red chest brightness develops over time. So the older the monkey, the more time for the redness to brighten (except for old adult when the brightness starts to fade).

- Red chest brightness may be related to reproductive hormones. So the brighter chest patch may signal mating readiness.
- Red chest brightness may be seen by other males as a sign of dominance for mating and may trigger the younger monkeys to stay away from a dominant male's mating territory.
- Brightness of the red chest patch may signal females that males are developmentally ready to mate/have maximum sperm production.

2.

(a)
- Males with brighter red chest patches mated more often/the males with less brightness to their red chest patch mated less often.
- The variation/standard deviation in number of mates was lowest when mating was lowest.
- Prime adults with the brightest red chest patch mated an average of 6 times more than did the old juvenile. (A point is awarded for any similar comparison between ages.)
- Variation/standard deviation increases from old juvenile to prime adult.

(b) 600% increase in mating for prime adult compared with old juvenile

Average number of matings for adolescent = 3

Average number of matings for prime adult = 7

$$\text{Percent change} = \frac{\text{New value} - \text{Old value}}{\text{Old value}} \times 100$$

$$\text{Percent change} = \frac{7-3}{7} \times 100 = 133(+/-1)\% \text{ increase}$$

(c)
- The prime adult
- The prime adult had the most mating events/32 compared with the others/next highest mating events of 10.
- The prime adult had more than twice the mating events of any other age of gelada monkey.

(d) One point is awarded for the suggested reason and one point for explaining the reason.
- Size of the monkey
- The size of the monkey could be part of the selection process. Older males are larger. The old juvenile would be the smallest/smaller size not attractive to females.
- The mating sound of the monkeys
- Mating sounds may be learned/innate sound enhanced as the monkey ages. The females may be responding to the sounds. The old juveniles may not have had time to master more complex sounds.
- The pheromones/hormones/scents of the monkeys may attract the females for mating.
- The pheromones/hormones/scents of the monkeys may change with age. The lower levels in the old juvenile may not attract as many mates.

- Behaviours/mating rituals of the monkeys
- The behaviours/mating rituals of the monkeys may be learned by observation. The old juveniles may not have had time to observe and perfect behaviours that attract mates.

(e)

- Organisms/gelada monkeys produced more offspring than could survive.
- Variation existed in the offspring/offspring born to the gelada monkeys.
- Some of the traits possessed by the monkeys were best for survival/most fit.
- Natural selection leads to the survival of the fittest/best adapted to the environment.
- Offspring that did not survive may not have been strong enough to find the breast/get food from the mother/resources.
- Offspring that survived outcompeted those that did not for food/resources.
- The surviving offspring will pass their fit traits on to future generations.

3.

(a)

- Cell membrane
- Regulates what can enter and exit the cell/separates the interior from the exterior environment

(b)

- Prokaryotic cell
- No nuclear membrane/no mitochondria/no membrane-bound organelles/circular DNA

(c)

- Naked DNA/not packaged (no histones/proteins)/not associated with protein
- Circular in structure
- No introns
- Not inside of a nucleus/found in nucleoid region

4.

(a)

- Structure I is the protein coat/capsid
- Structure II is the nucleic acid/DNA or RNA

(b) One point is awarded for the correct answer and one for the correct calculation.

- 5,000X/magnification
- The size bar is actually 5 cm long.

$$5 \text{ cm} = 50 \text{ mm and } 50 \text{ mm} = 50,000 \text{ } \mu m$$

The size bar label states 10 μm.

$$50,000 \text{ } \mu m \div 10 \text{ } \mu m = 5,000X$$

(c)

- Virus
- No organelles/membrane-bound structures
- Small size (10 μm)
- Only nucleic acid and protein cap/capsule/coat present

(d)

- Prokaryotic DNA is circular, and eukaryotic DNA is linear.
- Prokaryotic DNA is not packaged/associated with proteins/histones. Eukaryotic DNA has these characteristics.
- Prokaryotic DNA is not inside of any nucleus/nuclear membrane. Eukaryotic DNA is inside the nucleus/nuclear membrane.
- Prokaryotic DNA lacks introns, and eukaryotic DNA has introns.
- Prokaryotic DNA is much shorter/contains fewer genes than eukaryotic DNA.

(e)

- Presence of a cell wall
- Presence of chloroplast
- Presence of a large vacuole
- Square/rectangular shape
- Lack of centrioles
- Presence of a cell plate during mitosis/cell division

5.

(a) Award a point for each correctly labelled *and* annotated structure.

Structures present:

- Phospholipid molecule—basic molecule for structure of cell membrane/possesses hydrophilic and hydrophobic properties
- Phospholipid bilayer—serves as selective barrier to cell/regulates what can enter and exit the cell

Structures that could be added:

- Cholesterol—buffers the fluidity of the cell/keeps it from being too fluid
- Integral/intrinsic protein—spans membrane and can be enzyme/part of glycoprotein
- Peripheral protein—on outside of membrane (either side), for enzyme/recognition
- Channel protein—spans membrane with a pore to allow material to pass through that could not any other way (charged or large particles)
- Protein pump —actively moves molecules against their concentration gradient
- Glycoprotein—functions for cell recognition/immune function/hormone recognition
- Gated ion channel—allows selective movement of molecules by opening and closing pore

(b)

- Facilitated diffusion
- Active transport/pumping by proteins
- Endocytosis
- Simple diffusion
- Osmosis (water)

(c)

- Needed so charged molecules/hydrophilic molecules can pass through since they cannot span the hydrophobic region of the membrane/tails
- Needed for larger molecules to pass through that would not fit through the cell membrane

Section B

6.

    (a) A point is awarded for each structure that is correctly drawn and labelled.

- Ovary/ovaries
- Ovules/follicles
- Fallopian tubes/oviducts
- Uterus
- Cervix
- Endometrium
- Vagina

    (b)

- HCG is produced by the foetus.
- HCG triggers processes that keep the corpus luteum intact/prevents the degeneration of the corpus luteum.
- HCG is a hormone/chemical messenger.
- The corpus luteum would disintegrate/break down if HCG were not present.
- The corpus luteum releases progesterone.
- Progesterone maintains the endometrium.
- HCG can be detected in the urine of pregnant women and is the basis of a home pregnancy test.
- HCG is first released by the blastocyst when it implants in the uterus.
- If HCG levels are not high enough, the endometrium will be shed/miscarriage will result.
- HCG is important during early pregnancy because later in pregnancy, the placenta secretes progesterone.
- Progesterone stimulates vascularization/capillary formation of the endometrium to bring needed nutrients to the foetus.

    (c)

- Oogenesis is the creation of mature ova/eggs.
- Oogenesis occurs in the ovaries.
- Oogenesis begins before birth.
- Oogenesis produces one functional egg from each developing/germinal follicle.
- Three polar bodies are produced from each developing/germinal follicle/primary oocyte.
- The first step of development involves mitosis.
- Primary oocytes are produced by mitosis.
- Primary oocytes divide in meiosis I to produce a secondary oocyte and a polar body.
- First division is unequal, and the secondary oocyte is much larger than the polar body.
- The secondary oocyte divides by meiosis II to produce a mature ovum and polar body.
- Meiosis I produces 2 haploid cells from a diploid cell.
- Meiosis II produces 4 haploid cells (3 polar bodies and 1 secondary oocyte).
- The first polar body/polar body from meiosis I will divide to produce two polar bodies.
- Production of the secondary oocyte occurs prior to birth.

- The second meiotic division/creation of mature ovum does not occur unless the secondary oocyte is fertilized.
- Normally one mature egg is produced each month/per menstrual cycle.
- Oogenesis continues until menopause.
- One point is awarded for correctly displaying a diagram of the process.

7. (a) Award a point for each correctly drawn and labelled structure.
- Outer membrane
- Inner membrane (with folds)
- Cristae (folds)
- Matrix
- Circular DNA
- Ribosomes (70S)

(b)
- The Krebs cycle occurs in the matrix.
- The Krebs cycle is an aerobic process.
- Pyruvate enters the matrix and is decarboxylated/removal of CO_2.
- Acetyl-CoA is produced from decarboxylation/removal of CO_2 from pyruvate.
- Pyruvate to acetyl-CoA is the linking reaction.
- Acetyl-CoA joins with a 4-carbon compound (oxaloacetate).
- A 6-carbon compound/citric acid is decarboxylated/CO_2 removed to form a 5-carbon compound.
- A 5-carbon compound is decarboxylated/CO_2 removed to form a 4-carbon compound.
- A 4-carbon compound (oxaloacetate) starts the cycle over again by joining with another 2-carbon acetyl-CoA.
- 2 ATP are produced from pyruvate (1 from each acetyl-CoA that enters).
- 1 $FADH_2$ and 3 NADH (per turn) are produced.
- 6 CO_2 per pyruvate/6 for both pyruvates entering Krebs cycle are produced.
- $FADH_2$ and NADH will be used in the ETC/oxidized/donate hydrogens and electrons.

(c)
- Both processes produce ATP.
- Both processes involve glycolysis.
- Aerobic respiration uses oxygen, and anaerobic does not.
- Aerobic involves the mitochondria, and anaerobic does not.
- Aerobic produces more ATP (36–38 ATP per glucose) than does anaerobic (2 ATP per glucose).
- Aerobic has one pathway, and anaerobic has two/can be lactic acid or alcoholic.
- Aerobic was most likely last to evolve since anaerobic does not require oxygen and early Earth lacked oxygen/prokaryotic cells lack mitochondria and are believed to be the first organisms to have evolved.

8.
(a)
- Macromolecules are too large to be carried in the bloodstream, so they must be digested.

- Enzymes digest macromolecules into monomers.
- Monomers need to be produced/digestion needs to occur to extract nutrients from large macromolecules.
- Digestion of macromolecules releases monomers that can be rearranged into new macromolecules useful to the digesting organism.
- Carbohydrates are digested into monosaccharides/proteins into amino acids/lipids into fatty acids and glycerol.
- Digestion releases energy stored in macromolecules that can be used by the body.
- Digestion can be extracellular or intracellular.

(b)

- The villi are folded to increase surface area.
- The villi are only 1 cell thick to allow for quick absorption of nutrients.
- Cells of the villi are columnar/rectangular in shape.
- Cells of the villi possess microvilli to increase surface area even further.
- Capillaries reach into each villus.
- Capillaries absorb digested material and bring the nutrients into the circulatory system.
- A lacteal reaches into each villus.
- The lacteal absorbs fats/glycerol and fatty acids/lipids.
- The lacteal is packed with white blood cells for immunity/part of the immune system.
- Goblet cells are dispersed in the villi.
- Goblet cells secrete mucus to keep digested material moving.

(c)

- Enzymes are biological catalysts/speed up the rate of reactions.
- Enzymes lower the activation energy by stressing/rearranging bonds of the substrate.
- Enzymes cannot function/slow rate of reactions at temperatures too low due to the decreased movement of molecules/less collisions between molecules/enzyme and substrate don't meet as often.
- As the temperature increases, the rate of enzyme activity increases.
- At a temperature that is too high, an enzyme will begin to denature.
- Denatured enzymes lose their 3-D structure/structure.
- Enzymes catalyse reactions at the active site.
- Loss of the enzyme's 3-D structure changes the shape of the active site inhibiting its ability to bind with the substrate.
- Each enzyme has an optimal temperature at which it functions.
- Denaturation is permanent.
- A point is awarded for correctly drawing the shape of the effect of temperature on enzyme function/rate.

9.

(a)

- Antibiotics stop the metabolic processes of bacterial cells.
- Bacteria are exposed to the antibiotic.
- Surviving bacteria have mutated their DNA/plasmid to develop resistance to the antibiotic.

- Changes in the DNA base sequences lead to mutations that can allow resistance.
- Resistance can develop from an overuse of antibiotics, exposing bacteria to them/ not taking all the antibiotic and thereby allowing some bacteria to survive and mutate.
- Bacteria that develop resistance pass it to all their offspring.
- Bacteria can share resistance by transferring DNA/plasmid for resistance by sexual reproduction/conjugation.
- Antibiotic resistance is responsible for many deaths each year.
- A point is awarded for naming any antibiotic-resistant bacterium (e.g., *Mycobacterium tuberculosis*).
- Antibiotics kill good bacteria/non-pathogenic bacteria as well as bad/pathogenic bacteria.

(b)

- Evolution is change in organisms as seen over time.
- Natural selection selects for those organisms most fit/best adapted to the environment.
- Organisms not suited/adapted/naturally selected for the environment will not survive in the environment.
- Those organisms most suited/adapted to the environment will survive.
- The surviving organisms will pass on their fit/best genes for survival to their offspring.
- Traits best suited/naturally selected for in the environment will be the dominant traits seen in the population.
- Variation should exist in the population in order for survival of organisms in case selection changes/the environment changes.
- Populations with no variation will have a difficult time/may not be able to evolve in changing environments.

(c)

- Base substitutions involve a change in a nucleotide in DNA.
- The change in the nucleotide will be present in mRNA after transcription.
- Translation of the altered base may or may not result in a new amino acid being brought to the ribosome/being part of the newly formed polypeptide/protein.
- Some base changes do not affect the protein formed since some amino acids are coded for by more than one codon/triple base on mRNA. The change in base sequence may still code for the same amino acid.
- If the base change results in a new amino acid being brought to the ribosome/ incorporated into the polypeptide/protein, it may alter the protein's structure.
- A change in the protein structure may inhibit the function of the protein.
- Sickle-cell anaemia is caused by a single base mutation changing the structure of haemoglobin/red blood cell shape.
- The base A is replace by T. GAG is transformed to GTG.
- Glutamate is replaced for valine. Glutamate is coded for by the codon CAC, and valine is coded for by the codon GTG.
- Sickle-cell anaemia results from sickle-shaped/misshapen red blood cells.
- Sickle cells cannot carry oxygen as well/tend to clump in the bloodstream.
- Sickle cell is a point mutation/involves only one base change.

Paper 3

ANSWERS EXPLAINED

Option A

A1. (a) Between 12 P.M. (noon) and 3 P.M. (+/− 0.5hr).

  (b)

- Lower levels of grazing occur during hotter times of the day/12 P.M. (noon) to 3 P.M.
- Grazing during hot times of the day takes more energy/more heating of body tissue during hot times of day.
- Early morning and late at night, the environment is cooler so less heating of body tissue.
- Most grazing occurs early in the morning and late at night.
- May be more moisture on the grasses to facilitate digestion in the early morning and evening hours.

  (c)

- The graph shows a steady decrease in grazing from 7 A.M. until 12 P.M. (noon).
- The graph levels off for grazing activity between 12 P.M. (noon) and 3 P.M./grazing is similar/less than 10 occurrences between 12 P.M. (noon) and 3 P.M.
- Graph shows steady increase in grazing levels from 3 P.M. until 7 P.M.
- 7 A.M. and 7 P.M. grazing occurrences are very similar/close in values.

  (d)

- Plasticity of the nervous system refers to the changing of neural pathways within the nervous system.
- Neurons that are not used are degraded/neural pruning occurs.
- Synapses that are not active are not maintained/are removed from neural pathways.
- Learning can change the neural pathways in order to rewire the thought process.
- Brain injury can change neural pathways/areas of brain may take on role that was previously carried out by the damaged area of the brain.
- Recovery of brain function following stroke is evidence of neural plasticity.

A2. (a) Photoreceptors detect the presence of light/detect images.

  (b) *A*—cerebellum; *B*—pituitary gland; *C*—medulla/medulla oblongata; *D*—cerebral cortex

  (c)

- Rods and cones are photoreceptors/absorb light.
- Rods see color, and cones see black-and-white images.
- Rods are sensitive to low light, and cones are sensitive to bright light.
- Rods are used mostly for night and peripheral vision, while cones are used for detailed central vision.
- Many rods connect to one bipolar cell, while each cone connects to an individual bipolar cell.
- Rods are more numerous (120 million), and cones are less numerous (7–8 million).
- Rods are distributed throughout the retina, and cones are most concentrated in the fovea.
- There are one type of rod and three types of cones (red, green and blue).

(d)
- Receives sensory input from sensory receptors
- Highest level of mental and behavioural processing
- Memory
- Problem solving/goal setting
- Reading/writing
- Math/calculations
- Creativity/art/dance/musical ability

A3.
- The action potential arrives at the terminal button.
- The action potential initiates opening of calcium channels.
- Calcium diffuses into the presynaptic membrane/button.
- Calcium activates the exocytosis of the neurotransmitter.
- Neurotransmitter (NT) is released from vesicles into the synapse/exocytosis.
- If excitatory, the NT binds to sodium-gated ion channel receptor/ligand.
- Sodium channels open and postsynaptic cell depolarizes/reaches threshold.
- If inhibitory, the NT binds to potassium channels and opens potassium channels.
- Potassium diffuses out of the postsynaptic cell into the synapse.
- Postsynaptic cell hyperpolarizes/moves farther away from resting potential/ -70 mV.
- Many synapses may intervene with one neuron and the total change in membrane due to the combined effects of the NT influence depolarization/state of cell/summation.

HL Extension Questions:
(a)
- The pupil reflex is controlled by the brain stem/basic brain stem reflex.
- The pupil reflex controls constriction of the pupil/iris muscles.
- Brain stem damage would interfere with the pupil dilation reflex.
- Brain stem damage could interfere with breathing.
- Pupil reflex is a quick way to determine brain stem damage to evaluate the need for a respirator.

(b) Ethology is the study of animal behaviour/behavioural patterns in natural settings/normal niche of the animals.

Option B

B1. (a) 1440 seconds (+/– 10)
(b)
- Carbon dioxide production is level from 0 to 400 seconds.
- Carbon dioxide production increases from time 400 to time 1440.
- Carbon dioxide production levels off at 1440/same from 1440–1600.
- Highest levels are between 1440–1600.

(c)
- Carbon dioxide (alcoholic fermentation)
- Ethanol (alcoholic fermentation)

- Lactate (lactic acid fermentation)
- ATP (alcoholic and lactic acid fermentation)

(d)
- All the sugar/carbohydrate is used up in the environment.
- Another variable became limited in the environment/temperature changed/pH changed.

B2. (a) Methanogenic bacteria/bacteria/prokaryotes/acid-forming bacteria

(b)
- Viral vectors/bacteriophages are used by placing the rDNA into the protein capsid for delivery to bacterial cell.
- The virus injects rDNA into the host bacterial cell.
- A gene gun/biolistics delivers a pellet coated with rDNA.
- Heat shock by cooling and heating causes the bacteria cells to take up rDNA from the environment.
- Calcium carbonate causes pores in membrane of host cell through which rDNA enters host cell.
- Liposomes are produced to deliver the rDNA by endocytosis of the vesicle.

(c)
- *Marinobacter* bacteria are used.
- *Marinobacter* produce enzymes that degrade hydrocarbon rings.
- The addition of *Marinobacter* biofilms to ocean oil spills facilitates the breakdown of the hydrocarbon rings found in oil.
- *Pseudomonas* bacteria produce surfactants/materials that can break up oil spills/result in more surface area for decomposition of the oil.
- Physical removal of the oil aids in the cleanup/efficiency of the bioremediation.
- Using living organisms to clean up oil spills is a type of bioremediation.

(d)
- Production of toxins/disease
- Clogging of pipes/infrastructure
- Blocking light from lower water levels
- Consuming available nutrients

B3.
- Uses a closed system
- Microorganisms and nutrients are combined and monitored as desired product is continually removed.
- Population is maintained at exponential growth phase.
- Ongoing process of continual product production and collection
- Continual monitoring of several variables/pH, temperature, oxygen levels
- From fermentation using bacteria or yeast/microorganisms
- Used to produce a usable product

HL Extension Questions
(a)
- Metabolites present in blood
- Metabolites present in urine

(b)

- Microarrays are used to identify the presence of specific DNA gene sequences.
- Microarrays can be used to test for the presence of genes that predispose an individual to a genetic disease/identify presence of viral disease.
- Microarrays test for the presence of the gene transcript/mRNA.
- PCR is used to create microarrays.
- Reverse transcriptase creates DNA from a single-stranded mRNA.
- Single-stranded complementary DNA (cDNA) is produced by removal of the RNA strand from the newly formed cDNA strand.
- cDNA strands are attached to microarray plates/glass plates/silicon chips.
- Attached cDNA strands are referred to as probes.
- Target cDNA is fixed with fluorescent markers.
- Complementary target cDNA strands bind with cDNA probes.
- Noncomplementary target cDNA strands wash off of the microarray plate.
- The level of fluorescence determines the level of expression of the target gene/target cDNA.

Option C

C1. (a) After a depth of 20 (+/–5) feet, net primary production dips below 0.

(b)

- Highest net primary productivity occurs at depths lower than 10 feet (+/– 5).
- Net primary productivity rapidly drops after 10 feet.
- Net primary productivity levels off slightly between 20–30 feet.
- Net primary productivity drops between 30–40 feet more rapidly than between 20–30 feet.
- Greater depths show lower net primary productivity/no primary productivity.

(c)

- Less/no light available at lower depths
- Temperature drops at greater depths, slows metabolic activities

(d)

- Algae
- Bacteria/prokaryotes
- Plants/aquatic grasses

C2. (a)

- Food/nutrients
- Light
- Space/territory
- Water

(b) A: 200 kJm^{-2}yr^{-1}; B: 20 kJm^{-2}yr^{-1}; C: 2 kJm^{-2}yr^{-1}

(c)

- Microplastic debris consists of very small, nondigestible plastic waste.
- Aquatic organisms consume the plastic with food/mistake plastic for food.
- Plastic becomes lodged in digestive tract/blocks digestion.
- Organisms die from malnutrition due to blockage/filled digestive tract with microplastics.

- Microplastic damages the digestive tract as it travels the tract/bleeding occurs internally.
- Microplastic can come from food containers/plastic bottles/bags/broken down plastic materials.
- Adult birds regurgitate microplastics to young offspring damaging/destroying young.

(d)

- Captive breeding involves the breeding of organisms outside their natural habitat.
- Captive breeding focuses on endangered species.
- Breeding organisms in zoological parks are in contained environments.
- Organisms are reintroduced to their natural environment.
- Monitoring after reintroduction is important to ensure successful population growth.
- Captive breeding helps in the production and survival of offspring/increases number of surviving offspring, helping to prevent extinction.
- Captive breeding provides greater control over the environment of the organism.
- One point is awarded for naming an example, such as the California condor.

C3.

- Alien species can disrupt ecosystems since these species have no natural predators.
- Alien species can outcompete native species for available food/niches.
- Native species lack behaviours to avoid predation by invasive species.
- Native species could become extinct.
- For example, the cane toad was brought in to Australia to feed on cane grub that were destroying the sugar cane crops/one point is awarded for describing any other alien species introduction.
- The cane toad did not feed on sugar cane but fed on many native organisms.
- The cane toad overpopulated the area and greatly affected biodiversity/limited the ability of other organisms to survive.

HL Extension Questions

(a)

- Both top-down and bottom-up control types affect population sizes.
- Top-down limiting factors involve increases in the number of organisms at higher trophic levels, and bottom-up limiting factors involve decreases in the number of organisms at the producer level of the food chain.
- Combinations to top-down and bottom-up limiting factors maintain stable populations at each trophic level.

(b)

- Nitrogen is essential for the formation of proteins and nucleic acids.
- Most available nitrogen exists as atmospheric gas.
- Microbes/nitrogen-fixing bacteria convert atmospheric gas to usable nitrogen compounds/nitrogen fixation.
- Nitrogen-fixing bacteria in the soil produce ammonia from atmospheric nitrogen.
- *Rhizobium* bacteria live in the roots of plants and fix nitrogen into ammonia for the plant and receive nutrients/mutualistic relationship.

- Industrial processes can produce ammonia from atmospheric nitrogen/Haver-Bosch process.
- Decaying organic material releases ammonia into ecosystems/putrefaction/rotting organic matter.
- Ammonia is oxidized into nitrate following with the oxidation of nitrites into nitrates.
- *Nitrosomonas* microbes convert ammonia to nitrites.
- *Nitrobacter* microbes convert nitrites to nitrates.
- *Pseudomonas denitrificans* remove nitrates from the environment and convert them back to atmospheric nitrogen.
- One point is awarded for a correctly drawn and annotated nitrogen cycle diagram.

Option D

D1. (a) Organic spinach

(b)

- Inorganic boron: 5 mg/100 g and organic boron: 20 mg/100 g.
- $20 - 10 = 10/10 = 1 \times 100 = 100\%$ increase

(c) Iron

(d)

- Lack of vitamin D leads to rickets/soft bones
- Poor calcium use by the body
- Skeletal deformities due to soft bones
- Vitamin D can be synthesized in the skin with UV rays/sunlight exposure
- Animal-based diets provide vitamin D
- Lack of vitamin D has been linked to many disorders/cancers

D2. (a) Hypothalamus

(b)

- Anorexia involves a lack of nutrients due to an eating disorder.
- Anorexia is a psychological issue where the individual feels fat even though he or she is extremely malnourished/distorted body image.
- Anorexia is associated with dramatic weight loss.
- The body will begin to digests its own muscle for energy/heart muscle affected.
- Anorexia can lead to heart failure and death.
- Thinning hair occurs.
- Anaemia occurs.
- In females, menses stops/no menstrual cycle.
- Weak bones occur.
- Anorexia is more common in females than in males.
- Nervous system issues/changes in brain chemistry/personality (moody/irritable).

(c)

- Fibre includes indigestible plant material/passes through digestive system undigested.
- Fibre adds bulk to food, giving a feeling of fullness and helping maintain body weight/less food is consumed.
- Fibre keeps food moving at the proper rate through the digestive system/helps with control of constipation.

- Fibre consumption reduces obesity and obesity-related diseases/diabetes/coronary heart disease.
- Fibre helps regulate blood sugar.
- Fibre helps lower LDL/bad cholesterol levels.

(d)

- Lower the pH in the stomach to help digest macromolecules
- Lower the pH in the stomach to aid in destroying pathogens/immune system
- Excrete pepsinogen/pepsin to digest proteins
- Excrete mucus to protect stomach lining

D3.

- Villi are folds in the digestive tract that increase surface area.
- Increased surface area allows more space for absorption of nutrients/monomers.
- Villi are only 1 cell thick for rapid absorption across a thin layer of tissue.
- Villi possess microvilli to increase surface area even further.
- Villi have lacteals reaching into them to absorb fats/immune function/packed with white blood cells.
- Capillaries reach into each villus to absorb nutrients from the villi.
- Nutrients absorbed by the villi are taken to the liver for processing/storage.
- Protein channels and pumps are located in the villi to aid in the transport of nutrients into the blood capillaries.
- One point is awarded for including a correctly drawn villus.

HL Extension Questions:

(a)

- Growth—GH
- Reproduction—FSH and LH
- Milk production and excretion—prolactin and oxytocin
- Uterine contractions—oxytocin
- Osmoregulation/homeostasis—ADH
- Endorphins
- Metabolism—TSH

(b)

- Milk is produced and released from mammary glands.
- Prolactin and oxytocin both stimulate mammary glands.
- Prolactin is responsible for milk production.
- Oxytocin is responsible for milk excretion.
- Prolactin and oxytocin are released from the pituitary gland.
- The anterior pituitary produces prolactin.
- The posterior pituitary produces oxytocin.
- The pituitary gland is under the control of the hypothalamus/hypothalamus-releasing hormones control the release of prolactin and oxytocin.
- Nursing infants stimulate the release of oxytocin and prolactin.
- The release of milk is known as the "let down" reflex and can be initiated by hearing an infant crying/seeing the baby/suckling the baby/nipple stimulation.

APPENDIX

Answers to Featured Questions

TOPIC 1 (Page 29)

(D) **(X 5.000)** Convert millimetres to micrometres (40 millimetres $\times$ 1,000 = 40,000 micrometres) to determine actual size of drawing in micrometres. Divide the actual size of the drawing by the actual size of the organism to determine magnification (40,000 micrometres/8 micrometres = X 5,000).

TOPIC 2 (Page 78)

(D) The polar nature of water allows it to dissolve both positively and negatively charged molecules.

TOPIC 3 (Page 120)

(C) **(CAC)** The mRNA complementary codon to the nucleotides on DNA consisting of the sequence CAC is GUG. The tRNA complementary anticodon to the mRNA GUG is CAC.

TOPIC 4 (Page 159)

(B) Methane and water vapour contribute to the greenhouse gases. Nitrous oxide (not nitrogen) is also a greenhouse gas.

TOPIC 5 (Page 174)

(C) They must be members of the same genus to be members of the same species. Some organisms do have the same species *name* but the genus is different. In that case, they are not in the same species.

TOPIC 6 (Page 201)

(B) Arteries withstand high pressure and veins withstand low pressure. Don't forget that although the pulmonary artery carries deoxygenated blood, most arteries carry oxygenated blood, and although the pulmonary vein carries oxygenated blood, most veins carry deoxygenated blood, making C incorrect!

TOPIC 7 (Page 239)

(C)   Helicase unwinds the DNA strand. DNA polymerase adds the complementary bases.

TOPIC 8 (Page 264)

(C)   Protons are pumped into the intermembrane in order to undergo facilitated diffusion back into the matrix through ATP synthetase. This process is known as chemiosmosis.

TOPIC 9 (Page 288)

(D)   All are adaptations for plants that live in arid environments (xerophytes).

TOPIC 10 (Page 313)

(A)   Linked genes are genes that are located on the same chromosome.

TOPIC 11 (Page 355)

(C)   The embryo produces HCG, which stimulates the corpus luteum to continue to produce oestrogen and progesterone in order to maintain the endometrium.